LICHEN
PHYSIOLOGY
AND
CELL BIOLOGY

LICHEN PHYSIOLOGY
AND
CELL BIOLOGY

Edited by

D. H. Brown

University of Bristol
Bristol, England

PLENUM PRESS • NEW YORK AND LONDON

Library of Congress Cataloging in Publication Data

International Conference on Recent Advances in Lichen Physiology (1st:
   1984: Bristol, Avon)
   Lichen physiology and cell biology.

   "Proceedings of the First International Conference on Recent Advances in
Lichen Physiology, held April 16–18, 1984, in Bristol, England"—T.p. verso.
   1. Lichens—Physiology—Congresses. 2. Lichens—Cytology—Congresses.
I. Brown, D. H. II. Title. III. Series.
QK581.I58   1984   589.1   85-24452
ISBN 0-306-42200-X

Proceedings of the First International Conference on Recent Advances in Lichen
Physiology, held April 16–18, 1984, in Bristol, England

© 1985 Plenum Press, New York
A Division of Plenum Publishing Corporation
233 Spring Street, New York, N.Y. 10013

Printed in the United States of America

PREFACE

It is currently impossible to grow lichens under controlled conditions in the laboratory in sufficient quantity for physiological experiments. Lichen growth is slow and conditions which might accelerate the process tend to favour either the algal or fungal partner, resulting in the breakdown of balance symbiosis. Lichen physiologists are therefore forced to use field-grown material with all the problems associated with the unknown influences of unpredictable and unreproducible climatic conditions. Study of major biochemical topics, such as the nature of the carbohydrate and nitrogenous compounds passing between the symbionts, is less influenced by climatic conditions than the intrinsic nature of the symbionts and many advances have been made in these areas.

Recently, the challenge of using field-grown plant material, the physiological status of which is intimately linked to environmental conditions, has proved to be a stimulus rather than a hindrance to a number of research groups. The occurrence of lichens in extreme habitats has prompted a number of field and laboratory studies with material from such diverse localities as the cold deserts of Antarctica and the temperate rain forests of the New Zealand bush. A comparative approach, using contrasted species or habitats from a particular geographical region has yielded much information and an appreciation of the variety of physiological adaptations which may exist. The close linkage between morphology and physiology is now being directly demonstrated, as is the relevance of ultrastructural information. Photosynthetic data is also now being analysed in terms of both the possible mechanisms involved in acclimation to seasonally changing conditions and the generation of predictive models. The use of isozyme patterns has become a new tool with which to demonstrate inter- and intra- species differences. A variety of other techniques have been recently introduced into the study of lichen physiology.

Lichens are notorious for their sensitivity to gaseous pollutants but most investigations have been at the ecological

rather than the physiological level. On the other hand,
physiological studies on mineral uptake by lichens has been
stimulated by their use as local and regional monitors of heavy
metal pollution patterns. A more sophisticated appreciation has
developed of the influence of season, age and the nature of the
mineral source on the accumulation of specific elements by
lichens; the precise role of the cell wall in this process is also
being investigated.

Many investigators of lichen physiology use these plants as a
single organism without regard to their composite nature. Because
the isolation and culture of symbionts rapidly modifies their
physiology, there is still limited information on the rôle
individual symbionts play in particular lichen species. Nitrogen
metabolism is one field in which the behaviour of particular
symbionts has been specifically characterised in the intact
lichen. The processes involved in the recognition of suitable
symbionts in order to initiate the early developmental stages of
the lichen thallus is another topic which emphasises the
subtleties and specificity of the symbiotic process. It is only
recently that attempts have been made to understand how the
balance between symbionts is maintained in the intact lichen
thallus.

With the present vitality and variety in lichen physiology it
is perhaps surprising that there has never been an International
Conference devoted exclusively to this subject. This deficiency
was rectified by a meeting I organised on behalf of the British
Lichen Society and International Association for Lichenology at
Badock Hall, University of Bristol from 15th to 18th April 1984.
Speakers were invited who were actively involved in research on
the lichen physiological topics mentioned above. The present
volume represents the proceedings of the meeting and includes the
majority of lectures presented.

One communication problem has been that lichen physiologists
are relatively thinly scattered around the world. The
international nature of this meeting was achieved by generous
grants from the Tansley Trust of the New Phytologist, the Nuffield
Foundation, The Royal Society and the British Council, towards the
travel expenses on many speakers, for which we were extremely
grateful. The success of the meeting was further stimulated by
hospitality received from the British Lichen Society and sherry
receptions provided by John Harvey and Sons Ltd (followed by a
conference dinner in the cellars of their wine museum) and the
University of Bristol. D.H. Brown, M. Galun, T.G.A. Green,
K.A. Kershaw, R. Lallement, O.L. Lange, T.H. Nash III,
D.H.S. Richardson, M.R.D. Seaward, D.C. Smith and C. Vicente
kindly acted as Chairmen for sessions and Prof. D.C. Smith
provided a stimulating final special lecture. I am most grateful

to R.P. Beckett, R.M. Brown and D.J. Hill for suggestions and discussions about the organisation of the meeting and the former for typing all the abstracts. Prof. A.E. Walsby kindly provided support and many facilities from the Botany Department, University of Bristol. The present volume was skilfully and patiently typed by Miss Lena Clarke, Mrs Pat Griffin and Miss Frances Wiggins. Dr. R. Andrews of Plenum Press provided much valuable assistance towards the production of this book.

I am especially grateful to my wife, Rosalie, for her active support and assistance throughout the meeting, her tireless help in reading and correcting both manuscripts and proofs and her major contribution to compiling the index.

D.H. Brown

Editorial note on terminology and nomenclature

For many years the term "phycobiont" has been considered a suitable, general, description for the photosynthetic partner(s) of lichens. This term, emphasising the algal nature of the photosynthetic partner, became inappropriate when the prokaryotic "blue-green algae" were reclassified as "cyanobacteria". V. Ahmadjian proposed (International Lichenological Newsletter (1982) 15: p. 19) that "photobiont" should now be used to describe all photosynthetic partners in lichens, "cyanobiont" used when cyanobacteria are involved and "phycobiont" confined to cases when only eukaryotic algae are present. This terminology has been adopted here.

As far as possible, the nomenclature used here follows the "Checklist of British lichen-forming, lichenicolous and allied fungi" by D.L. Hawksworth, P.W. James & B.J. Coppins (The Lichenologist (1980) 12: 1-115). Authorities are only cited when they do not appear in this checklist. When referring to previous publications, names are generally quoted as in the original article. Hawksworth et al. did not accept the designation of Cladina as a genus separated from Cladonia, although many workers now prefer this segregation (see T. Ahti (1984) "The status of Cladina as a genus segregated from Cladonia." Beihefte zur Nova Hedwigia 79: 25-61). Cladonia is used here, except where comparisons between these groups have been made.

CONTENTS

ix

SOME ASPECTS OF CARBOHYDRATE METABOLISM IN LICHENS

J.D. MacFarlane and K.A. Kershaw

Department of Biology, McMaster University
Hamilton, Ontario, Canada

INTRODUCTION

There has been considerable interest over the last 15 years in the carbon metabolism of a range of lichen species and particularly in the movement of the products of photosynthesis from the algal component across to the fungal symbiont. This work has largely been developed by Smith and his co-workers and has been summarised in a series of reviews by Smith (1974, 1975, 1980). Smith (1980) outlined the general features of photosynthate transport as follows:

1) Transport is substantial and a high proportion of all carbon fixed by the alga passes to the fungus.

2) Fixed carbon moves predominantly as a single type of molecule; glucose in lichens with blue-green algal (cyanobacterial) symbionts; ribitol, sorbitol or erythritol in lichens with green algal symbionts.

3) This rapid and massive carbohydrate efflux ceases quite rapidly as soon as the algae are isolated from the lichen and may be scarcely detectable 2 to 3 hours later.

The identification of the transported molecule was determined by isotope trapping where, for example, ^{14}C-glucose was accumulated in the medium when ^{12}C-glucose was supplied in excess to saturate the glucose uptake sites. Subsequent scans of paper chromatograms then allowed identification and quantification of the mobile compounds as well as the final alcohol-soluble form that appeared in the mycobiont. Thus, one-way paper chromatography has played a central and fundamental role in our current understanding of the carbon metabolism of lichens.

1

However, over the last decade, there has been a rapid development
of high performance liquid chromatography (HPLC) which allows
extremely accurate detection and measurement of a wide range of
compounds (Wight and van Niekerk, 1983; Conrad and Palmer, 1976). The
potential importance of HPLC in studies of lichen carbon metabolism
and particularly the mechanism of transfer of fixed carbon to the
mycobiont is explored in this paper.

METHODS

    Peltigera praetextata, P. aphthosa, Cladonia rangiferina, Usnea
fulvoreagans and Evernia mesomorpha were stored dry in bench-top
growth chambers until required. Storage conditions approximated the
day-length, light intensity and temperature levels occurring in the
field at the time of collection. Air-dry thalli (c. 10% of their oven
dry weight moisture content) were immersed in 100 ml. of double
distilled and deionised (Type 1) water at 25°C for 2, 10, 30 and 60
minutes. This ecologically rather excessive treatment was contrasted
in P. aphthosa with a simulated rainfall period of 30 minutes achieved
by gently misting a suspended thallus with a water aerosol at intervals and collecting the thallus run-off. Each experiment was
replicated three times. In addition, the total alcohol-soluble carbohydrate content of each species was assessed by extraction of
replicates in hot ethanol for 1 hour followed by overnight extraction

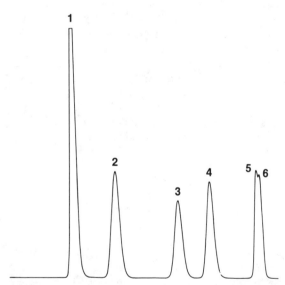

Fig. 1.   HPLC pattern of carbohydrate standards emphasising the
          difficulty of separation of mannitol and arabitol.
          1 = Trehalose; 2 = Glucose; 3 = Arabinose; 4 = Ribitol;
          5 = Mannitol; 6 = Arabitol.

in cold ethanol (MacFarlane and Kershaw, 1982). After evaporation of the supernatant to dryness, the residue was taken up in 1 ml of Type 1 water, filtered, using a Whatman GC filter paper, and finally centrifuged at 20,000 g. for 20 minutes prior to analysis.

Carbohydrate separation was achieved using a Biorad Aminex HPX-87C, 300 mm x 7.8 mm column, with water as the eluant at 0.5 ml min^{-1} and 85°C. Fifty microlitre samples were used throughout and peak measurement employed a refractive index detector coupled to a Varian 5010 HPLC, linked with a 4270 Varian microprocessor for data storage and peak integration. An example of the separation of 5 standards is given in Fig. 1, which emphasises the difficulty of clear separation of arabitol from mannitol using water as an eluant. Full separation has been subsequently achieved using 25% acetonitrile as the mobile phase.

Peak identification was by peak retention time matched against a range of standards (cf. Fig. 1), co-chromatography, colorimetric tests on sugar alcohols and hydrolysis of the trehalose fraction.

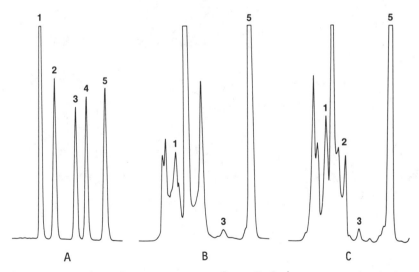

Fig. 2. The carbohydrate pattern from <u>Peltigera praetextata</u>.
A: 10 mmol. standards at attenuation 8. 1 = Trehalose; 2 = Glucose; 3 = Arabinose; 4 = Ribitol; 5 = Mannitol.
B: Alcohol-soluble extract from the control replicates.
C: The carbohydrates appearing in solution after 2 minutes thallus rehydration at 25°C, at attenuation 8.

RESULTS

    The alcohol-soluble carbohydrates in P. praetextata and the
leachates after 30 minutes immersion in Type 1 water are contrasted in
Fig. 2A & B. Four peaks have so far been identified, trehalose,
glucose, arabinose and mannitol. The initial unknown peak(s)
certainly comprise short-chain oligosaccharides which require further
analysis but, interestingly, the largest carbohydrate pool is an
additional unknown di- or tri-saccharide with a retention time close
to, but quite distinct from, glucose. Equally significant is the
close correspondence of the pattern of carbohydrates leached from the
thallus with the oligosaccharides, the unknown di-/tri-saccharide,
glucose, arabinose and mannitol all appearing in solution.

    In P. aphthosa there is a more complex pattern of carbohydrates
which not only reflect the dual thallus algal components and hence the
presence of ribitol as well as glucose but also a sequence of
oligosaccharides, a very large pool of trehalose and the unknown di-/
tri-saccharide (Fig. 3A). The total concentration of carbohydrates is
also large. During rehydration, either as an immersion treatment
(Fig. 3B) or as simulated rain (Fig. 3C), a very similar carbohydrate

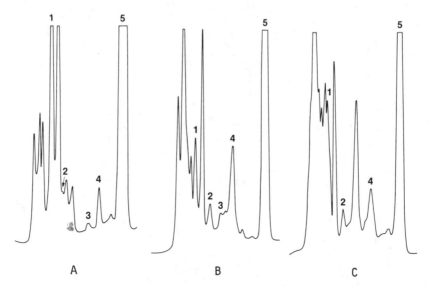

          A   B   C

Fig. 3.   The carbohydrate pattern from Peltigera aphthosa.
          A: Alcohol-soluble extract from the control replicates at
              attenuation 64.
          B: The carbohydrates appearing in solution after 2 minutes
              thallus rehydration at 25°C, at attenuation 16.
          C: The carbohydrates appearing in solution after 8 hours of
              simulated rainfall at 25°C, at attenuation 8.

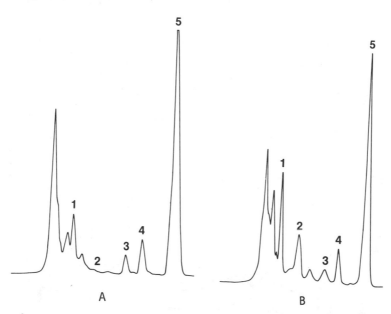

Fig. 4. The carbohydrate pattern from <u>Cladonia rangiferina</u>.
         A: Alcohol-soluble extract from the control replicates at
             attenuation 32.
         B: The carbohydrates appearing in solution after 2 minutes
             thallus rehydration at 25°C, at attenuation 8.

pattern appeared in the medium, although the amount of carbohydrate
differed between the two treatments (see below). The alcohol-soluble
extracts from <u>C. rangiferina</u> (Fig. 4A) again shows a basic pattern of
oligosaccharides, simple sugars and sugar-alcohols with ribitol
present as a characteristic of the green component of the lichen. The
comparable pattern obtained during a 2 minute rehydration period (Fig.
4B) again closely matched the pattern of alcohol-soluble carbohydrates.
There is an identical pattern of alcohol-soluble/rehydration carbohydrate losses in <u>E. mesomorpha</u> and <u>U. fulvoreagans</u>.

     A summary of the time course of carbohydrate release is given in
Fig. 5A & B. With the exception of <u>P. aphthosa</u> the absolute amount of
carbohydrate released over a period of 2 minutes to 1 hour remained
constant. In <u>P. aphthosa</u> there was an initially rapid and then more
gradual increase in carbohydrates in the medium throughout the
experimental period. The absolute maximum amounts released by the 5
species ranged from c. 17 mg carbohydrate per gram dry weight of
lichen in <u>P. praetextata</u>, c. 12 mg g^{-1} in <u>P. aphthosa</u> to c. 2 mg g^{-1}
in <u>U. fulvoreagans</u>, <u>C. rangiferina</u> and <u>E. mesomorpha</u>. When expressed
as a proportion of the total alcohol-soluble carbohydrate content of
each species, there remained a massive (c. 24%) efflux of carbohydrate
in <u>P. praetextata</u>, c. 20% in <u>Evernia</u> and <u>Usnea</u>, but significantly less

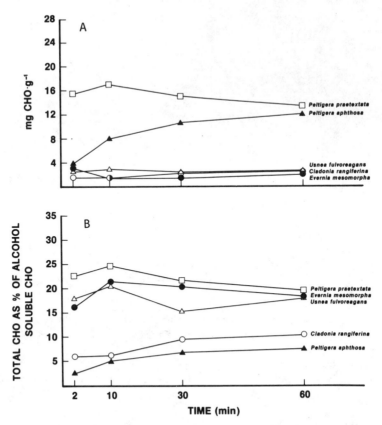

Fig. 5.   The time course of total carbohydrate loss during the first
          hour after rehydration in <u>Peltigera praetextata</u>, <u>P. aphthosa</u>,
          <u>Cladonia rangiferina</u>, <u>Usnea fulvoreagans</u> and <u>Evernia</u>
          <u>mesomorpha</u>.   The data are expressed as
          (A) absolute quantities, mg carbohydrate per gram dry weight;
          (B) percentage of the total alcohol soluble carbohydrates.

(up to 10%) in <u>C. rangiferina</u> and in <u>P. aphthosa</u> (c. 7.5%).

DISCUSSION

        It is very apparent from this data that HPLC provides an
extremely powerful research tool for examination of the carbohydrate
metabolism of lichens.   In <u>Peltigera</u> the standard model developed by
Drew and Smith (1967a,b), Hill and Smith (1972) and Richardson et al.
(1968) suggests that a single substance, glucose, is important in the
carbohydrate balance and transfer between alga and fungus.
Subsequently this is converted to mannitol within the fungal hyphae
and stored.   Hill (1972) suggested a second compound, a "glucan", was
involved which was then hydrolysed extracellularly to form glucose.

This glucose was then taken up by the fungus and converted to mannitol. All of this early work relied on one-way paper chromatography for the separation and identification of the carbohydrates involved. The ethanol-soluble, non-mobile peak at the origin of the paper chromatogram was additionally identified as "sugar-phosphates" presumably precursors of glucose (Drew and Smith, 1967a,b) or as polysaccharide and sucrose (Richardson et al., 1968). Similar evidence supported the importance of ribitol in lichens with green algal symbionts.

The HPLC data here, with its capability for the exact separation of a wide range of oligosaccharides, monosaccharides and sugar-alcohols, now throws considerable doubt on the simple carbohydrate model that we have previously accepted. The pattern of carbohydrates, which at this time is still only partially identified, is not only considerably more complex than has been suspected but also contains sizeable pools of carbohydrates such as trehalose, which have not previously been reported. The substantial pool of the undetermined di-/tri-saccharide potentially represents a further unrecorded but important intermediate, since preliminary studies using $^{14}CO_2$ indicate that it is indeed heavily labelled.

As a consequence of these results the potential environmental control of glucose transport in a range of Peltigera species and Collema furfuraceum (Tysiaczny and Kershaw, 1979; MacFarlane and Kershaw, 1982) should be re-examined. Initially, it was accepted that the primary peak that remained at the origin of the paper chromatograph was indeed sugar-phosphates as reported by Drew and Smith (1967a,b). Accordingly, the ratio of the mannitol peak size to the remaining peak areas would provide an accurate summary of the proportion of the $^{14}CO_2$ pulse-label appearing in each of the bionts. This assumption now appears to be incorrect and the potential control of metabolite movement between symbionts by the degree of thallus hydration requires a much more detailed examination, once each specific carbohydrate is identified as algal or fungal.

It is even more surprising to find that, in all five species examined, the ethanol-soluble fraction is almost perfectly paralleled in the carbohydrates detected in the supernatant two minutes after thallus rehydration. Specific pool sizes vary somewhat from species to species, particular carbohydrates specific to certain algal genera are present in some species and not in others, but generally the soluble carbohydrates rapidly appear in solution, presumably prior to full membrane reconfiguration in the first two minutes following thallus rehydration. The concept of carbohydrate transfer being as a single molecule, therefore, requires careful re-examination in a range of species. The complexity of the carbohydrate pattern also demands the establishment of which compounds are algal, which fungal and their specific identity. Finally the sequence of $^{14}CO_2$ labelling, and hence the carbon pathway, requires defining. We are endeavouring to proceed along these lines.

ACKNOWLEDGMENTS

    We wish to thank Dr. D.S. Coxson for shipments of Peltigera and
Usnea to us at appropriate intervals and The National Science and
Engineering Research Council of Canada for research support during
this work.

REFERENCES

Conrad, E.C., and Palmer, J.K., 1976, Rapid analysis of carbohydrates
    by high-pressure liquid chromatography, Food Technology, 10: 84-92.
Drew, E.A., and Smith, D.C., 1967a, Studies in the physiology of
    lichens. VII. The physiology of the Nostoc symbiont of
    Peltigera polydactyla compared with cultured and free-living
    forms, New Phytologist, 66: 379-388.
Drew, E.A., and Smith, D.C., 1967b, Studies in the physiology of
    lichens. VIII. Movement of glucose from alga to fungus during
    photosynthesis in the thallus of Peltigera polydactyla, New
    Phytologist, 66: 389-400.
Hill, D., 1972, The movement of carbohydrate from the alga to the
    fungus in the lichen Peltigera polydactyla, New Phytologist, 71:
    31-39.
Hill, D.J., and Smith, D.C., 1972, Lichen physiology. XII. The
    inhibition technique, New Phytologist, 71: 15-30.
MacFarlane, J.D., and Kershaw, K.A., 1982, Physiological-environmen-
    tal interactions in lichens. XIV. The environmental control of
    glucose movement from alga to fungus in Peltigera polydactyla,
    P. rufescens and Collema furfuraceum, New Phytologist, 91: 93-101.
Richardson, D.H.S., Hill, D.J., and Smith, D.C., 1968, Lichen
    physiology. XI. The role of the alga in determining the
    patterns of carbohydrate movement between lichen symbionts, New
    Phytologist, 67: 469-486.
Smith, D.C., 1974, Transport from symbiotic algae and symbiotic
    chloroplasts to host cells, Symposium of the Society for
    Experimental Biology, 28: 437-508.
Smith, D.C., 1975, Symbiosis and the biology of lichenised fungi.
    Symposium of the Society for Experimental Biology, 29: 373-405.
Smith, D.C., 1980, Mechanisms of nutrient movement between the
    lichen symbionts, in: "Cellular Interactions in Symbiosis and
    Parasitism," C.B. Cook, P.W. Pappas, and E.D. Rudolph, eds, pp.
    197-227, Ohio State University Press.
Tysiaczny, M.J., and Kershaw, K.A., 1979, Physiological-environmental
    interactions in lichens. VII. The environmental control of
    glucose movement from alga to fungus in Peltigera canina v.
    praetextata Hue, New Phytologist, 83: 137-146.
Wight, A.W., and van Niekerk, P.J., 1983, Determination of reducing
    sugars, sucrose and inulin in chicory root by high-performance
    liquid chromatography, Journal of Agricultural and Food
    Chemistry, 31: 282-285.

PHOTOSYNTHETIC PARAMETERS IN <u>RAMALINA DURIAEI</u>, <u>IN VIVO</u>, STUDIED BY

PHOTOACOUSTICS

R. Ronen[a], O. Canaani[b], J. Garty[a], D. Cahen[c], S. Malkin[b]
and M. Galun[a]

[a] Department of Botany, Tel Aviv University, Tel Aviv,
    Israel
[b] Department of Biochemistry and [c] Department of Structural
    Chemistry, Weizmann Institute of Science, Rehovoth,
    Israel

INTRODUCTION

Sulfur dioxide is a well-documented atmospheric pollutant and
many studies have dealt with the effect of SO_2 on lichens. Laboratory
studies have demonstrated the effect of SO_2 on respiration, photosynthesis and chlorophyll in lichens (Pearson and Skye, 1965; Rao and
Le Blanc, 1966; Hill, 1971, 1974; Showman, 1972; Baddeley et al.,
1973; Richardson and Puckett, 1973; Türk et al., 1974; Hällgren and
Huss, 1975; Eversman, 1978; Beekley and Hoffman, 1981; Malhotra and
Khan, 1983). Effects of SO_2 on these physiological parameters have
also been observed in lichens fumigated in nature (Eversman, 1978;
Moser et al., 1980) and transplanted to polluted areas (Brodo, 1961;
LeBlanc and Rao, 1966; Schönbeck, 1969; Kauppi, 1976; Ferry and
Coppins, 1979). These parameters have been measured in lichens by
different methods such as: Infra Red Gas Analysis (IRGA) (Lange, 1965;
Türk et al., 1974; Larson and Kershaw, 1975); ^{14}C incorporation (Hill,
1971; Puckett et al., 1973; Hällgren and Huss, 1975; Malhotra and
Khan, 1983); oxygen electrode (Baddeley et al., 1971); differential
respirometry (Showman, 1972) and microfluorometry (Kauppi, 1980).

In this article we introduce a novel technique, photoacoustic
spectroscopy (PAS), for determining the photosynthetic activity of the
lichen <u>Ramalina duriaei</u> and assessing its applicability in measuring
SO_2 damage to lichens.

The principles of the PAS technique and the adaptation of this

technique for photosynthetic measurements have been described by Cahen
et al. (1980) and Malkin and Cahen (1979). Photoacoustics has been
used for studying the photosynthetic activity of isolated, broken
chloroplasts (Lasser-Ross et al., 1980), the development of the photosynthetic apparatus in greening leaves (Canaani et al., 1982), the
photosynthetic activities in whole leaves of higher plants (Bults et
al., 1982; Poulet et al., 1983) and in cyanobacteria (Carpentier et
al., 1984). O'Hara et al. (1983) also applied it to the determination

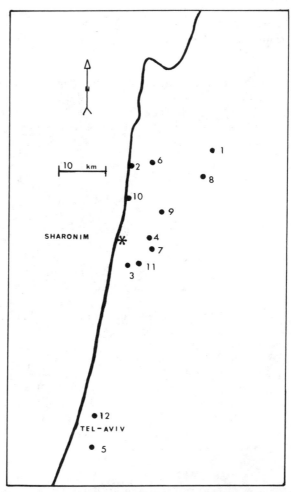

Fig. 1.   Map of the twelve monitoring stations - 1. HaZorea;
          2. Nahsholim; 3. Hedera; 4. Pardes Hana; 5. Tel Aviv
          University (old campus); 6. Ofer; 7. Shemurat Allon; 8. En
          haShofet; 9. Givat Ada; 10. Ma'ayan Mikhael; 11. Bet Eliezer;
          12. Kefar haYarok Junction.

of the _in vivo_ absorption and photosynthetic properties of the lichen _Acarospora schleicheri_.

In this study photosynthetic parameters of R. duriaei in natural conditions, after exposure to air pollution and after treatment with NaHSO3 in the laboratory, were examined by PAS.

MATERIALS AND METHODS

The study was carried out with the fruticose lichen _Ramalina duriaei_ (De Not.) Jatta growing on carob trees (_Ceratonia siliqua_ L.) near HaZorea (Esdraelon Valley, N.E. Israel). This area is known to be clean in terms of air pollution (Garty and Fuchs, 1982; Fuchs and Garty, 1983; Ronen et al., 1983; Galun et al., 1984). The lichen samples were collected with the twigs on which they grew and were used for measurements within three days. Prior to use the lichens were illuminated overnight in the laboratory in a humid atmosphere. PAS measurements were made on cut discs (10 mm diam.) of the lichen.

Twigs covered with R. duriaei were transplanted from HaZorea to eleven monitoring stations (Fig. 1). They were exposed for one year (December 1981–December 1982) and then taken to the laboratory for immediate measurements. Ten samples were examined from each of the eleven stations. Samples from HaZorea, but from transplanted twigs, served as control material.

Determination of Chlorophyll Degradation

The R. duriaei thallus discs were first checked for photosynthetic activity by PAS and the same discs were then extracted in dimethyl sulphoxide (DMSO) and the degree of chlorophyll degradation estimated by using the parameter OD435/OD415, as described by Ronen and Galun (1984).

Bisulphite Treatment

Discs of R. duriaei were immersed in a freshly prepared buffer containing 40 mM MES (2-(N Morpholino) ethane sulphonic acid), 5 mM NaHCO3, pH 5.7, with added 5 mM NaHSO3. After treatment the discs were washed for one min. in buffer, blotted and used for photoacoustic measurements. For each disc the inhibiting effect of NaHSO3 was calculated relative to control measurements.

Photoacoustic Instrument

The instrument is depicted in Fig. 2. Light from a 450 W Xenon-lamp passed through a Bausch and Lomb monochromator and was modulated by a mechanical chopper. The light was directed onto the sample with a randomised bifurcated light guide allowing simultaneous irradiation with modulated and direct current (d.c.) saturating light; the latter

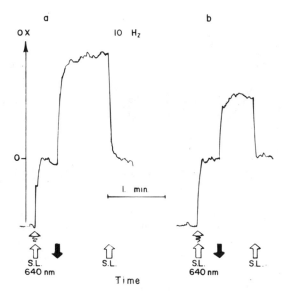

Fig. 4.   Oxygen evolution quantum yield obtained at 10 Hz. Lights as
in Fig. 3.   When S.L. is added to the modulated light it
gives the zero level of O_2 evolution.
a) Control = untreated lichen; b) Lichen treated with $NaHSO_3$
for one hour.

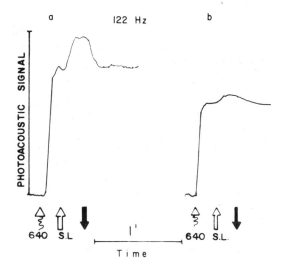

Fig. 5.   Photoacoustic signal obtained at 122 Hz. Lights as in Fig. 3.
The addition of the S.L. results in a maximum photothermal
signal. The drop of the signal when switching off S.L., as
related to the total signal, is the percentage of photochemical energy storage.
a) Control = untreated lichen; b) Lichen treated with $NaHSO_3$
for one hour.

The fluorescence signal passed through a light guide and a series of blocking filters to a photodetector, the signal from which was amplified and recorded.

RESULTS AND DISCUSSION

The frequency dependence of the oxygen evolution quantum yield was measured to determine the conditions necessary to obtain the maximal oxygen evolution signal. For R. duriaei it was found (Fig. 3) that a maximal signal was obtained around 10 Hz beyond which the signal decreased until at about 100 Hz no oxygen evolution could be detected. This is in contrast to leaves where the signal is damped at a much slower rate. From this observed rate of damping it can be calculated that the average diffusion path of oxygen in vivo in the algal cells is four times greater than in an intact leaf (Canaani et al., 1984).

Field Studies

Table 1 shows the values obtained with the thalli of R. duriaei that were transplanted to the monitoring stations for one year. The highest values for the oxygen evolution quantum yield and the OD 435/ OD415 parameter, measuring chlorophyll degradation (Ronen and Galun, 1984), were obtained at the control station (HaZorea) and the lowest

Table 1: The OD 435/OD 415 and Aox/Apt values of R. duriaei Samples Suspended at the Monitoring Stations (Dec. 1981 – Dec. 1982)

Location	$\bar{x} \pm$ s.d. (10 samples)	Aox/Apt $\bar{x} \pm$ s.d. (the same 10 samples)
HaZorea	1.43 $\pm$ 0.05	1.71 $\pm$ 0.68
Nahsholim	1.40 $\pm$ 0.03	1.39 $\pm$ 0.92
Hadera	1.34 $\pm$ 0.11	1.23 $\pm$ 0.64
Pardes Hana	1.32 $\pm$ 0.10	1.43 $\pm$ 0.60
Tel Aviv University (old campus)	1.32 $\pm$ 0.20	1.23 $\pm$ 0.61
Ofer	1.28 $\pm$ 0.20	1.47 $\pm$ 0.71
Shemurat Allon	1.24 $\pm$ 0.15	0.74 $\pm$ 0.51
En haShofet	1.23 $\pm$ 0.18	0.98 $\pm$ 0.54
Givat Ada	1.19 $\pm$ 0.18	0.90 $\pm$ 0.70
Ma'agan Mikhael	1.12 $\pm$ 0.25	0.65 $\pm$ 0.68
Bet Eliezer	1.07 $\pm$ 0.31	0.75 $\pm$ 0.69
Kefar haYarok Junction	0.92 $\pm$ 0.24	0.35 $\pm$ 0.58

$r = 0.895$ $\qquad$ $p < 0.01$

from samples transplanted to the urban road intersection (Kefar
haYarok Junction). There was a high degree of correlation between the
observed oxygen evolution quantum yield and chlorophyll degradation
($r = 0.895$, $p < 0.01$). The values of oxygen quantum yield decreased
from 1.71 (control samples) to 0.35 (samples transferred to Kefar
HaYarok Junction) and the parameter for chlorophyll degradation from
1.43 to 0.92 respectively. The parameter value for total degradation
of chlorophyll a to phaeophytin a in vitro was 0.56 (Ronen and Galun,
1984). Table 1 shows that, when the value of this parameter is
smaller than 1.28, there is a drop in the photosynthetic activity of
the lichen. According to the calibration curve obtained with mixtures
of chlorophyll and phaeophytin in different proportions, this value of
1.28 indicates the degradation of 10% of the chlorophyll.

Our results could indicate that the lichens are able to continue
their photosynthetic activities without severe damage up to a degradation of 10% of the chlorophyll. In the urban station (Kefar haYarok
Junction), 50% of the chlorophyll was degraded and there was almost no
photosynthetic activity. Together with the drop in the value of the
photosynthetic activities at the different monitoring stations, there
is an increase in the value of the standard deviation, showing that
the population of the lichens at the different monitoring stations
become more heterogenous due to the effect of pollution. The two
parameters, the oxygen evolution quantum yield and the parameter for

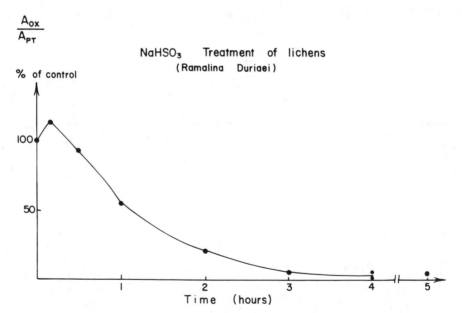

Fig. 6.   Oxygen evolution quantum field (Aox/Apt) as function of
           incubation time in 5 mM $NaHSO_3$ in the presence of light (8
           W/m^{-2}, 400-700 nm). Lights for PAS measured as in Fig. 3.

chlorophyll degradation (OD435/OD415), are both sensitive and accurate and together give a very rapid indication of the state of the lichen photosynthetic activities.

Laboratory Studies

The effect of bisulphite treatment on R. duriaei under laboratory conditions was also studied by PAS. The kinetics of photosynthesis inhibition by bisulphite (Fig. 6) show that inhibition is not immediate; there is initially some stimulation above the control level. This stimulation was also found in other research works (Puckett et al., 1973; Hällgren and Huss, 1975; Beekley and Hoffman, 1981). The quantum yield of oxygen evolution was reduced by about 40% after a one hour treatment and the quantum yield of energy storage decreased by about 75%. After a treatment of 4 hours, oxygen evolution and energy storage were totally inhibited. The oxygen evolution quantum yield, reflecting mostly rate-limiting reactions in the non-cyclic electron flow, was damaged to a lesser extent than the quantum yield for the energy storage. It seems, therefore, that other reactions, besides the linear electron transport chain, which store photochemical energy were strongly affected by the bisulphite treatment.

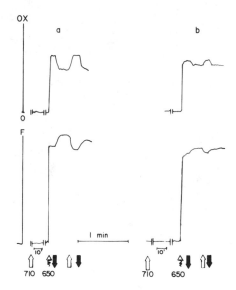

Fig. 7. Emerson enhancement as expressed by changes in O_2 evolution (upper trace) and fluorescence (lower trace) by addition of 710 nm (18 W/m^2) background to 650 nm (10.8 W/m^2) modulated light. Lichen was preilluminated for 10' by 710 nm light prior to measurement.
a) Control = untreated lichen; b) Lichen treated with 5 mM NaHSO$_3$ for one hour.

Emerson enhancement and chlorophyll fluorescence were also observed in R. duriaei after treatment with bisulphite (5 mM) for one hour. Comparison with control measurements (Fig. 7) shows that there is a drop in the value of the Emerson enhancement from 1.20 to 1.03, as well as a drop in fluorescence.

With these different PAS parameters, it is very convenient to calculate the spectrum of absorbance. The intensity (I) of the modulated light is measured between 400 nm and 750 nm at intervals of 10 nm; the photothermal maximal signal (Apt) (proportional to the energy absorbed) is measured at the same wavelengths. The ratio (Apt x λ)/I then yields a number proportional to the percentage of the emitted quanta that have been absorbed, that is proportional to the absorbance. The spectrum obtained is presented in Fig. 8, along with the diffuse reflectance spectrum of the same lichen measured by means of a Cary 15 spectrophotometer equipped with two integrating spheres. The two spectra were normalized at 550 nm and appear very similar, even though absorbance and reflectance do not measure the same parameter. However, a part of the light may have been reflected at the surface of the lichen or scattered when passing through the hyphae

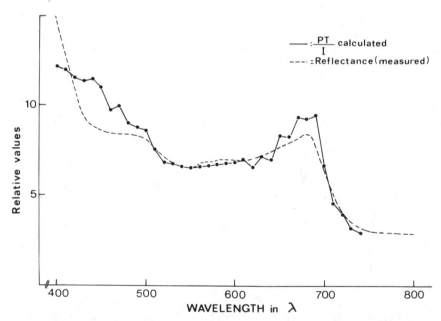

Fig. 8.   Absorbance spectrum (Apt/I) between 400-750 nm (●—●) compared with diffuse reflectance (----) measured in a Cary 15 spectrophotometer with the use of two integrating spheres. The two curves were normalized at 550 nm. Apt = photothermal signal; I = intensity of the light.

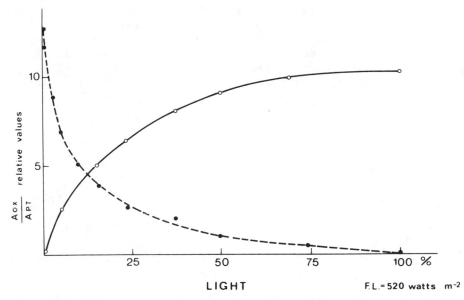

Fig. 9.   Saturation curve of the rate of oxygen evolution. Quantum
           yield of oxygen evolution is first measured with a constant
           modulated light (640 nm, 10.8 W/m^2) but at different
           intensities of background light. The curve obtained ($\bullet$---$\bullet$)
           is integrated and yields the rate saturation curve of oxygen
           evolution ($\circ$——$\circ$).

of the thallus. The difference between the spectra in the red region
is certainly due to this phenomenon.

     Another datum that can easily be obtained is a saturation curve
of the rate of oxygen evolution against light intensity (Fig. 9).
This curve is obtained by measuring the quantum yield of oxygen
evolution with a constant modulated light, but at different
intensities of background light. By plotting the results of the
quantum yield as a function of the background light intensities a
curve is obtained which, when integrated, yields the rate saturation
curve (Poulet et al., 1983).

     The advantages of the photoacoustic techniques are obvious.
First, it permits the checking of gross photosynthesis of the lichen.
Respiration does not interfere with the acoustic signals because it is
a much slower process (Bults et al., 1982). Secondly, measurements
are accurate, repetitive and rapid (a sample can be checked within a
few minutes). Thirdly, we simultaneously obtain information on the
state of the photosystems (oxygen evolution and energy storage quantum
yields), the state of the pigments in vivo (absorption and quantum

yield spectra) and electron flow (rate saturation curves). Comparison
of these parameters in control lichens with lichens that have been
exposed to air pollution or treated with pollutants in the laboratory,
enable qualitative and quantitative evaluation of the damage by
pollutants. Comparisons can also be made between sensitive and
tolerant lichens for understanding resistance to air pollution.

REFERENCES

Baddeley, M. S., Ferry, B. W., and Finegan, E. J., 1971, A new method
    of measuring lichen respiration : response of selected species to
    temperature, pH and sulfur dioxide, The Lichenologist, 5 :
    18-25.
Baddeley, M. S., Ferry, B. W., and Finegan, E. J., 1973, Sulphur
    dioxide and respiration in lichens, in : "Air Pollution and
    Lichens", B. W. Ferry, M. S. Baddeley, and D. L. Hawksworth, eds,
    pp. 299-313, Athlone Press, London.
Beekley, P. K., and Hoffman, G. R., 1981, Effects of sulphur dioxide
    fumigation on photosynthesis, respiration and chlorophyll content
    of selected lichens, The Bryologist, 84 : 379-390.
Brodo, I. M., 1961, Transplant experiments with corticolous lichens
    using a new technique, Ecology, 42 : 838-841.
Bults, G., Horwitz, B. A., Malkin, S., and Cahen, D., 1982,
    Photoacoustic measurements of photosynthetic activities in whole
    leaves. Photochemistry and gas exchange, Biochimica et
    Biophysica Acta, 679 : 452-465.
Cahen, D., Bults, G., Garty, H., and Malkin, S., 1980, Photoacoustics
    in life sciences, Journal of Biochemical and Biophysical Methods,
    3 : 293-310.
Canaani, O., Cahen, D., and Malkin, S., 1982, Use of photoacoustic
    methods in probing development of the photosynthetic apparatus in
    greening leaves. in : "Cell function and Differentiation. Part
    B," G. Akoyunoglou, ed., pp. 299-308, Alan R. Liss Inc., New
    York.
Canaani, O., Ronen, R., Garty, J., Cahen, D., Malkin, S., and Galun,
    M., 1984, Photoacoustic study of the green alga Trebouxia in the
    lichen Ramalina duriaei in vivo, Photosynthesis Research, 5: 297-
    306.
Carpentier, R., Larue, B., and LeBlanc, R. M., 1984, Photoacoustic
    spectroscopy of Anacystis nidulans. III. Detection of
    photosynthetic activities, Archives of Biochemistry and
    Biophysics, 228 : 534-543.
Eversman, S., 1978, Effects of low level SO_2 on Usnea hirta and
    Parmelia chlorochroa. The Bryologist, 81 : 368-377.
Ferry, B. W., and Coppins, B. J., 1979, Lichen transplant experiments
    and air pollution studies, The Lichenologist, 11 : 63-74.
Fuchs, C., and Garty, J., 1983, The content of some elements in the
    lichen Ramalia duriaei (De Not.) Jatta in air quality
    biomonitoring stations, Environment and Experimental Botany, 23 :
    29-43.

Galun, M., Garty, J., and Ronen, R., 1984, Lichens as bioindicators of air pollution, Webbia, 38 : 371-383.

Garty, J., and Fuchs, C., 1982, Heavy metals in the lichen Ramalina duriaei transplanted to biomonitoring stations. Water, Air and Soil Pollution, 17 : 175-183.

Hällgren, J. E., and Huss, K., 1975, Effects of SO_2 on photosynthesis and nitrogen fixation, Physiologia Plantarum, 34 : 171-176.

Hill, D. J., 1971, Experimental study of the effect of sulphite on lichens with reference to atmospheric pollution, New Phytologist, 70 : 831-836.

Hill, D. J., 1974, Some effects of sulphite on photosynthesis in lichens, New Phytologist, 73 : 1193-1205.

Kauppi, M., 1976, Fruticose lichen transplant technique for air pollution experiments, Flora, 165 : 407-414.

Kauppi, M., 1980, Fluorescence microscopy and microfluorometry for the examination of pollution damage in lichens, Annales Botanici Fennici, 17 : 163-173.

Lange, O. L., 1965, Der CO_2 Gaswechsel von Flechten nach Erwarmung im feuchten Zustand, Berichte der Deutschen Botanischen Gesellschaft, 78 : 441-454.

Larson, D. W., and Kershaw, K. A., 1975, Measurement of CO_2 exchange in lichens: a new method, Canadian Journal of Botany, 53 : 1535-1541.

Lasser-Ross, N., Malkin, S., and Cahen, D., 1980, Photoacoustic detection of photosynthetic activities in isolated broken chloroplasts, Biochimica et Biophysica Acta, 593 : 330-341.

LeBlanc, F., and Rao, D. N., 1966, Reaction de quelques lichens et mousses epiphytiques a l'anhydride sulfureux dans la region de Sudbury, Ontario, The Bryologist, 69 : 338-346.

Malhotra, S. S., and Khan, A. A., 1983, Sensitivity to SO_2 of various metabolic processes in an epiphytic lichen Evernia mesomorpha, Biochemie und Physiologie der Pflanzen, 178 : 121-130.

Malkin, S., and Cahen, D., 1979, Photoacoustic spectroscopy and radiant energy conversion : theory of the effect with special emphasis on photosynthesis, Photochemistry and Photobiology, 29 : 803-813.

Moser, T. J., Nash III, T. H., and Clark, W. D., 1980, Effects of a long-term field sulfur dioxide fumigation on Arctic caribou forage lichens, Canadian Journal of Botany, 58 : 2235-2240.

O'Hara, E. P., Tom, R. D., and Moore, T. A., 1983, Determination of the in vivo absorption and photosynthetic properties of the lichen Acarospora schleicheri using photoacoustic spectroscopy, Photochemistry and Photobiology, 38 : 709-715.

Pearson, L., and Skye, E., 1965, Air pollution affects pattern of photosynthesis in Parmelia sulcata, a corticolous lichen, Science, 148 : 1600-1602.

Poulet, P., Cahen, D., and Malkin, S., 1983, Photoacoustic detection of photosynthetic oxygen evolution from leaves: quantitative analysis by phase and amplitude measurements, Biochimica et Biophysica Acta, 724 : 433-446.

Puckett, K. J., Nieboer, E., Flora, W. P., and Richardson, D. H. S., 1973, Sulphur dioxide : its effect on photosynthetic ^{14}C fixation in lichens and suggested mechanisms of phototoxicity, New Phytologist, 72 : 141-154.

Rao, D. N., and LeBlanc, F., 1966, Effects of sulfur dioxide on the lichen alga, with special reference to chlorophyll, The Bryologist, 69 : 69-75.

Richardson, D. H. S., and Puckett, K. J., 1973, Sulphur dioxide and photosynthesis in lichens, in : "Air Pollution and Lichens," B. W. Ferry, M. S. Baddeley, and D. L. Hawksworth, eds, pp. 283-298, Athlone Press, London.

Ronen, R., Garty, J., and Galun, M., 1983, Air pollution monitored by lichens, in : "Developments in Ecology and Environmental Quality," Proc. Int. Meeting – The Israel Ecological Society., Balaban Int. Science Ser. Rehovoth/Philadelphia, 2 : 167-176.

Ronen, R., and Galun, M., 1984, Pigment extraction from lichens with dimethyl-sulfoxide (DMSO) and estimation of chlorophyll degradation, Environmental and Experimental Botany, 24 : 239-245.

Schönbeck, H., 1969, A method for determining the biological effects of air pollution by transplanted lichens, Staub-Reinhalt-Luft, 29 : 17-21.

Showman, R. E., 1972, Residual effects of sulphur dioxide on the net photosynthetic and respiratory rates of lichen thalli and cultured lichen symbionts, The Bryologist, 75 : 335-341.

Türk, R., Wirth, V., and Lange, O. L., 1974, CO_2-Gaswechsel Untersuchungen zur SO_2- Resistenz von Flechten, Oecologia, 15 : 33-64.

METHOD FOR FIELD MEASUREMENTS OF CO_2-EXCHANGE. THE DIURNAL CHANGES IN NET PHOTOSYNTHESIS AND PHOTOSYNTHETIC CAPACITY OF LICHENS UNDER MEDITERRANEAN CLIMATIC CONDITIONS

O.L. Lange[a], J.D. Tenhunen[b], P. Harley[c] and H. Walz[d]

[a] Lehrstuhl für Botanik II der Universität Würzburg, F.R.G.
[b] Systems Ecology Research Group, San Diego State University, San Diego, California, U.S.A.
[c] Department of Biology, University of Utah, Salt Lake City, Utah, U.S.A.
[d] Effeltrich, F.R.G.

INTRODUCTION

The determination and interpretation of patterns in photosynthetic primary production which are characteristic of different lichen species within their respective habitats may lead to a better understanding of habitat preferences and species distributions. However, the necessary photosynthetic CO_2 exchange measurements are difficult to accomplish in the field. Thus, only a few examples of diurnal courses of photosynthetic CO_2 uptake by poikilohydric organisms may be found in the ecophysiological literature. Recently mathematical models have been developed which have been used to simulate the net photosynthetic performance of lichens under natural conditions. Because such models are usually based upon physiological parameters obtained from laboratory experiments, their validity can only be established by comparing simulated rates of photosynthesis with rates actually measured in the field. Thus, the development of instrumentation and methods which allow accurate determination of CO_2 exchange of lichens under natural conditions is essential.

Two portable gas exchange measurement systems are described here which are used to determine photosynthesis and respiration rates of lichens under natural conditions by means of infrared gas analysis. Their function is illustrated with a series of examples drawn from experiments conducted with lichens under Mediterranean climate conditions in Portugal. (The instrumentation described is available from H. Walz Company, Eichenring 10-14, D-8521 Effeltrich, F.R.G.).

23

MEASUREMENT OF NATURAL CO_2 EXCHANGE OF LICHENS

To determine photosynthetic CO_2 exchange, lichen samples must be enclosed in a ventilated cuvette. If the period of enclosure is prolonged, the natural situation of the thalli is altered. Measurement of the natural response may no longer be possible due to changes in temperature, humidity and convection. One approach, which partly avoids alteration of the natural microhabitat conditions, has been to insert lichen samples for measurement into a climatically adjusted cuvette (a kind of small phytotron). Cuvette air temperature and humidity are controlled to correspond to ambient conditions in the surrounding natural environment (Lange et al., 1969).

During our work in the Negev Desert, we used several chambers of this type to investigate, for instance, the importance of moistening lichens by dew and atmospheric humidity (Lange et al., 1970) and to determine the influence of different exposure conditions on lichen photosynthetic CO_2 uptake (Kappen et al., 1980). In these investigations, the thalli were allowed to experience natural environmental conditions prior to measurement. They were moistened by dew in the early morning and samples were subsequently transferred to the cuvettes, where their CO_2 exchange was monitored. A period of about 12 to 20 minutes was required to obtain steady-state CO_2 exchange with the large gas volume of the pre-conditioned chambers. During the period of enclosure, the thalli were no longer moistened by rain or dew, and rates of drying were possibly changed due to altered convection. Since the rates of drying in the natural environment were also high, the high convection within the cuvettes may not have had a substantial effect. In other habitats, however, it is doubtful that the hydration conditions which are occurring naturally can be maintained during experiments of this type.

Recent technological developments, including production of new instrumentation for infrared CO_2 gas analysis, have permitted the construction of apparatus better suited to performing field determinations of net photosynthesis in lichens. We have built a "CO_2-porometer", which allows accurate gas-exchange rate determinations after samples have been enclosed for approximately 1 to 2 minutes in a small cuvette. In the periods between measurements, the material remains in situ and experiences its natural environment.

Description of the CO_2-Porometer

Detailed descriptions of the CO_2-porometer were presented by Schulze et al. (1982) and Lange (1984). It was primarily designed for photosynthesis and transpiration measurements of higher plant leaves. Therefore, the name "porometer", as introduced by Darwin and Pertz (1911) for the first instrument which allowed estimation of stomatal pore size by means of air flow measurement, was retained. For CO_2 exchange measurements with lichens (Lange et al., 1984b) the lid of

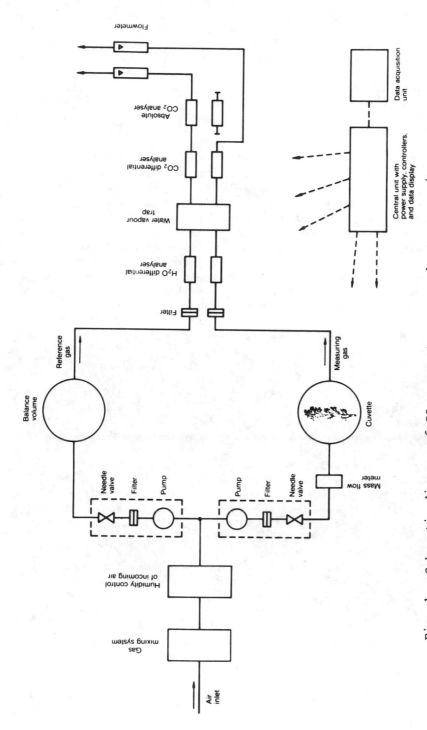

Fig. 1. Schematic diagram of CO₂ porometer gas exchange-measuring system. For details see text.

the porometer cuvette is opened to insert the sample. Thereafter, two foam cell rings between cuvette and lid ensure an airtight seal. The chamber is of small volume (c. 100 ml) and is covered by a thin polythene film. A fan rapidly circulates air within the cuvette. The cuvette is surrounded by a cylindrical radiation shield with a second fan for circulating ambient air through the cylinder along the outer walls of the cuvette. Thus, internal air temperature is maintained at almost ambient temperature even under conditions of high solar irradiance.

Figure 1 shows a schematic diagram of the complete porometer gas exchange measurement system. Air is taken from outside, passes through a buffer vessel, and is split into two streams. The measurement air stream is pumped through the porometer cuvette while the second stream is pumped directly to the gas analyzer as a reference. Subsequently, water vapour and CO_2 differentials between measuring and reference air streams are determined in a BINOS infrared gas analyser (Leybold Heraeus, Hanau, F.R.G.). Flow rates are measured with a mass flow meter. There is a unit for data display where environmental and meteorological data pertinent to the sample under study, such as air

Fig. 2   Porometer cuvette with enclosed lichen sample (<u>Parmelia</u> <u>reticulata</u>) near its natural habitat on the trunk of a pear tree (Quinta São Pedro, Portugal). A tray with another experimental sample is fixed to the trunk above the cuvette (arrow).

temperature and air humidity in the cuvette as well as incident photosynthetically active photon flux density, are monitored. The measurement system is portable and is powered in the field by a small
generator (or a car battery). The response time of the porometer is
extremely short. Steady-state conditions, after enclosure of lichen
material, are reached within 60 to 80 seconds. After this time a
reliable CO_2 exchange reading is possible.

Small thallus samples (c. 0.15 to 0.4 g dry weight) are cleaned
and inserted into fine wire mesh baskets (c. 40 mm diameter). These
samples are then returned to their original location, where the thalli
once again experience natural habitat conditions. The porometer
cuvette may be held by hand or otherwise supported close to the
thallus samples. Figure 2 shows such an arrangement for measurements
with lichen species growing on the branch of a tree. At regular
intervals, a basket with its lichen sample is transferred into the
porometer cuvette and its CO_2 exchange is determined. After the short
period of enclosure, the sample is returned to its natural position.
We feel that the description of lichen gas exchange behaviour obtained
by these procedures is representative of the undisturbed situation and
thus reliable information about photosynthetic primary production by
lichen species in their natural habitat is obtained.

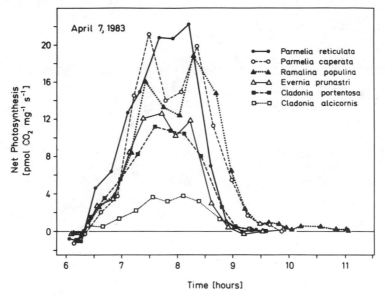

Fig. 3. Diurnal course of CO_2 exchange of different lichen species
         after fog and dew imbibition, examined under the same
         environmental conditions (Quinta São Pedro, Portugal).
         Abscissa: time of day; ordinate; net photosynthesis
         (positive) and respiration (negative).

Diurnal Courses of CO_2 Exchange

The following results demonstrate the different types of information which may be obtained using the methods described. A series of experiments were conducted to examine the CO_2 exchange behaviour of several lichen species growing in a Mediterranean type climate in Portugal (Quinta São Pedro Research Station, Sobreda, south of Lisbon). In Fig. 3 absolute rates of net photosynthesis of different species are compared under the same environmental conditions and exposed to the same light regime during measurement. On this day in spring (March 1983) clear weather was experienced and maximal air temperatures were about 20°C. The night before, however, was rather cool. This led to heavy fog and dew formation in the early morning and activation of metabolism in the moistened lichen thalli. Shortly after sunrise, all samples exhibited CO_2 uptake, which continued until they later dried out. Substantial species-specific variation in the dry weight-related rates of net photosynthesis occurred.

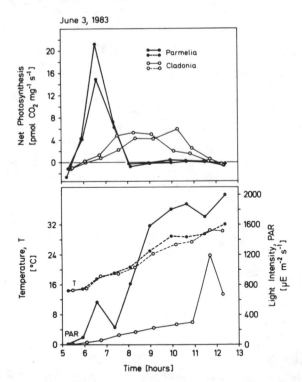

Fig. 4. Diurnal courses of CO_2 exchange of two samples of __Parmelia__ __reticulata__ and __Cladonia portentosa__ after heavy fog imbibition in their natural habitat (see Fig. 3); air temperature (T) and photosynthetically active radiation (PAR) incident on the lichens.

     In a second experiment, photosynthetic primary production of an
epiphytic lichen species growing on a free-standing pear tree was
compared with that of a terricolous species growing at the edge of an
evergreen scrub below a stand of <u>Lavandula</u> and <u>Cistus</u> shrubs.   These
two lichen habitats differ from each other with respect to light
intensity, wind speed and relative air humidity.   Figure 4 illustrates
the time courses of photosynthetic CO_2 uptake of both the epiphytic
<u>Parmelia reticulata</u> and the terricolous <u>Cladonia portentosa</u> after
moistening by heavy fog in the early morning.   After sunrise the fog
cleared.   Photosynthetically active radiation (PAR) at the <u>P.</u>
<u>reticulata</u> site increased rapidly and reached maximal values by noon.
In contrast, light intensity remained low until 1100 h in the <u>C.</u>
<u>portentosa</u> habitat.   <u>P. reticulata</u> exhibited a short period of rapid
CO_2 uptake between 0530 and 0800 h.   This was due to the relatively
high light intensity incident on the well-hydrated thalli positioned
to intercept the morning sun.   However, high irradiance as well as
exposure to wind led to rapid drying and photosynthetic rates
decreased to zero after only 3 hours.   Due to low light intensity and
a lower photosynthetic capacity (Fig. 3), much lower rates of CO_2
uptake occurred in <u>C. portentosa</u>.   The light compensation point was
only reached after 0600 h.   Thereafter, photosynthesis increased
slowly.   However, this lichen remained moist much longer, so that
maximal rates of photosynthesis occurred several hours later at a time

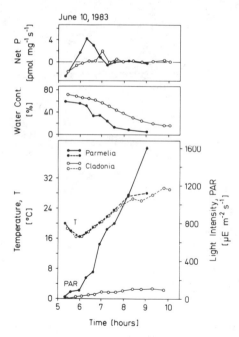

Fig. 5.   Diurnal courses of CO_2 exchange of <u>Parmelia reticulata</u> and
          <u>Cladonia portentosa</u> after slight dew imbibition; for air
          temperature and photosynthetically active radiation see Figs
          3 and 4.   Water content related to thallus dry weight.

when P. reticulata was already dry. Interestingly, total daily carbon
gain did not differ much between species.

The maximal rates of photosynthesis as well as the duration of
the periods with positive CO_2 uptake depend on the degree of initial
moistening of the lichen thalli. On the particular morning depicted
in Fig. 4, peak rates of CO_2 uptake of more than 20 pmol mg^{-1} dry
weight s^{-1} were attained with P. reticulata. One week later (Fig. 5)
much less dew formation occurred and P. reticulata exhibited a maximal
photosynthetic rate of only 4 pmol mg^{-1} s^{-1}, whereas CO_2 uptake of C.
portentosa was only slightly positive for a short period of about one
hour. The differences in radiation environment, which affect both the
time course of water content and the time at which maximal photosynthetic rate is attained, are apparent.

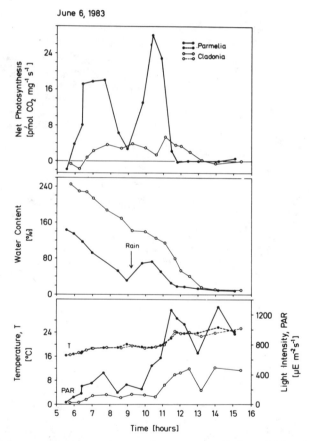

Fig. 6. Diurnal courses of CO_2 exchange of Parmelia reticulata and
Cladonia portentosa on a day with rain showers; for air
temperature, photosynthetically active radiation and water
content see Figs 3 and 4.

Although infrequent, the occurrence of rain during the summer in Mediterranean climate regions prolongs lichen photosynthetic activity or activates CO_2 uptake at other times of the day. Figure 6 depicts the time course of photosynthetic CO_2 uptake of the same two lichen species on a summer day when rain showers occurred. At sunrise, the water content of C. portentosa was higher than that of P. reticulata (lichen samples were briefly weighed after each gas exchange measurement, water content being related to thallus dry weight). Nevertheless, P. reticulata demonstrated more rapid CO_2 uptake than C. portentosa. When gathering clouds reduced available light incident on P. reticulata at 0900 h, its rate of CO_2 uptake decreased sharply to approximately the same level as observed for C. portentosa. Water content was at this time approaching a critical level at which CO_2 uptake would be expected to cease (compare Figs 5 & 6). However, a

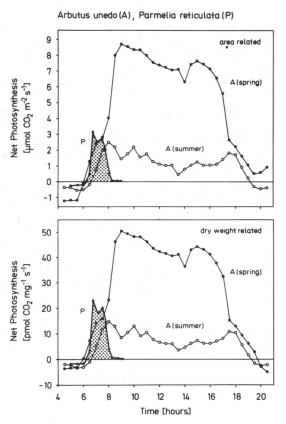

Fig. 7. Diurnal courses of area- or dry weight-related CO_2 exchange of Parmelia reticulata in comparison with Arbutus unedo (see Fig. 3). Response curve of the lichen after fog and dew imbibition in summer; typical response curves in spring and summer of the evergreen shrub.

rain shower (arrow in Fig. 6) rewetted the exposed <u>Parmelia</u>,
increasing the water content again to approximately 80%. For a short
period thereafter, maximal rates of net photosynthesis were observed
in this species. Subsequently the water content of the thallus was
reduced to about 18% before noon, and the moisture compensation point
for CO_2 reached. In terms of additional carbon fixed, <u>P. reticulata</u>
was much more able to profit from this rain event than was
<u>C. portentosa</u>.

Poikilohydric and homoiohydric plants are fundamentally different
with regard to their ability to control water relations. This
difference affects CO_2 uptake as illustrated in Fig. 7 for the diurnal
time courses of net photosynthesis obtained with <u>P. reticulata</u> and the
evergreen Mediterranean shrub <u>Arbutus unedo</u> growing in a nearby
evergreen scrub formation (Tenhunen et al., 1982). A typical time
course for CO_2 uptake for <u>P. reticulata</u>, obtained in early summer
after water imbibition from dew and fog, is superimposed on a characteristic diurnal time course of net photosynthesis for <u>A. unedo</u> during
spring and summer. Gas exchange rates are related to dry weight as
well as to area of the assimilating organs. In spring, photosynthetic
rates of the leaves of the shrub are much higher. Lichen peak CO_2
uptake rates are only one third to one half of those of <u>Arbutus</u>. In
contrast, during the dry summer, when soil water shortage and high
temperatures restrict the photosynthetic activity of the woody plants,
their maximal rates of net photosynthesis may be lower than those of
lichen. However, the time period of metabolic activity is always
much shorter for the lichen in comparison with the homoiohydric plant.

MEASUREMENT OF LICHEN CO_2 EXCHANGE UNDER CONTROLLED CONDITIONS IN THE
FIELD

In order to understand more fully the patterns of net photosynthesis which emerge from investigations of lichen CO_2 exchange
under natural conditions, experimental work must be carried out which
defines species-specific responses to individual environmental
factors. To conduct the required experiments in the field, we have
developed a "minicuvette system".

Description of the Minicuvette System

The minicuvette system is similar to the previously described
porometer. However, cuvette air temperature may be controlled to set
values by means of a Peltier-operated cooling and heating unit and an
electronic controller. Artificial illumination is provided for the
enclosed plant material with a lamp fixed to the top of an extension
tube which connects with the glass-covered lid of the cuvette. Light
intensity (maximally up to 2000 μE m^{-2} s^{-1} PAR) is changed with the
aid of neutral density filters inserted in the cylinder below the
light source. The humidity and CO_2 concentration of the air stream to
the chamber are controlled by means of a humidity control system and a

CO_2-mixing apparatus at the air inlet. A second infrared gas analyser is used to monitor the absolute CO_2 concentration of the air stream. Thus, photosynthetic measurements can be conducted at set conditions of temperature, light, humidity and ambient CO_2 concentration. Figure 8 illustrates the cuvette with inserted lichen samples.

Fig. 8.   Minicuvette (below) with inserted lichen sample (arrow). The lid of the cuvette with the extension tube and light source (above) is partly opened. (From Lange et al., 1984a).

Fig. 9.   Minicuvette measuring system assembled in the field.   Cuvette
          fixed on a tripod (1); gas analyser (2); water vapour trap
          (3); controlling unit and read-out system (4); line recorder
          (5).   (From Lange et al., 1984a).

This instrumentation is also portable and Fig. 9 shows the experimental set up as used for measurements in the field.   More detailed
information about the instrumentation are given by Lange and Tenhunen
(1984) and Lange et al. (1984a).

Factor Dependencies of Lichen Net Photosynthesis and Lichen Photosynthetic Capacity

     The data necessary for the estimation of those parameters used in
mathematical models of lichen photosynthesis can be obtained with the
minicuvette system.   An example is shown in Fig. 10, which depicts the
CO_2 dependence of net photosynthetic rate of P. reticulata (Portugal)
on CO_2 concentration, at constant temperature and saturating light
intensity.   The initial slope of the response curve may be used as a
measure of the overall carboxylation efficiency of the lichen, which
is affected by endogenous limitations on rates of biochemical
reactions as well as by resistances to diffusion of gases within the
thallus.   At approximately 1250 ppm external CO_2 concentration, the
carboxylation process is substrate saturated and maximal rates of net
CO_2 uptake are attained.   Values for the carboxylation efficiency and
maximal rates of net photosynthesis provide important input information for biochemically-oriented photosynthesis models (e.g.   Tenhunen
et al., 1978).   Similarly, temperature, water content and light
dependencies of net photosynthesis can be determined in the field with
the minicuvette system and analyzed for key characteristic parameters.

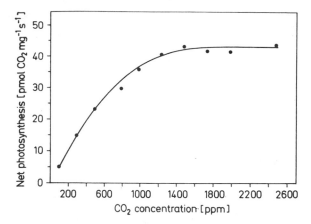

Fig. 10. CO$_2$ response curve of <u>Parmelia reticulata</u> at saturating
         light intensity (thalli fully water saturated, 20°C).
         Abscissa: CO$_2$ concentration external to the plant; ordinate:
         net photosynthetic CO$_2$ uptake. (From Lange et al., 1984a).

         Rapid determination of the photosynthetic capacity of a lichen is
possible with this instrumentation in the field and changes in
capacity can be monitored.   Thus, several types of useful information
may be gathered.   Seasonal changes in lichen photosynthetic capacity
are of significance when estimating annual lichen productivity and
lichen growth (Brown and Kershaw, 1984 and this volume).   Secondly,
since photosynthetic processes of lichens have been shown to respond
sensitively to SO$_2$, measurements of rates of CO$_2$ exchange and the
photosynthetic capacity of transplanted or exposed lichen samples have
been used in attempts to monitor air pollution.   Nevertheless, inter-
sample variation in lichen photosynthesis estimates has often been
very high in such studies.   Thus, to obtain relevant information, it
has been necessary to work with a large number of replicates in the
laboratory.   The described minicuvette systems provides another means
of establishing statistically significant trends in photosynthetic
response.   Repeated measurements may be conducted in the field on the
same individual samples over a longer period of time, while allowing
them to experience the natural environment between measurements.

         When lichens are studied under a high radiation load, the thallus
inevitably loses water.   During the course of drying, net photosynthesis changes dramatically.   Low, as well as very high, water
contents reduce the apparent carboxylation efficiency due to biochemical inactivation and/or due to increase in thallus diffusion
resistance respectively (Lange, 1980).   Despite the inability to
experimentally control thallus water content at predetermined levels,
reproducible measurements of lichen photosynthesis are obtained if we
work with saturating CO$_2$ concentrations and a high water content.

mathematical descriptions of lichen photosynthesis have been developed (Lange et al., 1977; Kappen et al., 1979; Link et al., this volume). As shown in Fig. 12, several steps are involved in achieving a satisfactory model to simulate lichen CO_2 exchange. The model is initially constructed with the aid of parameter values derived from individual factor response curves (cf. Figs 10 & 11). Ideally, these data are based on experimentation conducted in the natural habitat and with fresh material but quite often information from laboratory experiments must be used. Based on microclimatic observations, environmental information is provided stepwise to the model and diurnal time course of CO_2 uptake as well as photosynthetic production may be predicted. At first, we are not certain whether the simulated performances match the natural performance of the plant. Therefore, the model must be tested by determining the diurnal courses of lichen CO_2 exchange in the field and comparing measurement with prediction. Our experience suggests that we can expect to improve our model many times before it satisfactorily reproduces the real-world situation. Instrumentation has been described here which permits these kinds of investigations. We feel that through such experimentation, we can obtain information which will help us understand the physiological basis of lichen distribution and the behaviour of lichens in their natural habitats.

ACKNOWLEDGEMENTS

We wish to thank Dr. O. Schulz-Kampfhenkel for providing facilities at the Quinta São Pedro Research Station. Dipl.-Biol. W. Beyschlag provided data on gas exchange of Arbutus unedo. The research was supported by the Deutsche Forschungsgemeinschaft.

REFERENCES

Brown, D., and Kershaw, K.A., 1984, Photosynthetic capacity changes in Peltigera. 2. Contrasting season patterns of net photosynthesis in two populations of Peltigera rufescens, New Phytologist, 96: 447-457.

Darwin, F., and Pertz, D.F.M., 1911, On a new method of estimating the aperture of stomata, Proceedings of the Royal Society London, Ser. B., 84: 136-154.

Kappen, L., Lange, O.L., Schulze, E.-D., Evenari, M., and Buschbom, U., 1979, Ecophysiological investigations on lichens of the Negev Desert. VI. Annual course of the photosynthetic production of Ramalina maciformis (Del.) Bory, Flora, 168: 85-108.

Kappen, L., Lange, O.L., Schulze, E.-D., Buschbom, U., and Evenari, M., 1980, Ecophysiological investigations on lichens of the Negev Desert. VII. The influence of the habitat exposure on dew inhibition and photosynthetic productivity, Flora, 169: 216-229.

Lange, O.L., 1980, Moisture content and CO_2 exchange of lichens. I. Influence of temperature on moisture-dependent net photosynthesis and dark respiration in Ramalina maciformis, Oecologia, 45: 82-87.

Lange, O.L., 1984, "CO_2/H_2O-Porometer zur Messung von CO_2-Gaswechsel

und Transpiration an Pflanzen unter natürlichen Bedingungen,"
Heinz Walz Mess- und Regeltechnik, Effeltrich.

Lange, O.L., Beyschlag, W., Meyer, A., and Tenhunen, J.D., 1984a,
Determination of photosynthetic capacity of lichens in the field
- a method for measurement of light response curves at saturating
CO_2 concentration, Flora, 175: 283-293.

Lange, O.L., Geiger, I.L., and Schulze, E.-D., 1977, Ecophysiological
investigations on lichens of the Negev Desert. V. A model to
simulate net photosynthesis and respiration of Ramalina
maciformis, Oecologia, 28: 247-259.

Lange, O.L., Kilian, E., Meyer, A., and Tenhunen, J.D., 1984b,
Measurement of lichen photosynthesis in the field with a portable
steady-state CO_2-porometer, The Lichenologist, 16: 1-9.

Lange, O.L., Koch, W., and Schulze, E.-D., 1969, CO_2-Gaswechsel und
Wasserhaushalt von Pflanzen in der Negev-Wüste am Ende der
Trockenzeit, Berichte der Deutschen Botanischen Gesellschaft, 82:
39-61.

Lange, O.L., Schulze, E.-D., and Koch, W., 1970, Experimentell-
ökologische Untersuchungen an Flechten der Negev-Wüste. II.
CO_2- Gaswechsel und Wasserhaushalt von Ramalina maciformis (Del.)
Bory am natürlichen Standort während der sommerlichen Trockenperiode, Flora, 159: 38-62.

Lange, O.L., and Tenhunen, J.D., 1981, Moisture content and CO_2
exchange of lichens. II. Depression of net photosynthesis in
Ramalina maciformis at high water content is caused by increased
thallus carbon dioxide diffusion resistance, Oecologia, 51:
426-429.

Lange, O.L., and Tenhunen, J.D., 1984, "A minicuvette system for
measurement of CO_2 exchange and transpiration of plants under
controlled conditions in field and laboratory," Heinz Walz Mess-
und Regeltechnik, Effeltrich, F.R.G.

Schulze, E.-D., Hall, A.E., Lange, O.L., and Walz, H., 1982, A
portable steady-state porometer for measuring the carbon dioxide
and water vapour exchange of leaves under natural conditions,
Oecologia, 53: 141-145.

Tenhunen, J.D., Lange, O.L., and Jahner, D., 1982, The control by
atmospheric factors and water stress of midday stomatal closure
in Arbutus unedo growing in a natural macchia, Oecologia, 55:
165-169.

Tenhunen, J.D., Yocum, C., and Gates, D.M., 1976, Development of a
photosynthesis model with an emphasis on ecological applications.
I. Theory, Oecologia, 26: 89-100.

WATER RELATIONS AND NET PHOTOSYNTHESIS OF USNEA. A COMPARISON
BETWEEN USNEA FASCIATA (MARITIME ANTARCTIC) AND USNEA SULPHUREA
(CONTINENTAL ANTARCTIC)

L. Kappen

Botanisches Institut und Institut für Polarökologie
der Universität Kiel
Olshausenstrasse 40-60
D-2300 Kiel
West Germany

INTRODUCTION

Our knowledge of the ecophysiology of photosynthetic
production of Antarctic lichens is based on separate investigations
made with species from the maritime Antarctic or species from the
continental Antarctic, but no studies have yet been made comparing
the same or similar species in both areas. Many data have been
published from field and laboratory measurements on Signy Island,
although much less about lichens (Lindsay, 1978; Hooker, 1980a,b,c)
than about bryophytes (Baker, 1972; Collins, 1977; Collins and
Callaghan, 1980; Fenton, 1980). With respect to the continental
Antarctic there are more studies on lichens (Gannutz, 1970; Lange
and Kappen, 1972; Schofield and Ahmadjian, 1972; Kappen 1983a;
Kappen and Friedmann, 1983) than on mosses (Rastorfer, 1970;
Longton, 1974; Ino, 1983). Climate and ecological conditions are
very different in the two areas, particularly the winter conditions.
In the maritime Antarctic, as the author has seen on King George
Island, the abundance and variety of lichens is remarkable. In
continental Antarctica, even near the coast of northern Victoria
Land (Kappen, 1983b) lichens may be considered to be extremely
resistant representatives of a pioneer vegetation in a polar
desert. Consequently, the aim of this study was to evaluate
whether lichens show adaptive physiological differences or whether
they have the same physiological and ecological requirements in the
polar desert and the maritime Antarctic due to presumably
convergent habitat conditions.

41

This study is taken from the results of two expeditions. In 1981 lichen vegetation and ecology was studied at Birthday Ridge, northern Victoria Land (70° 48' S, 167° E. Kappen, 1983a) and field data about the microclimate and vegetation are being prepared for publication. Lichens from the maritime Antarctic were studied and collected on King George Island (62° 12' S, 58° 56' W) in 1983 when the author was on an expedition together with Prof. Dr. J. Redon, University of Valparaiso. Laboratory and field data will be published in detail elsewhere by Kappen and Redon. For the photo-synthetic measurements reported here the air dry thalli were transported to Kiel in closed dewar vessels and then kept in an illuminated chamber at c. +2°C.

Neuropogon is the only fruticose lichen subgenus that is very prominent in the vegetation of both the maritime and the continental Antarctic (Lamb, 1964). Usnea antarctica D.R. was present in both areas but was too rare in northern Victoria Land for experiments. Usnea fasciata Torr. is the most abundant species on King George Island. Interestingly, U. fasciata exists either as an erect fruticose thallus that is fixed to the rock and fertile, as is the case with most species of Neuropogon, or it occurs more

Table 1.   Cryptogamic Production in the Antarctic, where Usnea sp. Occur

---

Subantarctic

Cryptogam-tundra, South Georgia 54° S   1000-2000 g m^{-2}
(Lindsay, 1978)

Maritime Antarctic

Usnea Himantormia - "heath", King George Island
Fildes Bay 62° 12' S, 58° 56' W   1891.75 g m^{-2}

Usnea fasicata separated   1040.0 g m^{-2}
(Kappen and Redon, unpublished)

Continental Antarctic

Pure Usnea sulphurea vegetation on granite,
Birthday Ridge, Yule Bay, NVL, 70° 47' S, 167° E   372.2 g m^{-2}
Usnea sulphurea - bryophyte community,
Birthday Ridge, Yule Bay, NVL, 70° 47' S, 167° E   138.6 g m^{-2}
(Kappen, unpublished)

---

or less prostrate and only loosely attached to the mossy ground and then never bears fruiting bodies. On King George Island both forms grow together and are widespread on terraces and ridges. At Birthday Ridge, northern Victoria Land, U. sulphurea (Koenig) Th. Fr. is widespread and dominates lichen formations. U. sulphurea in northern Victoria Land can also be found in two forms: one is dark pigmented, the other is pale yellow. Generally the following text refers to the erect form of U. fasciata and the dark pigmented form of U. sulphurea. The contributions of Usnea to biomass formation is apparent from Table 1. For comparison, some other lichen species are included in the discussion.

RESULTS

Water uptake and effects of water content on net photosynthesis

Snowfall and rain are frequent during the warm season in the maritime Antarctic. For example, the annual precipitation on King George Island in 1977 was 522 mm, with more than 0.1 mm recorded on 193 days (Anon. 1977). The lichens remain soaked for periods longer than 24 hours, and dry periods are short. Maximum rain water-soaked field water contents were 164.2% of dry weight in

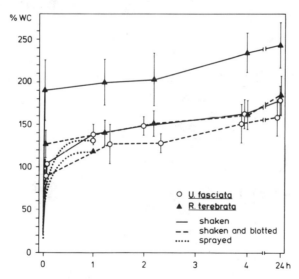

Fig. 1.   Water uptake of thalli immersed in liquid water for certain periods of time. The relative water content (% dry weight) was measured by weighing samples after having them shaken and blotted or only shaken. The increase of water content is also shown after the thalli were sprayed twice, the normal wetting procedure before gas exchange measurements.

U. fasciata and to 263% dry weight in the fenestrated thalli of
Ramalina terebrata Hook et Tayl (Kappen and Redon, in preparation).
Usnea sulphurea on the other hand grows in an area where
precipitation is much lower (200 mm or less?) and consists only of
snow. Under natural conditions the thalli never reach a high
degree of water saturation, being maximally 97% of dry weight
(Kappen, 1984).

     At first tests were made to discover which laboratory method
best reproduced the natural thallus water contents. Submersion in
distilled water was compared to spraying with water and the effect
of shaking and blotting the soaked thalli before weighing was
tested (Fig. 1). Within three minutes of submersion in water the
thalli had already reached 2/3 of their maximum water content and
after 24 hours of submersion water contents were at the maximum
level of 178% of dry weight in U. fasciata and 243% of dry weight
in R. terebrata. These values were found when the thalli had only
been shaken. Blotted thalli had lower maximum values of 158%
(U. fasciata) and 184% of dry weight (R. terebrata) and the maximum
values of sprayed thalli were still lower again. Thus, submersion
in water and shaking of the thalli best reproduced the natural
maximum rainwater-soaking. U. sulphurea from the continent took up
to a maximum of 160% of dry weight (Kappen, 1983a), similar to
U. fasciata but well above field levels.

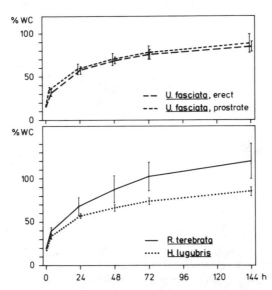

Fig. 2.   Water vapour uptake of the two forms of Usnea fasciata,
           Ramalina terebrata and Himantormia lugubris from the
           maritime Antarctic during a period of 144 hours in an
           almost water vapour saturated atmosphere at 5°C.

All the species were tested for their ability to absorb water
from humid atmospheres. Both the erect and prostrate form of
U. fasciata reached twice their initial water content (20% of dry
weight) within 6 hours, four times the initial water content after
24 hours, but after 6 days had still reached only 50% of the water
content reached if immersed in water (Fig. 2). Other maritime
Antarctic species which grow in the same area as U. fasciata (like
Himantormia lugubris (Hue) M. Lamb and R. terebrata) take up water
vapour to the same order of magnitude (Fig. 2). This ability
appears well adapted to high air humidity and fog which normally
occurs during about 15% of the summer season on King George Island.

In U. sulphurea from northern Victoria Land water vapour
uptake of initially dry thalli was measured in gas exchange
cuvettes (Siemens type) at 5°C and high relative humidity (Kappen,
1983a). The thalli reached 1/3 of full saturation in light and 1/2
of full saturation in the dark, very similar values to the above
mentioned species. Thus, this "desert" lichen was also able to
absorb water during the periods of higher humidity in hazy or
overcast conditions which were sometimes observed.

An important question is whether the rate of photosynthetic
CO_2 uptake is limited at high water contents (Lange, 1980; Brown
and Kershaw, 1984). Carbon dioxide exchange of the thalli was
measured by means of an infrared gas analyzing system (H. Walz,
Effeltrich, W. Germany). The thalli were placed, in a temperature
controlled plexiglas chamber, on an electronic microbalance
(Precisa 80A, PAG Zürich), allowing the weight to be continuously

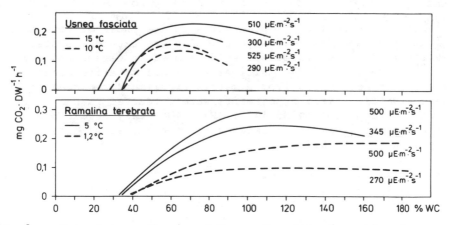

Fig. 3.   Net photosynthesis of Usnea fasciata (erect) and
          Ramalina terebrata at different water contents (% dry
          weight) at different quantum flux densities (indicated
          near the curves) and at different thallus temperatures
          (solid or broken lines).

recorded by a YEW Pen recorder Mod 3021, YOKOGAWA Electr., which
avoided interrupting measurements.

Carbon dioxide uptake in U. fasciata starts at water contents
around 20-30% of dry weight depending on light intensity (Fig. 3).
The optimum water content was about 70% of dry weight at 15°C and
60% dry weight at 10°C. Net photosynthesis diminished at water
contents greater than optimal at 10°C and 15°C, with the decrease
being less at higher light levels (Fig. 3). The depression did not
occur at all at 7°C and 2.5°C (Fig. 4). These measurements of net
photosynthesis versus water content in U. fasciata, at different
temperatures but constant quantum flux density, confirm Lange's
(1980) results with the hot desert species R. maciformis but the
depressions of net photosynthesis were in no instance as extreme as
was observed with other lichens by Brown and Kershaw (1984).

The same general pattern was also found with R. terebrata
where a slight depression takes place only at 4°C but not at 1.2°C
(Fig. 3). Consequently, the usually cold rain would have little
affect on the gas exchange of the thalli in the Antarctic. Water
vapour uptake always results in water contents which are less than
optimal or just optimal for net photosynthesis.

In its natural environment U. sulphurea receives water only
from snow. Snow can accumulate on the thalli only where aero-
dynamic conditions allow this. Snow falls mostly with strong winds
and must be blown onto the thalli; drifted snow crystals can reach
lichens in the lee of granite rocks at Birthday Ridge. The black
pigmented thalli are heated by the high irradiation and the fresh
snow melts rapidly. Experimentally it has been shown that CO_2
exchange in U. sulphurea starts at about 40% of dry weight, as

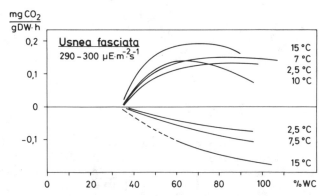

Fig. 4.   Net photosynthesis and dark respiration of Usnea fasciata
           (erect) at different water contents (% dry weight) and
           different thallus temperatures (as indicated).

in U. fasciata, and reaches maximum rates at a water content about
70% of dry weight. However, at 5°C and 250 μE m^{-2} s^{-1} even a water
content of 140% of dry weight has no significant detrimental effect
on net photosynthesis (Kappen, 1983a). Thus, the negative effect
of water soaking is ecologically not relevant for this species.

Temperature Effects

The effect of temperature on net photosynthesis of the two
forms of U. fasciata is shown in Fig. 5. The overall minimum net
photosynthesis is at -5°C. Optimum temperatures and upper compensation points increase with increasing quantum flux density. The
prostrate form appears much less productive than the erect form.
However, it must still be competitive with the erect form and underlying mosses since it looks healthy and forms a large proportion
of the total standing biomass of the lichen-moss heaths of King George
Island. However, one cannot exclude the possibility of the prostrate

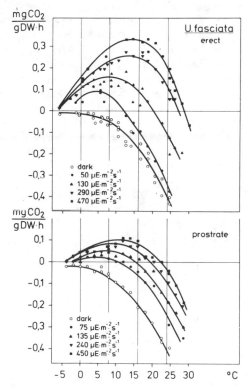

Fig. 5.   Net photosynthesis and dark respiration of the two forms
of Usnea fasciata at different thallus temperatures
(abscissa) and at different quantum flux densities (as
indicated).

form being more sensitive to transportation to the laboratory.
Lamb (1964) interpreted the "muscicole occurrence" of U. fasciata
as a consequence of intensive growth of the bryophytes around the
lichens so that they lost their attachment to the rocks. Our
results may indicate that these lichens also had a reduced vitality.

The CO_2 exchange rate of the erect U. fasciata from the
maritime Antarctic has been compared with the same, but always
sterile, growth form of U. sulphurea from northern Victoria Land
(Fig. 6). The photosynthetic rates of U. sulphurea were only half
those of U. fasciata. Dark respiration of U. sulphurea increased
more with increasing temperature than that of U. fasciata. This
may characterize the former species as being more adapted to cold
conditions (Billings, 1974) and the temperature dependence of net
photosynthesis also suggests this since, at the same light
intensity, optimum and maximum temperatures are lower than those of
U. fasciata (Fig. 7).

Usnea sulphurea grows in open places as a black pigmented
"sun" form and on the underside of rocks as a pale yellow "shade"
form. Photosynthetic rates on a chlorophyll basis of the "shade"
form are inferior to those of the "sun" form. The chlorophyll a/b
ratio is equal in both forms. However, chlorophyll content per
unit dry matter in the "shade" form is more than twice as much as
that of the "sun" form. Consequently, the photosynthetic rates on
a dry weight basis of the "shade" form are not so low and at lower
temperatures they are even higher than those of the "sun" form.
The differences between "sun" and "shade" forms are clearly shown
in their response to light intensity at different temperatures

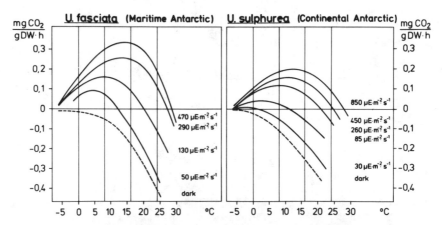

Fig. 6.   Net photosynthesis (solid lines) and dark respiration
           (broken lines) of Usnea fasciata and Usnea sulphurea at
           different temperatures (abscissa) and quantum flux
           densities (as indicated).

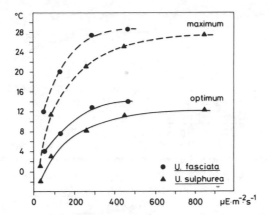

Fig. 7.   Maximum (---) and optimum (———) values of net photosynthesis of <u>Usnea fasciata</u> (circles) and <u>Usnea sulphurea</u>
(triangles) at different temperatures and quantum flux
densities.

(Fig. 8). At -1°C the net photosynthetic rate of the "shade" form
is always higher than that of the "sun" form; at 8°C both have the
same photosynthetic rates while at 15°C the "sun" form is superior
to the "shade" form. Thus, the overall adaptation of the "shade"

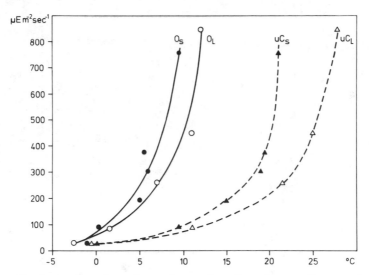

Fig. 8.   Maximum (---) and optimum (———) values of net photosynthesis of the "sun" form (O, $\triangle$) and the "shade" form
($\bullet$, $\blacktriangle$) of <u>Usnea sulphurea</u> at different quantum flux
densities and temperatures (from Kappen 1983a).

form to lower temperatures and low light intensities is apparent
(Kappen, 1983a).

Light Intensity

     Under optimal water conditions net photosynthesis in
U. fasciata is light saturated above 450 μE m^{-2} s^{-1}, while that of
the "sun" form of U. sulphurea is saturated at c. 800 μE m^{-2} s^{-1}
(Fig. 9) and with increasing temperature the maximum values of
photosynthesis increase. The light compensation point varies
significantly with temperature in each species. It is remarkable,
however, that the generally low temperature adapted U. sulphurea
had higher light compensation points than U. fasciata. This
obviously is an adaptation to the usually higher solar radiation of
its habitat in the continental Antarctic.

Ecology and Carbon Production

     Maximum photosynthesis and dark respiration of the continental
species is shifted to lower temperature optima compared with that

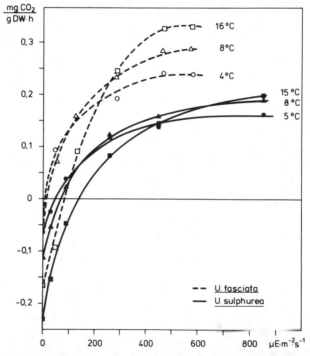

Fig. 9.   Light dependency of net photosynthesis at different
           temperatures (indicated near the curves) in Usnea
           fasciata (erect) and U. sulphurea ("sun" form).

of the maritime Antarctic. This adaptation of U. sulphurea to
colder conditions also, however, includes a much lower photosynthetic rate, probably representing a higher cost of resistance
to extreme frost and desiccation. There is no difference between
the species with respect to the moisture dependency of net
photosynthesis. Thus, U. sulphurea, in the dry continental
Antarctic, needs particular localities where sufficient moisture is
available.

The microclimatic conditions of both lichen habitats on a
sunny day (a) and a foggy day with precipitation (b) are presented
in Fig. 10. Although there is a seasonal difference between the
observations, the general character of both habitats is obvious and
may be taken as representative. Air temperatures and relative
humidity at Birthday Ridge are mostly lower than at Fildes Bay,
King George Island. Differences in maximum quantum flux rate are
probably a result of the different dates of the observation. The
thallus temperatures of U. sulphurea were not so high as those of
U. fasciata. However, temperatures above 20°C may not support
photosynthesis in the field because then the lichens are mostly
desiccated by the sun. Periods with thallus temperatures above 0°C
were usually longer for U. fasciata. The temperature difference
between the top layer and the inner canopy of U. fasciata possibly
reflects a local microclimate in the dense growth that the lichens
form on the top of the pebble ridges.

Although, on King George Island the lichens were moistened by
fog and rain, light limited photosynthesis under these conditions
(Fig. 10b). In general the microclimate of the continental site is
colder and drier. However, lack of heat and of moisture will, to a
certain extent, be compensated for by a higher amount and longer
duration of light. The lichens grow on big granite rocks but not
on the tops and open exposed parts of the rocks which, although
warmer, are too dry (Fig. 10D). The lichens are restricted to
fissures and niches in deeper parts between the rocks. The black
pigmentation increases the heat absorption of the thalli of
U. sulphurea, so that thalli were warm enough to rapidly melt light
snow. As a result the main factor controlling the occurrence of
lichens in the Antarctic is moisture availability and more
indirectly temperature ("micro-oases", Kappen and Friedmann,
1983).

Since estimates of the growth of maritime Antarctic lichens
are available (Hooker, 1980c), it is interesting to estimate the
carbon production of U. sulphurea under natural conditions in
northern Victoria Land. During one third (58h) of the one week
period of observation (168h) lichens were wet and capable of
carrying out photosynthesis. Summing up the hourly temperature
related rates of CO_2 uptake the lichens would gain a total of
5.761 mg CO_2 g dry weight^{-1} (Kappen, 1984).

The productive season in northern Victoria Land is estimated to be 5 months. It is proposed that it can be sub-divided into one month, like the example week, 2 months when half as productive and 2 months when one quarter as productive, yielding 22.8 mg and 11.6 mg CO_2 g dry weight^{-1} respectively. The rest of the year is too dark and too cold for any CO_2 exchange so that the annual total will be 57.0 mg CO_2 g dry weight^{-1} (= 15.5 mg C g dry weight^{-1}).

Carbon, nitrogen (CN analyser Carlo Erba) and ash content (dry combustion 550°C) were determined in order to have a basis to which the carbon production would be related (Table 2). Both Usnea species had a C content of around 41% of dry weight. This value is a little higher than that of the desert lichen R. maciformis (35.4% of dry weight, Lange et al., 1970). The total N-content of Usnea sulphurea was in the range of the lowest reported values in lichens (0.3%: Cladia retipora, Sphaerophorus tener; Green et al., 1980). Usnea fasciata is in the upper part of the range of lichens with green phycobionts (e.g. Siphula; Green et al., 1980). The C/N ratio of U. fasciata was double that of U. sulphurea. This may be indicative of poorer nutrient conditions in the continental Antarctic in contrast to the maritime Antarctic. The ash content of the thalli appears low for lichens and is lowest in U. sulphurea, possibly an effect of the low dust content of the Antarctic atmosphere. Further analyses of Antarctic lichens could be indicative of their species specific and ecological differences.

The above estimated annual yield of 15.5 mg C g dry weight^{-1} year^{-1} represents a 3.8% annual gain if the carbon content of the dry matter is 41%. Large thalli of U. sulphurea attain a dry weight of 500 mg. Such thalli would have needed 204 years to grow from a young state with a dry biomass of 0.25 mg, at a constant annual rate of 3.8% (compound interest calculation: $K_n = K_0 + (1 + q/100)^n$. The assumption is made that there were no climatic changes during this period and that the thallus does not lose biomass by respiration or erosion. The initial phase, until the thallus reaches 0.25 mg dry weight, is estimated to cover 100 years, so these thalli were about 300 years old.

By comparison, the thalli of U. fasciata from Signy Island investigated by Hooker (1980c) had dry weights up to 8.6 g. This biomass was produced within about 500 years and U. antarctica had reached a dry weight of 1.0 g within 200 years. This is a good demonstration of the difference in productivity between the maritime Antarctic species and the continental Antarctic species of Usnea.

ACKNOWLEDGEMENTS

The expeditions to northern Victoria Land and to King George Island were supported by the Deutsche Forschungsgemeinschaft

(Schwerpunkt Antarktisforschung). The author is gratefully
indebted also to Mr. H. Mempel, Institut für Meereskunde, Kiel, for
the C,N-analysis and to Dr. M. Bölter for the determination of ash
content. Mrs. C. Lasalle skillfully carried out the CO_2 exchange
measurements and prepared the graphs. Dr. T.G.A. Green, while at
the Botanical Institute of Kiel, read and revised the English
version and Mrs. Ch. Strauss typed the original manuscript.

REFERENCES

Anon., 1977, Boletin Meterologico de la Fuerza Aérea de Chile,
    Centro Meteorologico Presidente Frei ("BMF AC-report").
Baker, J.H., 1972, The rate of production and decomposition of
    Chorisodontium aciphyllum (Hook f. and Wils.), Broth., British
    Antarctic Survey Bulletin, 27:123-129.
Billings, W.D., 1974, Arctic and alpine vegetation: plant
    adaptations to cold summer climates, in: "Arctic and alpine
    environments," Ives, J.D. and Barry, R.G. eds, pp.403-443, Methuen.
Brown, D., and Kershaw, K.A., 1984, Photosynthetic capacity changes
    in Peltigera. 2. Contrasting season patterns of net
    photosynthesis in two populations of Peltigera rufescens, New
    Phytologist, 96:447-457.
Collins, N.J., 1977, The growth of mosses in two contrasting
    communities in the maritime Antarctic: Measurement and
    prediction of net annual production, in : "Adaptations within
    Antarctic ecosystems," Llano, G.A. ed., pp.921-933, Proc.
    3rd S.C.A.R. Symposium, Smithsonian Institution Washington D.C.
Collins, N.J., and Callaghan, T.V., 1980, Predicted patterns of
    photosynthetic production in maritime Antarctic mosses, Annals
    of Botany, 45: 601-620.
Fenton, J.H.C., 1980, The rate of peat accumulation in Antarctic
    moss banks, Journal of Ecology, 68:211-228.
Gannutz, T.P., 1970, Photosynthesis and respiration of plants in
    the Antarctic peninsula area, Antarctic Journal of the US, 5:
    49-51.
Green, T.G.A., Horstmann, J., Bonnet, H., Wilkins, A.,
    and Silvester, W.B., 1980, Nitrogen fixation by members of the
    Stictaceae (Lichenes) of New Zealand, New Phytologist,
    84:339-348.
Hooker, T.N., 1980a, Factors affecting the growth of antarctic
    crustose lichens, British Antarctic Survey Bulletin, 50:1-19.
Hooker, T.N., 1980b, Growth and production of Cladonia rangiferina
    and Sphaerophorus globosus on Signy Island, South Orkney
    Islands, British Antarctic Survey Bulletin, 50:27-34.
Hooker, T.N., 1980c, Growth and production of Usnea antarctica and
    U. fasciata on Signy Island, South Orkney Islands, British
    Antarctic Survey Bulletin, 50:35-49.
Ino, Y., 1983, Photosynthesis and primary production in moss
    community at Syowa Station, Antarctica, Japanese Journal of
    Ecology, 33:427-433.

Kappen, L., 1983a, Ecology and physiology of the Antarctic
     fruticose lichen Usnea sulphurea (Koenig) Th. Fries, Polar
     Biology, 1:249–255.
Kappen, L., 1983b, Anpassungen von Pflanzen an kalte
     Extremstandorte, Berichte der Deutschen Botanischen
     Gesellschaft, 96:87–101.
Kappen, L., 1984, Vegetation and ecology of ice-free areas of
     northern Victoria Land, Antarctica. II. Ecological
     conditions in typical microhabitats of lichens at Birthday
     Ridge. Polar Biology, (submitted).
Kappen, L., and Friedmann, E.I., 1983, Kryptoendolithische Flechten
     als Beispiel einer Anpassung an extrem trocken-kalte
     Klimabedingungen, Verhandlungen der Gesellschaft für Ökologie
     (Mainz 1981), 10:517–519.
Lamb, I.M., 1964, Antarctic lichens, 1. The genera Usnea,
     Ramalina, Himantormia, Alectoria, Cornicularia, British
     Antarctic Survey Scientific Report, 38:1–34.
Lange, O.L., 1980, Moisture content and CO_2 exchange of lichens.
     I. Influence of temperature on moisture-dependent net
     photosynthesis and dark respiration in Ramalina maciformis,
     Oecologia, 45:82–87.
Lange, O.L., and Kappen, L., 1972, Photosynthesis of lichens from
     Antarctica, Antarctic Research Series, 20:83–95.
Lange, O.L., Schulze, E.-D., and Koch, W., 1970, Experimentell-
     ökologische Untersuchungen an Flechten der Negev Wüste: II.
     CO_2 Gaswechsel und Wasserhaushalt von Ramalina maciformis
     (Del). Bory, Flora, 159:38–62.
Lindsay, D.C., 1978, The role of lichens in Antarctic ecosystems,
     The Bryologist, 81:268–276.
Longton, R.E., 1974, Microclimate and biomass in communities of the
     Bryum association on Ross Island, continental Antarctica, The
     Bryologist, 77:109–127.
Rastorfer, J.R., 1970, Effects of light intensity and temperature
     on photosynthesis and respiration of two east Antarctic
     mosses, Bryum argenteum and Bryum antarcticum, The Bryologist,
     73:544–556.
Schofield, E., and Ahmadjian, V., 1972, Field observation and
     laboratory studies of some Antarctic cold desert cryptogams,
     Antarctic Research Series, 20:97–142.

PHOTOSYNTHESIS, WATER RELATIONS AND THALLUS STRUCTURE OF STICTACEAE LICHENS

T.G.A. Green[a], W.P. Snelgar[b], A.L. Wilkins[c]

[a] Biological Sciences, [c] Chemistry Department
   Waikato University
   Hamilton, New Zealand
[b] Department of Scientific and Industrial Research
   Mt Albert, Private Bag
   Auckland, New Zealand

INTRODUCTION

The net photosynthetic rate (NP) of any autotrophic plant can be affected by a variety of external and internal factors. Extensive work with higher plants would suggest photosynthetically active photon flux density (PPFD) as a dominant factor, together with CO_2 concentration, temperature and diffusion resistance of the photosynthesising structure. Also, probably because of the strong interest in stomatal regulation, there has been considerable emphasis on leaf and canopy resistances. In recent years the adaptive significance of different photosynthetic pathways (C3, C4, CAM) has been extensively researched and, with this, water-use efficiency has become a central issue.

The NP of lichens will be also under the control of these factors but it is interesting that there are considerable differences between lichen and higher plant research in the relative importance accorded each factor. In lichens, the relationship between thallus water content (WC) and NP has become the most studied aspect. This is not only a result of the poikilohydric nature of lichens but also a reflection of the fascination of lichenologists with the ability of lichens to tolerate environmental extremes, particularly temperature and desiccation (Ahmadjian and Hale, 1973). With the arrival of an easier system of CO_2 measurement, the infrared gas analyser (IRGA), research on the CO_2 exchange of lichens, particularly the NP:WC relationship, increased extensively. Some of the earliest work was the field studies of <u>Ramalina maciformis</u> in the Negev Desert by Lange and associates (Lange et al., 1970). The importance of transient

factors such as dew and high relative humidity for lichen NP was
clearly demonstrated. A model linking NP with environmental parameters was developed (Lange et al., 1977). This work, which was
unusual in that comparisons were made between lichens and higher
plants (Lange et al., 1975), set high standards for others to follow.
It is unfortunate that the challenge has not been taken up and, in
general, there have been few measurements of NP in the field. The
present result of over a decade of IRGA studies is a literature
dominated by NP:WC response curves with only a minimum of supporting
ecological data (see Lange and Matthes (1981) for a data summary).

It has become quite clear that there is remarkable diversity in
the NP:WC response curve both between lichen species (see Snelgar et
al. (1980) for examples) and within a single species (Kershaw, 1972
and much subsequent work). From one of the earliest reports (Kershaw,
1972) these differences have been interpreted with respect to the
ecology of the lichens. These interpretations have been extended to
include apparent acclimation of the NP:WC response to seasonal
influences particularly through studies of Kershaw, Larson and co-
workers. (See Larson and Kershaw, 1975; Kershaw, 1977; Kershaw and
Smith, 1978; Larson, 1980; and many others listed in Lange and
Matthes, 1981). Interpretation of the NP:WC curves has been confined
to comparisons of response shape and key features such as WC at
NP_{max}, NP_{max} and respiration rate. Very little information is
actually available on what was producing the shifts in the curves
although there was broad agreement that NP limitation at low WC was
biochemical through decreased water potential (Cowan et al., 1979) and
an unwritten, but often inferred, assumption that other changes were
also physiological.

Until the pioneering work of Larson and associates there was very
little information on the management of water by lichens. It is now
clear that lichens can modify their water relations in response to
external water demand by thallus shape and water storage (Larson and
Kershaw, 1976; Larson, 1979; Snelgar and Green, 1981a). However, even
now, there are few lichens where the significance of the NP:WC response
is known in terms of residence times at different WC in the field.

An alternative approach is to analyse the NP:WC response in terms
of lichen internal factors. This article is a synthesis of published
and unpublished results from a series of studies on the large, foliose,
Stictaceae lichens of New Zealand. The overall aim was to obtain
information on the physiology and structure of the lichens which would
allow a more soundly based interpretation of the NP:WC response. An
important experimental procedure was the application of diffusion
resistance techniques that had been extensively developed in higher
plant studies. It proved possible to not only obtain this better
interpretation, but also to develop a further concept that the lichen
thallus, in the Stictaceae, represented an adaptive solution to the,
often opposing, problems of CO_2 exchange and optimising water relations.

PHOTOSYNTHETIC PERFORMANCE OF LICHENS

CO_2 Exchange and CO_2 Concentration

Surprisingly this remained an overlooked aspect of lichen photosynthesis for many years. Increased interest has followed the finding
by Snelgar and Green (1980) that lichens may possess very low CO_2 compensation values (Γ). Lichens appear to have a similar response of NP
to CO_2 concentration to that of higher plants. A typical saturation
curve (Fig. 1) is found, with the actual saturating CO_2 concentration
depending on thallus WC.

In some cases a biphasic response has been found (Coxson and
Kershaw, 1984). Such results are extremely difficult to interpret
using normal diffusion theory since a varying resistance is suggested.
It is possible, as for respiratory studies considered later, that the
CO_2 response form may be sensitive to methodology since biphasic
curves have been produced only by the discrete sampling system.
However Link et al. (1984), in a comparison of several methodologies,
obtained only the standard saturation curve. It is clear that further
investigations are required.

Investigations into the effect of oxygen concentration on NP
indicate that photorespiratory rates differ between lichens (Snelgar
and Green, 1980, 1981b). Further investigations of the response of NP
to CO_2 concentration at 2% and 21% oxygen reveal a complex situation
(Fig. 2). Lichens appear to mimic the responses of C3, C4 and C3/C4
intermediate higher plants (Brown, 1980; Kennedy et al., 1980). In
Fig. 2 Pseudocyphellaria homoeophylla behaves as a typical C3 plant
with marked sensitivity of NP, carboxylation and Γ to oxygen. Pseudocyphellaria billardieri is similar to C4 plants with no oxygen effect,
whilst Peltigera dolichorhiza behaves as a C3/C4 intermediate

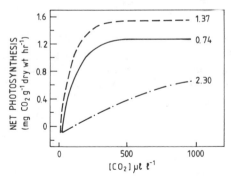

Fig. 1. Response of net photosynthesis of Sticta latifrons to CO_2
concentration at different thallus water contents (given as g
H_2O gDW^{-1} by each line). Modified from Snelgar (1981); Green
and Snelgar, (1981a).

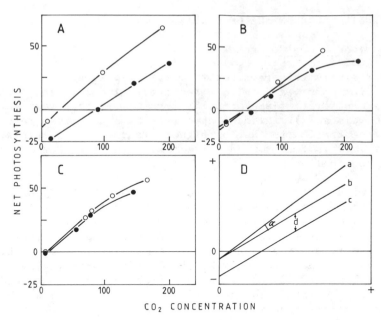

Fig. 2.   The effect of different oxygen concentrations on the
          relationship between net photosynthesis and CO_2 concentra-
          tion.
          Response at < 1% O_2 = O; response at 21% O_2 = ●.
          A: Pseudocyphellaria homoeophylla, typical C3 response;
          B: Peltigera dolichorhiza, C3/C4 intermediate;
          C: Pseudocyphellaria billardieri, typical C4 response.
          D: summary of the components of the interaction between net
             photosynthesis, oxygen level and CO_2 concentration; α =
             the effect of oxygen on carboxylation efficiency, d = the
             rate of photorespiratory release of CO_2. Net
             photosynthetic rates in μgCO_2 $m^{-2}s^{-1}$, CO_2 concentrations
             in μlCO_2 l^{-1}. After Snelgar (1981).

with carboxylation (α) but not Γ sensitive to oxygen. In higher
plants the C4 pattern can only be produced when CO_2 pumping occurs,
whilst the C3/C4 intermediate represents increased PEP carboxylase
activity.   There is no evidence from lichen [13] values for CO_2 uptake
predominantly through PEP carboxylase as in the C4 pathway
(Shomar-Ilan et al., 1979).   However evidence does exist of CO_2
pumping systems in algae (Badger et al., 1980) with variable Γ.   It
appears that CO_2 pumps operating at the cell level may be the
explanation of the different oxygen effects in lichens.

CO₂ Exchange and Thallus Water Content (WC)

     A typical CO_2 exchange response to WC is shown in Fig. 3.

Curiously this type of curve has been the subject of considerable
controversy over two aspects: the dark respiration response and the NP
response at high WC. Two general patterns of dark respiratory
response exist. First, a constant respiration rate over the majority
of WC (solid line Fig. 3), a result typical of flow-through or
non-sampling IRGA systems. Second, a linear or more complex response
of respiration to increasing WC (dashed line Fig. 3), typical of
discrete-sampling IRGA methodology (Lange and Matthes, 1981). In the
absence of a controlled comparison between the two methodologies it is
not yet possible to be certain whether the pattern is merely
coincidental.

At high WC there has been controversy over the degree and causes
of NP depression. It is now accepted that the NP depression (dashed
line Fig. 3) is the consequence of increased diffusion resistances
(Snelgar et al., 1981b; Lange and Tenhunen, 1981). The original
controversy was fuelled by several methodology factors. There were
differences in how thallus wetting was carried out, i.e. the lichen
could be blotted or simply shaken, treatments producing quite
different WC (Lange and Tenhunen, 1981). In many cases the use of
non-saturating PPFD affected the NP depression as reported in the
majority of investigations before 1980. It is important that when NP
response to a single factor is being investigated, only that factor is
limiting NP; an experimental requirement that has been met in far too
few lichen investigations.

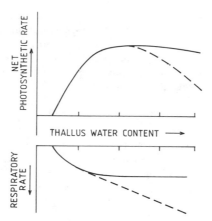

Fig. 3.   Typical relationship between net photosynthesis (upper
          curves), respiration (lower curves) and thallus water
          content.
          For net photosynthesis the dashed line shows depression at
          high water contents; for respiration the solid and dashed
          lines are those expected from gas flow and discrete injection
          systems respectively.

DIFFUSION PATHWAYS IN LICHENS

Introduction

The exchange of CO_2 and water vapour between lichen thalli and the atmosphere takes place by diffusion. The diffusion rate depends on both the concentration gradient between lichen and atmosphere and the physical structure of the diffusion pathway. The effect of the latter can be described in terms of a 'resistance' value which, for any particular diffusion pathway, is constant but will differ depending on the substance diffusing, e.g. resistance to water vapour is about 0.6 of the resistance to CO_2 (Jarvis, 1971). If the morphological parameters of the pathway are changed then the resistance will also be altered. Concentration gradient, diffusion rate and resistance are linked in a manner analogous to Ohm's Law in electrical circuits (potential or gradient equals flux multiplied by resistance).

Lange (1980) proposed a resistance model for CO_2 diffusion through a lichen thallus although he acknowledged that no direct information then existed either for the values of the resistances or for the diffusion pathway locations. Collins and Farrar (1978) had, in fact, measured the total resistance to CO_2 diffusion of <u>Xanthoria</u> <u>parietina</u>. Although the value obtained was technically in error, their conclusion that it was greater than values obtained for higher plants still appears correct. Values for diffusion resistances in the Stictaceae have been published (Snelgar et al., 1981a; Green and Snelgar, 1981b) and these are discussed in this article.

Although there has been continuing discussion about possible diffusion pathways in lichen thalli there has been very little direct investigation of this topic. Some lichen structures have been suggested to limit diffusion, in particular the cortex (Scott, 1960; Hill, 1976). Other structures, such as the cyphellae and pseudo-cyphellae of the Stictaceae, have been allocated important roles as diffusion pathways (Ahmadjian and Hale, 1973; Hale, 1974; Henssen and Jahns, 1974; Rundel et al., 1979).

Two complementary experimental approaches exist which can provide useful information about lichen thallus gas exchange. First, by studying parts of the thallus in isolation the actual sites of gas exchange can be directly identified. Second, the diffusion pathways in the thallus can be analysed in terms of diffusion resistances by either experimental determination of resistances or by calculating them from the dimensions of the diffusion pathway. Both approaches have been used for higher plants where, by comparison of magnitudes and locations, the relative importance of resistances can be assessed (Jarvis, 1971). The application of both approaches to lichens are discussed here.

Water Vapour Diffusion Pathways

In the leaves of higher plants water vapour and CO_2 follow the same diffusion pathways, the stomata, so that the measurement of the resistance of one can be used in calculations concerning the other. This situation does not occur in lichens. Resistances to water vapour loss by lichen thalli are very low (<1 s cm^{-1}) at thallus water contents equal to or greater than those for optimal photosynthesis (Harris, 1976; Green and Snelgar, 1982a). At lower thallus water contents the resistance to water loss is determined by the water potential of the thallus (Jones and Norton, 1979; Green and Snelgar, 1982a; Fig. 4). The use of a diffusion porometer has shown that water loss is equal from both upper and lower surfaces of the thalli (Green and Snelgar, 1982a). Lichens behave much like a moistened filter paper (Smith, 1962) so that, because of their low resistances to water loss, boundary layer resistances and therefore thallus shape become important in the water relations of the thalli (Larson and Kershaw, 1976; Larson, 1979).

CO_2 Diffusion Pathways

Green et al. (1981) used a split incubation chamber to demonstrate that CO_2 diffusion in Sticta latifrons was entirely through the cyphellate lower cortex whilst, in Pseudocyphellaria amphisticta, it was through the pseudocyphellate upper cortex. This provided strong circumstantial evidence that the cyphellae/

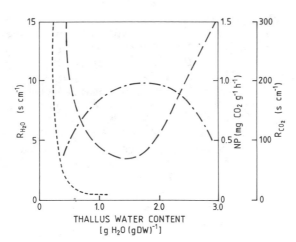

Fig. 4.   Summary of the relationship between total water diffusion resistance (rH_2O), total CO_2 diffusion resistance (rCO_2), net photosynthesis (NP) and thallus water content for Sticta latifrons.
rH_2O:----- ; rCO_2:— — — ; NP:—·—·—·— Modified from Green and Snelgar (1982a).

Table 1. CO_2 Diffusion Resistances (s cm^{-1}) in Stictaceae Thalli [1]

Lichen	Total CO_2 resistance	Medulla resistance	Pore[2] resistance	Biochemical resistance
Sticta latifrons	20	2.2	16.3	3.3
Pseudocyphellaria amphisticta	37	3.5	29.0	7.0

[1] values from Green and Snelgar (1981b), Snelgar et al. (1981b).
[2] cyphellae or pseudocyphellae

pseudocyphellae were gas exchange pores and allowed any models of CO_2 diffusion to omit the non—pored surface. It should be noted that this contrasts with the results obtained for water vapour diffusion.

Total CO_2 diffusion resistance (rCO_2) can be determined by standard higher plant techniques as described in Snelgar et al. (1981b). A typical result for S. latifrons is shown in Fig. 4 with rCO_2 being high at both low and high WC and reaching a minimum value at the optimum WC for photosynthesis. Values for a range of species are given in Snelgar et al. (1981b).

The subdivision of rCO_2 for a lichen thallus is difficult. In higher plants rCO_2 can be subdivided into stomatal and mesophyll components by using the stomatal resistance for water vapour to calculate the internal CO_2 concentration of the leaf. This is clearly not possible for lichens since CO_2 and water vapour are not diffusing along the same pathways (Fig. 4, Green and Snelgar, 1982a). At present, subdivision of rCO_2 can be done only by two methods. First, by subdividing rCO_2 into transport and biochemical resistances by manipulation of the relationship between NP and CO_2 concentration, a technique that has been open to criticism (Green and Snelgar, 1981b). Second, the resistances can be calculated from the morphology of the pathways using equations developed for higher plants. The latter method has been applied to Stictaceae lichens by Snelgar et al. (1981a). The important resistances to CO_2 diffusion were found to be in the fungal medulla and in the pore systems (cyphellae or pseudo-cyphellae), with the latter being up to eight-fold larger and the dominant diffusion resistance (Table 1). CO_2 diffusion must take place through an air-filled pathway since even the calculated resistance for the cortex alone (around 400 s cm^{-1}) is many times larger than total CO_2 diffusion resistances found for lichens at the optimum WC for photosynthesis (around 20-50 s cm^{-1}).

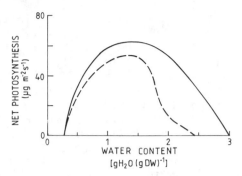

Fig. 5.   Net photosynthetic rate for top-dried (– – – –) and bottom-
          dried (————) thalli of <u>Sticta latifrons</u> with respect to
          thallus water content.
          Vertical axis is net photosynthesis in $\mu g\ m^{-2}\ s^{-1}$; horizontal
          axis is thallus water content in $gH_2O\ gDW^{-1}$. Modified from
          Green et al. (1981).

Photosynthetic Depression at High Thallus Water Content

     Depressed photosynthetic rates at high thallus water contents
result from increased rCO_2 (Snelgar et al., 1981b). Complete or
partial blockage of diffusion pathways by surface liquid water is the
generally accepted explanation since the diffusivity of CO_2 is 10,000
times slower in water than in air. A considerable body of evidence
now exists to support this explanation.

     Green and Snelgar (1981b) subdivided rCO_2 into biochemical and
transport (diffusion) components. The increase at high WC was
entirely due to increased transport resistance, a conclusion also
reached by Lange and Tenhunen (1981) from more limited evidence. The

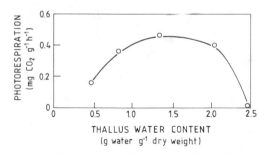

Fig. 6.   Photorespiration rate of <u>Sticta latifrons</u> at several thallus
          water contents.
          Photorespiration was measured as the difference between net
          photosynthesis at 1% and 21% oxygen. Modified from Snelgar
          and Green (1981b).

rôle of external water films in producing the increased resistance is
indicated by three lines of evidence. The first, and simplest, is the
finding that blotting of the soaked thallus of a Stictaceae lichen
will return its WC to optimum values for NP (Green and Snelgar,
1982a). Such blotting will only remove liquid water from the lichen.
Second, the change in NP as WC declines in thalli of S. latifrons
being dried by air currents directed onto either the upper or lower
thallus surface is shown in Fig. 5. Maximum NP is reached more
quickly, and at higher WC, on thalli subjected to bottom drying. This
is a consequence of the earlier removal of the water film from the
lower surface since CO_2 exchange in this lichen is entirely through
that surface (Green et al., 1981). Another important implication of
this result is that measurements of NP on a single thallus may not be
reproducible, even if made at identical WC, because of differences in
water location. Third, the external location of the resistance
increase is indicated by the loss of the oxygen effect on NP (Fig. 6);
a result most easily explained by increased internal recycling of
photorespired CO_2 due to the presence of a high surface resistance
(Snelgar and Green, 1981b).

The effect of a water film on diffusive resistance can be
calculated and a comparison of the effect of thin (1 or 10 µm thick)
water films in hepatics and lichens is given in Table 2. Resistances
are greatly increased by water films where CO_2 exchange occurs through
pore systems. This is a consequence of the small proportion of the
total surface area actually involved in gas exchange and the effect is

Table 2.   Effect of Surface Water Film on Total CO_2 Diffusion
           Resistance (rCO_2)

Thallus type	Species	At optimum WC	Additional Resistance	
			+ 10 µm H_2O	+ 1 µm H_2O
PORES	Marchantia[1]	75	+ 480	–
	Sticta[2]	25	–	+ 4700
SOLID	Monoclea[1]	119	+ 63	–

TOTAL   RESISTANCE ($s\ cm^{-1}$)

[1]   from Green and Snelgar (1982b).
[2]   from Snelgar et al. (1981a) for Sticta latifrons.

more marked in <u>S. latifrons</u> because pore diameters are so much smaller
in lichens than in hepatics (3.6 μm and 140 μm respectively). In
contrast, these water films had little effect on rCO_2 of the solid
hepatic, <u>Monoclea forsteri</u>, because CO_2 exchange occurred over the
whole surface. It follows that NP of lichens with a continuous
cortex, particularly fruticose species, would not be expected to be as
sensitive to water films.

Adaptive Significance of Stictaceae Thallus Structure

Cyphellae and pseudocyphellae. The Stictaceae lichens have
higher rCO_2 values than higher plants and would be expected to have a
linear relationship between NP and CO_2 concentration at normal ambient
CO_2 levels. However, at optimum WC for photosynthesis, this does not
appear to be true and biochemical limitations dominate (Green and
Snelgar, 1981b). Examination of published $NP:CO_2$ concentration curves
confirms that, although lichen photosynthesis is not CO_2 saturated
until above ambient levels, biochemical limitation is present well
below ambient CO_2 levels (Green and Snelgar, 1981a; Link et al.,
1984). This effectively means that, although CO_2 diffusion resistances of cyphellae and pseudocyphellae are high, they have little
effect on rates of photosynthesis. What, therefore, is the adaptive
significance of these pore systems? Three possible advantages are
proposed:

1. Recycling of respired CO_2. In contrast to the photosynthetic
organs of bryophytes and higher plants, a large proportion of the
lichen thallus is non-photosynthetic mycobiont. Therefore in the
light there is present, as well as phycobiont photorespiration,
continuing fungal respiration. The presence of high surface CO_2
diffusion resistances (the cortices and the pore systems), together
with a low internal medullar diffusion resistance, would encourage
recycling of respired or photorespired CO_2 inside the lichen thallus.
This recycling is extremely effective being 70% at zero external CO_2
concentration (Snelgar and Green, 1981b) and 100% at ambient CO_2
levels (Green et al., 1981).

2. Phycobiont microclimate. The high diffusion resistance of
the pore systems would effectively maintain a moist internal atmosphere and insulate the phycobiont cells from the external drying
atmosphere.

3. Surface Water Storage. Pore systems confine the CO_2 exchange
to a small proportion of the lichen thallus surface. Liquid water can
then be stored over the remainder of the surface without affecting CO_2
diffusion. In terms of water relations there is no disadvantage in
external water storage since, at high WC, lichens have no appreciable
diffusion resistance to water loss except that generated by the
boundary layer. Table 3 shows that more than 60% of total stored
water may be external and that the quantity can increase with environ-

extraction of the two main groups of lichen substances suggest that
such a protected diffusion pathway might exist (Fig. 7). The yellow
pulvinic lactones provide a granular filling to voids in the medulla,
whilst the colourless triterpenoids coat a central-medullary band of
hyphae and also hyphae in the vicinity of the pseudocyphellae. The
impression is gained of a water-storage region (pulvinic-lactone
regions) and a water-proofed CO_2 diffusion pathway from pseudocyphellae to phycobiont (triterpenoids).

SUMMARY

The application of standard techniques from higher plant photosynthesis have allowed a better understanding of the diffusion
pathways and their resistances in Stictaceae lichens. The thallus of
these lichens appears to show an unexpected degree of structural
complexity as a result of adaptations to allow both CO_2 diffusion and
water storage. Lichens have no known system of water translocation
and, therefore, an important method of meeting increased desiccation
stress is by increasing water storage capacity (Larson, 1979; Snelgar
and Green, 1981a). The low radiation, wet, evergreen forests allow
this option to be developed to its full in the form of the

Fig. 7.   Scanning electron micrographs of <u>Pseudocyphellaria colensoi</u>
          thalli that have been subjected to different extraction
          procedures.
          A: Transverse section of untreated thallus. Note the large
              quantities of lichen substances in the medulla which
              obscure the hyphae and cause excessive charging in the
              electron beam. Scale bar is 200 μm.
          B: Transverse section of thallus that has been extracted in
              acetone. All lichen substances have been removed
              revealing a medulla with large voids and a band of hyphae
              running through the middle of the medulla. Note
              pseudocyphella in lower cortex and what are apparently
              blue-green secondary phycobionts filaments in the lower
              medulla and unicellular primary phycobiont below upper
              cortex. Scale bar is 200 μm.
          C: Transverse section of medulla and pseudocyphella of a
              thallus that has been treated with hexane to selectively
              remove the triterpenoids but not the pulvinic-lactones.
              Note how the band of hyphae has been revealed in the
              middle of the medulla (compare A) as well as hyphae at
              the pseudocyphella and between the pseudocyphella and the
              hyphal band. Scale bar is 100 μm.
          D: Close up of the central medullary band of hyphae in C.
              Note how clean the hyphae are and the granular
              pulvinic-lactone above and below the band. Scale bar is
              20 μm.

cyphellate and pseudocyphellate thalli of the Stictaceae, which allow
the majority of the thallus surface to be available for storage. The
major limitation of the system is the possibility of pore blockage by
excess surface water. This risk is minimised by the pores usually
being in the lower cortex, often being raised clear of the tomentum,
rimmed, and impregnated with substances that appear to be hydrophobic,
even in Sticta spp. (Henssen and Jahns, 1974). It is suggested that,
as in bryophytes (Green and Snelgar, 1982b), the adaptations are
primarily to improve water relations and that improvements in CO_2
exchange, e.g. respiratory CO_2 recycling, are secondary. The adaptations described apply only to foliose Stictaceae lichens. The same
principles can, and should, be applied to other major morphologies,
particularly fruticose lichens. It is predicted that in those cases,
as well, water relations optimisation will be a satisfactory
explanation for much of the thallus structure.

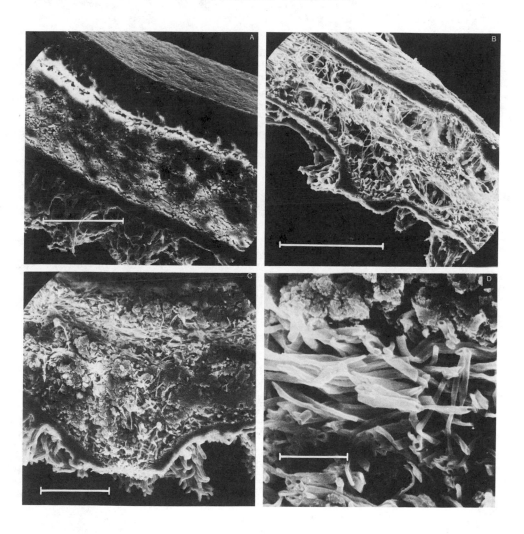

ACKNOWLEDGEMENTS

This article is a summary of several years of experimentation and many
lichenologists have helped with comments and discussion. Particular
thanks are due to Dr D.H. Brown, Professor Dr O.L. Lange, Dr M.C.F.
Proctor and Professor Dr L. Kappen. The University Grants Committee
supported the work with grants for equipment. The article was written
whilst T.G.A. Green was the holder of a research fellowship from the
Alexander von Humboldt - Stiftung, Bonn, West Germany. Dr C.K. Beltz,
School of Science, Waikato University, is thanked for assistance in
the preparation of the scanning electron micrographs, and Mr F. Bailey
for preparing the figures.

REFERENCES

Ahmadjian, V.A., and Hale, M.E., 1973, "The Lichens," Academic Press,
    New York and London.
Badger, M.R., Kaplan, A., and Berry, J.A., 1980, Internal inorganic
    carbon pool of Chlamydomonas reinhardtii. Evidence for a carbon
    dioxide - concentrating mechanism, Plant Physiology, 66: 407-413.
Brown, R.H., 1980, Photosynthesis of grass species differing in
    carbon dioxide fixation pathways IV. Analysis of reduced oxygen
    responses in Panicum milioides and Panicum schenckii, Plant
    Physiology, 65: 346-349.
Collins, C.A., and Farrar, J.F., 1978, Structural resistances to mass
    transfer in the lichen Xanthoria parietina, New Phytologist, 81:
    71-83.
Cowan, D.A., Green, T.G.A., and Wilson, A.T., 1979, Lichen metabolism
    I. The use of tritium labelled water in studies of anhydrobiotic
    metabolism in Ramalina celastri and Peltigera polydactyla, New
    Phytologist, 82: 489-503.
Coxson, D.S., and Kershaw, K.A., 1984, Low temperature acclimation of
    net photosynthesis in the crustaceous lichen Caloplaca
    trachyphylla (Tuck.) A. Zahlbr., Canadian Journal of Botany, 62:
    86-95.
Gates, D.M., and Papian, E.L., 1971, "Atlas of energy budgets of plant
    leaves," Academic Press, New York.
Goebel, K., 1926, Morphologische and biologische Studien VII. Ein
    Beitrag zur Biologie der Flechten, Annales Jardin Botanique
    Buitenzorg, 36: 1-83.
Green, T.G.A., and Snelgar, W.P., 1981a, Carbon dioxide exchange in
    lichens: Relationship between net photosynthetic rate and CO_2
    concentration, Plant Physiology, 68: 199-201.
Green, T.G.A., and Snelgar, W.P. 1981b, Carbon dioxide exchange in
    lichens: Partition of total CO_2 resistances at different thallus
    water contents into transport and carboxylation components,
    Physiologia Plantarum, 52: 411-416.
Green, T.G.A., and Snelgar, W.P., 1982a, Carbon dioxide exchange in
    lichens: relationship between the diffusive resistance of carbon
    dioxide and water vapour, The Lichenologist, 14: 255-260.

Green, T.G.A., and Snelgar, W.P., 1982b, A comparison of
    photosynthesis in two thalloid liverworts, Oecologia, 54:
    275-280.
Green, T.G.A., Snelgar, W.P., and Brown, D.H., 1981, Carbon dioxide
    exchange in lichens. CO_2 exchange through the cyphellate lower
    cortex of Sticta latifrons Rich., New Phytologist, 88: 421-426.
Hale, M.E., 1974, "The Biology of Lichens," Arnold, London.
Harris, G.P., 1976, Water content and productivity of lichens, in:
    "Water and Plant Life - Problems and Modern Approaches,"
    O.L. Lange, L. Kappen and E-D. Schulze, eds, pp. 452-468,
    Springer Verlag, Berlin.
Henssen, A., and Jahns, H.M., 1974, "Lichens, Eine Einfuhrung in die
    Flechtenkunde," George Thieme Verlag, Stuttgart.
Hill, D.J., 1976, The physiology of lichen symbioses, in:
    "Lichenology: Progress and Problems," D.H. Brown, D.L. Hawksworth
    and R.H. Bailey, eds, pp. 457-496, Academic Press, London.
Jarvis, P.G., 1971, The estimation of resistances to carbon dioxide
    transfer, in: "Plant Photosynthetic Production Manual of
    Methods," Z. Sestak, J. Catsky and P.G. Jarvis, eds, pp.
    566-631, W. Junk, The Hague.
Jones, G.J., and Norton, T.A., 1979, Internal factors controlling the
    rate of evaporation from fronds of some intertidal algae, New
    Phytologist, 83: 771-781.
Kennedy, R.A., Eastburn, J.L., and Jensen, K.G., 1980, C3-C4 photo-
    synthesis in the genus Mollugo: structure, physiology and
    evolution of intermediate characteristics, American Journal of
    Botany, 67: 1207-1217.
Kershaw, K.A., 1972, The relationship between moisture content and
    net assimilation rate of lichen thalli and its ecological
    significance, Canadian Journal of Botany, 50: 543-555.
Kershaw, K.A., 1977, Physiological-environmental interactions in
    lichens II. The pattern of net photosynthesis acclimation in
    Peltigera canina (L.) Willd. var. praetextata (Floerke in Somm.)
    Hue. and P. polydactyla (Neck.) Hoffm., New Phytologist, 79:
    377-390.
Kershaw, K.A., and Smith, M.M., 1978, Studies on lichen dominated
    systems XXI. The control of seasonal rates of net photo-
    synthesis by moisture, light and temperature in Stereocaulon
    paschale, Canadian Journal of Botany, 56: 2825-2830.
Lange, O.L., 1980, Moisture content and CO_2 exchange of lichens I.
    Influence of temperature on moisture-dependent net photosynthesis
    and dark respiration in Ramalina maciformis, Oecologia, 45:
    82-87.
Lange, O.L., Geiger, I.L., and Schulze, E-D., 1977, Ecophysiological
    investigations on lichens of the Negev Desert V. A model to
    simulate net photosynthesis and respiration of Ramalina
    maciformis, Oecologia, 28: 247-259.
Lange, O.L., and Matthes, U., 1981, Moisture dependent CO_2 exchange
    of lichens, Photosynthetica, 15: 555-574.
Lange, O.L., Schulze, E-D., Kappen, L., Buschbom, U., and Evenari, M.,

1975, Photosynthesis of desert plants as influenced by internal and external factors, in: "Perspectives of Biophysical Ecology," D.M. Gates and R.B. Schmerl, eds, pp. 121–143, Springer Verlag, New York.

Lange, O.L., Schulze, E-D., and Koch, W., 1970, Experimentell-ökologische Untersuchungen an Flechten der Negev-Wüste II. CO_2-Gaswechsel und Waserhaushalt von Ramalina maciformis (Del.) Bory am natürlichen Standort während der sommerlichen Trockenperiode, Flora, 159: 38–62.

Lange, O.L., and Tenhunen, J.D., 1981, Moisture content and CO_2 exchange of lichens II. Depression of net photosynthesis in Ramalina maciformis at high water content is caused by increased thallus carbon dioxide diffusion resistance, Oecologia, 51: 426–429.

Larson, D.W., 1979, Lichen water relations under drying conditions, New Phytologist, 82: 713–731.

Larson, D.W., 1980, Seasonal change in the pattern of net CO_2 exchange in Umbilicaria lichens, New Phytologist, 84: 349–369.

Larson, D.W., and Kershaw, K.A., 1975, Acclimation in arctic lichens, Nature, 254: 421–423.

Larson, D.W., and Kershaw, K.A., 1976, Studies on lichen dominated systems XVIII. Morphological control of evaporation in lichens, Canadian Journal of Botany, 54: 2061–2073.

Link, S.O., Moser, T.J., and Nash, T.H. III., 1984, Relationships among initial rate, $^{14}CO_2$ techniques with respect to lichen photosynthesis CO_2 dependencies, Photosynthetica, 18: 90–99.

Rhoades, F.M., 1977, Growth rates of the lichen Lobaria oregana as determined from sequential photographs, Canadian Journal of Botany, 55: 2226–2233.

Rundel, P.W., Bratt, G.G., and Lange, D.L., 1979, Habitat ecology and physiological response of Sticta filix and Pseudocyphellaria delisei from Tasmania, The Bryologist, 82: 171–180.

Scott, G.D., 1960, Studies of the lichen symbiosis, I. The relationship between nutrition and moisture content in the maintenance of the symbiotic state, New Phytologist, 59: 374–381.

Shomer-Ilan, A., Nissenbaum, A., Galun, M., and Waisel, Y., 1979, Effect of water regime on carbon isotope composition of lichens, Plant Physiologist, 63: 201–205.

Smith, D.C., 1962, The biology of lichen thalli, Biological Reviews, 37: 537–570.

Snelgar, W.P., 1981, "The Ecophysiology of New Zealand Forest Lichens with Special Reference to Carbon Dioxide Exchange," Ph.D. Dissertation. Waikato University, Hamilton, New Zealand.

Snelgar, W.P., Brown, D.H., and Green, T.G.A., 1980, A provisional survey of the interaction between net photosynthetic rate, respiratory rate and thallus water content in some New Zealand cryptogams, New Zealand Journal of Botany, 18: 247–256.

Snelgar, W.P., and Green, T.G.A., 1980, Carbon dioxide exchange in lichens: low carbon dioxide compensation levels and lack of apparent photorespiratory activity in some lichens, The

Bryologist, 83: 505-507.

Snelgar, W.P., and Green, T.G.A., 1981a, Ecologically linked
variations in morphology, acetylene reduction and water relations
in Pseudocyphellaria dissimilis, New Phytologist, 87: 403-411.

Snelgar, W.P., and Green, T.G.A., 1981, Carbon dioxide exchange in
lichens: Apparent photorespiration and possible role of CO_2
refixation in some members of the Stictaceae (Lichenes), Journal
of Experimental Botany, 32: 661-668.

Snelgar, W.P., and Green, T.G.A., 1982, Growth rates of Stictaceae
lichens in New Zealand beech forests, The Bryologist, 85:
301-306.

Snelgar, W.P., Green, T.G.A., and Beltz, C.K., 1981a, Carbon dioxide
exchange in lichens: Estimation of internal thallus CO_2 transport
resistances, Physiologia Plantarum, 52: 417-422.

Snelgar, W.P., Green, T.G.A., and Wilkins, A.L., 1981b, Carbon dioxide
exchange in lichens: resistances to CO_2 uptake at different
thallus water contents, New Phytologist, 88: 353-361.

Note added in proof: The names of many New Zealand lichens have been
corrected in D.J. Galloway's recent, excellent, "Flora of New Zealand
Lichens" (1985, P.D. Hasselberg, New Zealand Government Printer,
Wellington, New Zealand). Reference to Pseudocyphellaria amphisticta
in this article should, therefore, be corrected to P. lividofusca.

CO_2 EXCHANGE IN LICHENS: TOWARDS A MECHANISTIC MODEL

S.O. Link[a], M.F. Driscoll[b], and T.H. Nash III[c]

[a] Department of Range Science, Utah State University
Logan, UT 84322, U.S.A.
[b] Department of Mathematics and [c] Botany-Microbiology
Arizona State University, Tempe, AZ 85287, U.S.A.

INTRODUCTION

Photosynthesis, a process by which radiant energy is converted to chemical energy, is, of course, the essential foundation for most forms of life. To measure photosynthetic rate in atmospheric environments, one normally focuses on CO_2 influx, which is partially balanced by respiratory CO_2 efflux, and consequently the measurement of CO_2 exchange in the light is called net photosynthesis (P_n). Ideally, to measure the absolute rate of CO_2 incorporation, the enzyme activity of ribulose-1,5-biphosphate carboxylase, the enzyme controlling the incorporation of CO_2 should be measured <u>in vivo</u>. However, in practice it is much easier to measure CO_2 exchange, particularly when dealing with a whole plant or intact portion thereof.

Photosynthetic rate is controlled by a variety of environmental factors, such as temperature, CO_2 concentration, light and water status. To understand how these factors affect photosynthesis requires a multivariate approach which may best be accomplished by use of a mechanistic model. Such models are usually based on controlled, laboratory experimentation, where unbiased parameter estimates may be obtained (Reed et al., 1976). Once constructed, the model may be used to explore the multivariate relationships of environmental variables and, when compared with measured data, allow for assessment of potential interactions among the environmental variables. Ultimately, the appropriate test for any physiological model is the degree to which field performance can be predicted.

In passing from the external air to the carboxylase sites, CO_2 molecules encounter a set of resistances. In fact, photosynthetic rate is appropriately expressed as CO_2 flux adjusted for these resistances. With vascular plants a major component of these resistances is controlled by the degree to which the stomates (through which gas exchange occurs) are open. Stomatal conductance is estimated indirectly by measuring transpiration rates (H_2O efflux) and these estimates are incorporated into many vascular plant photosynthetic models. Lichens, in contrast to vascular plants, possess no stomates and consequently water efflux does not convey parallel information, particularly as part of the water originates from intercellular spaces. In constructing a CO_2 exchange model for lichens, the relationship to water must, therefore, be defined in a different manner compared to that used in vascular plant models, although other relationships, such as those to light, may be similar to those used in vascular plant models.

Until now, lichen CO_2 exchange models have been largely empirically defined (Lange et al., 1977; Paterson et al., 1983). In the present case we attempt to develop mechanistic, univariate relationships of lichen CO_2 exchange to the lichen's internal water content, light intensity and thallus temperature. Subsequently these relationships are combined into multivariate models which are then used to predict CO_2 exchange in the field. Much of the inspiration for the development of our model is drawn from previous vascular plant modelling of Tenhunen et al. (1980) and Reed et al. (1976). For a more detailed exposition of the model and its application see Link (1983).

Methods

Laboratory experimental procedures follow those of Link et al. (1984) and Link and Nash (1984a,b,c). The gas exchange system used is a modification of the closed system developed by Larson and Kershaw (1975). In the case of the multivariate model relative total resistances are estimated from a matrix experiment reported in Link and Nash (1984c). Field experiments were conducted in the Chiricahua Mountains, Arizona (31°55'N, 109°15'W). The study organism was Parmelia praesignis Nyl. Procedures for the latter studies are discussed in Link (1983).

All computations were performed on an IBM 3081 computer using the Statistical Analysis System (SAS) software package (Helwig and Council, 1981). Non-linear regressions for parameter estimation were performed with the NLIN procedure using the method of Marquardt (1970). Parameter estimates are reported with associated asymptotic standard errors (Bard, 1974).

The Basic Model Equation

Since net photosynthesis, P_n, is taken here to be the difference between CO_2 influx, P_c, at (21% O_2) and CO_2 efflux, R_t, $(P_n = P_c - R_t)$ and since CO_2 efflux, R_t, is the total of photorespiration, R_p, and dark respiration, R_d, $(R_t = R_p + R_d)$ we can express net photosynthesis as $P_n = P_c - R_p - R_d$. However, photorespiration is difficult to measure independently, so we model net photosynthesis, P_n, as

$$P_n = P_{cp} - R_d, \tag{1}$$

where P_{cp} is defined by $P_{cp} = P_c - R_p$ (Reed et al., 1976) and is called the carboxylation rate.

UNIVARIATE RELATIONSHIPS TO WATER CONTENT

Internal water content has an overriding influence on gas exchange processes in lichens. When, for example, water content, W, is less than that amount required for maximal carboxylation, W_{max}, rates are reduced by internal resistances (Collins and Farrar, 1978; Green and Snelgar, 1981a). At levels above W_{max}, the partial filling of air spaces by liquid water results in physical resistance to CO_2 flux (Lange, 1980; Lange and Tenhunen, 1981; Snelgar et al., 1981a; Green and Snelgar, 1981b).

Since internal CO_2 concentrations cannot be estimated by the standard techniques used for vascular plants, the effect of water content, W, on carboxylation rate, P_{cp}, and on dark respiration, R_d, (but especially on P_{cp}) remains somewhat elusive. The use of classical diffusion formulations and geometrical considerations has at least led to some recent progress in the definition of various CO_2 resistance components (Collins and Farrar, 1978; Lange, 1980; Lange and Tenhunen, 1981; Green et al., 1981; Green and Snelgar, 1981a, 1982; Snelgar et al., 1981a,b).

Modelling Carboxylation Rate and Water Content

To describe the dependence of P_{cp} on W, we develop a relation between P_{cp} and apparent carboxylation CO_2 concentration, denoted by C_{ac}. We refer to C_{ac} as an apparent concentration because, as a lichen thallus dries to levels at which diffusion resistance no longer dominates, increased carboxylation resistance plays an important role in reducing gas exchange rates (Bewley, 1979).

Our model for P_{cp} on W is first given in general terms, and is then briefly described in a manner which pertains more closely to the mechanics of gas exchange.

The Model in Terms of P_n

At given light and temperature levels, net photosynthesis is conveniently described as a polynomial function of W, as in

$$P_n = b_1W + b_2W^2 + \dots . \tag{2}$$

The coefficients in this relationship are estimated by statistical regression analysis of an empirical matrix data set (Link and Nash, 1984c). From the fitted equation, we can compute the maximum net photosynthesis level, $P_{n,max}$, and the water level, W_{max}, at which this maximal activity is achieved. When $W \geq W_{max}$, the CO_2 compensation point is $\Gamma = 2.6$ mmol m^{-3} (see Link et al., 1984). When $W < W_{max}$, we have an apparent compensation point of $\Gamma = 2.6\ P_n/P_{n,max}$. That is,

$$\Gamma = 2.6\ P_n/P_{n,max}\ \text{(for } W < W_{max}) \text{ or } \Gamma = 2.6\ \text{(for } W \geq W_{max}). \tag{3}$$

C_{ac} is a function of photosynthesis, Γ, and external CO_2 concentration C_a. We assume that photosynthetic enzymes are maximally active when $W \geq W_{max}$ but that carboxylation rates are reduced due to increased resistance to CO_2 diffusion into the thallus. These assumptions suggest that $C_{ac} = \Gamma$ when $C_a = \Gamma$. At lower levels of W, enzyme activity decreases, reducing the carboxylation rate, so we assume that, as W goes to zero, C_{ac} goes to zero, as do P_n (by equation 2) and Γ (by equation 3). These results suggest

$$C_{ac} = (0.80\ P_n/P_{n,max})\ (C_a - \Gamma) + \Gamma, \tag{4}$$

along with (2) and (3), as a model for apparent carboxylation concentration as a function of water content.

Since experimental CO_2 concentrations were at atmospheric levels, the relationship between C_{ac} and carboxylation rate is taken to be linear, so that

$$P_{cp} = k_fC_{ac}, \tag{5}$$

where k_f is a marginal scaling factor. Equations (2) through (5) provide our model for the univariate dependence of P_{cp} on W.

Model in Terms of Resistance

We observe here that net photosynthesis and resistance to photosynthesis are inversely related (in a complementary, not a reciprocal, sense). See, for example, the graphs given by Green and Snelgar (1981a, 1982) and Snelgar et al. (1981b). This inverse relationship is present over a full range of hydration conditions, and may be given as

$$R_t = 0.80\ (P_{n,max} - P_n)/P_{n,max} + 0.20. \tag{6}$$

The constants in this equation were chosen as follows: first, so that $P_n = 0$ would correspond to $R_t = 1$ (and hence our reference to R_t as a relative total resistance) and, second, so that $P_n = P_{n,max}$ would correspond to $R_t = 0.20$ (a minimal value suggested by graphs in Snelgar et al. (1981b) in which minimal resistance levels ranged from 0.10 to 0.30). The complementary nature of this relationship can be emphasized by rewriting (6) in the form

$$(R_t - 0.20)/0.80 + (P_n/P_{n,max}) = 1. \qquad (7)$$

The relationship (7) is, obviously, a simplification of a complex phenomenon, and no attempt is made in it to distinguish between the various components of resistance to photosynthesis. Also, (7) is not as specific a statement as (6). Nonetheless, (7) would seem to be of fundamental importance in understanding the mechanics of gas exchange in lichens. We illustrate by using (7) to rewrite (3) as

$$\Gamma = 2.6 \ (1 - (R_t - 0.20)/0.80) \ \text{(for } W < W_{max})$$
$$\text{or } \Gamma = 2.6 \ \text{(for } W \geq W_{max}),$$

and again to rewrite (4) as

$$C_{ac} = (1 - R_t) \ (C_a - \Gamma) + \Gamma.$$

Modelling Dark Respiration and Water Content

The dependence of dark respiration, R_d, on water content, W, is taken to be

$$R_d = k_s \ W/(W + W_{half}), \qquad (8)$$

where k_s is another scaling factor and W_{half} is the value of W for which dark respiration is half-maximal. Details behind this model may be found in Link and Nash (1984c).

Univariate Results for Water Content

Four sets of thallus tips were used to obtain the upper curve in Fig. 1, which related P_n to W by a regression fit of equation (2). The fitted curve accurately reflects the rapid rise in P_n at low W, the broad range of near optimal W (with $P_{n,max} = 0.00106$ mmol m^{-2} s^{-1} at $W = 1.41$), and the drop in activity at high W. The fit is highly significant ($F = 125$, $p = 0.0001$, $R^2 = 0.92$).

Figure 1 also shows R_d as a saturation function of W, the model equation being equation (8). The fit was again highly significant ($F = 125$, $p = 0.0001$) with $W_{half} = 1.21$ and maximal $R_d = 0.0007$ mmol $m^{-2}s^{-1}$. Since $P_{cp} = P_n + R_d$, Fig. 1 implicitly shows carboxylation rate P_{cp} as a function of W.

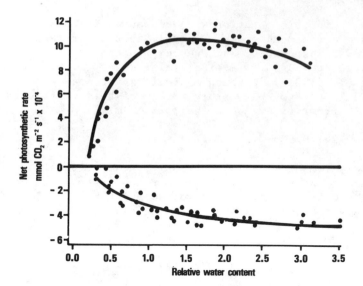

Fig. 1.   The relationship between CO_2 exchange and thallus water
content in the light (0.24 mmol $m^{-2}s^{-1}$PAR) or in the dark at
16°C.

Figure 2 supports the linear relationships assumed between P_{cp}
and C_{ac}. The concentration of C_{ac} points between 10 and 12 mmol m^{-3}
correspond to the points in Fig. 1 which are at maximal levels. The
correspondence between $P_{cp} = P_n + R_d$ and the ordinate (P_n) of Fig. 2
is direct, it being assumed that respiration in the light equals
respiration measured in the dark, R_d, under the same temperature
condition.

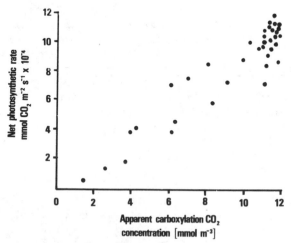

Fig. 2.   Relationship between P_n and computed apparent carboxylation
CO_2 concentration at 0.24 mmol $m^{-2}s^{-1}$PAR and at 16°C.

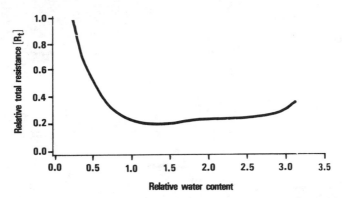

Fig. 3.   The total relative resistance as a function of thallus
relative water content at 0.24 mmol $m^{-2}s^{-1}$PAR and at $16°C$.

Figure 3 shows total relative resistance, R_t, against water
content, W. The complementary nature of R_t and P_n is strikingly clear
from comparison of Fig. 3 with the upper part of Fig. 1.

UNIVARIATE RELATIONSHIPS TO LIGHT

The environmental condition which most obviously affects photosynthesis is the level of light (or, more correctly, photosynthetically active radiation). The relationships of P_{cp} and R_d to light (I)
are similar to those for vascular plants.

Modelling Carboxylation Rate and Light

A general formulation commonly used to describe P_{cp} as a function
of I is

$$P_{cp} = P_{cp,max} I /(I_{half} + I^n)^{1/n}, \tag{9}$$

where $P_{cp,max}$ is the maximal carboxylation rate and I_{half} is that
light level which yields a half-maximal carboxylation rate. In
particular, setting n = 1 in (9) gives the Michaelis-Menten
formulation; similarly, n = 2 yields the Smith equation.

For modelling purposes, it is convenient to rewrite (9) by
dividing both numerator and denominator on the right-hand side by
I_{half}. We obtain

$$P_{cp} = k_g I/(1 + c I^n)^{1/n}, \tag{10}$$

in which the scaling factor k_g and the parameter c must be estimated.
Since k_g is the derivative of (9) at I = 0, it is sometimes called the
initial or maximal light-use efficiency. Once it and c have been

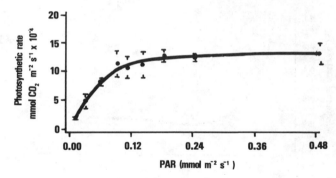

Fig. 4.   Relationship between the estimated carboxylation rate and
           light at 16°C, 21% O_2 and optimal water content.

estimated, we can compute estimated values of $I_{half} = (1/c)^{1/n}$ and
$P_{cp,max} = k_g I_{half}$.

Modelling Dark Respiration and Light

No model is needed here because, after all, dark respiration
occurs in the dark. But it would be more precise to say that dark
respiration, R_d, is that amount of respiration which occurs
independent of light, and that we assumed that an equivalent amount of
mitochondrial respiration is occurring when light is present.

Univariate Results for Light

To fit equation (9), we investigated values of n between 1 and 3,
using the F-test for lack-of-fit (Neter and Wasserman, 1974) as a
criterion for choosing n. The best value for n using this criterion
was n = 1.95, although all values of n except those near 1 passed this
fit test. With n = 2, the Smith equation was used.

The resulting fit, given as Fig. 4, shows saturation at roughly
0.16 mmol $m^{-2}s^{-1}$, and is highly significant (F = 601, p = 0.0001, R^2 =
0.95) with a highly insignificant lack-of-fit (F = 1.48, p = 0.7).
For this fit, $P_{cp,max}$ = 0.00131 mmol $m^{-2}s^{-1}$ and I_{half} = 0.072 mmol
$m^{-2}s^{-1}$. The maximal light-use efficiency is 0.018 mmol CO_2 mmol
photon^{-1}. (The additional hatch marks in the figure depict a 95
per cent confidence band on mean P_{cp} response to light.)

UNIVARIATE RELATIONSHIPS TO TEMPERATURE

The dependence of carboxylation rate, P_{cp}, and dark respiration,
R_d, on atmospheric temperature, T, are, for lichens, again similar to
those for vascular plants.

Modelling Carboxylation Rate and Temperature

To describe the relationship bewtween P_{cp} and T, we use the Arrhenius equation

$$P_{cp} = \frac{k_h \, T_k \, \exp \, (-\Delta H \neq /RT_k)}{1 + \exp \, (\Delta S/R - \Delta H_1/RT_k)} \tag{11}$$

Here, k_h is a marginal scaling factor, $\Delta H \neq$ is the energy of activation of the enzyme catalysed photosynthetic reaction, ΔS is the entropy of the denaturation equilibrium, ΔH_1 is the energy of activation of the denaturation equilibrium, T_k is the thallus temperature in degrees Kelvin and R is the universal gas constant.

Modelling Dark Respiration and Temperature

The model for R_d as a function of T is another Arrhenius equation, namely

$$R_d = k_t \, \exp \, (-E/RT_k). \tag{12}$$

In (12), k_t is a scaling factor and E is the energy of activation.

Univariate Results for Temperature

Figure 5 shows the results of fitting the Arrhenius equation (11). The resultant fit shows typical responses of enzymes to temperature, with P_{cp} gradually increasing at low T to an optimal level at about T = 24°C, and then decreasing at still higher temperatures. The fit was highly significant (F = 8.44, p = 0.00001,

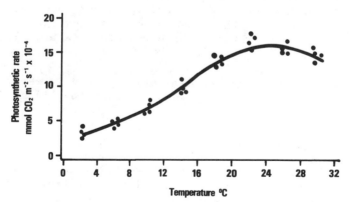

Fig. 5.   The relationship between the estimated carboxylation rate and temperature at 0.24 mmol m⁻²s⁻¹, 21% O_2 and optimal water content.

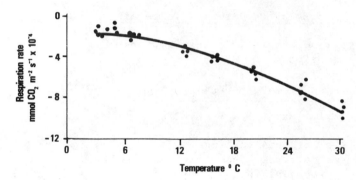

Fig. 6.   The relationship between dark respiration and temperature at
           optimal water content.

$R^2 = 0.97$). Estimated values of the parameters in equation (11) were
$\Delta H\neq = 70066$ J mol^{-1}, $\Delta S = 498$ J mol^{-1}°C^{-1}, $\Delta H_1 = 147990$ J mol^{-1},
and $k_h = 2.33 \times 10^{-7}$. The results are at optimal thallus water
content.

     The fitted relationship of R_d to T, again for optimal thallus
water content, is given in Fig. 6. Dark respiration increased (in
absolute value) as temperature increased. The estimated parameter
values for equation (12) were $E = 45053$ J mol^{-1} and $k_t = 54114$. This
relationship is also highly significant ($F = 693$, $p = 0.00001$,
$R^2 = 0.96$).

A MULTIVARIATE GAS EXCHANGE MODEL

     The previous sections have concerned univariate relationships.
The goal of the present section is to combine these several
relationships into models which describe gas exchange in lichens as
simultaneously dependent on W, I, and T, rather than describing these
dependencies marginally, that is, rather than describing one
dependency (on T, say) while the other factors (W and I) are held
fixed.

     Initially, we continue to deal with P_{cp} and R_d separately. The
two models are then combined by means of equation (1) to describe P_n
multivariately.

Modelling Carboxylation Rate

     Equations (2) through (5) postulate an implicit relationship
between P_{cp} and W, equation (10) postulates a relationship between P_{cp}
and I, and equation (11) postulates one between P_{cp} and T. Let us
denote these postulated marginal relationships by

     $P_{cp} = k_f \, f^*(W)$, $= k_g \, g(I)$, and $= k_h \, h(T)$,

the * on f*(W) being a reminder that this relationship is only
implicitly described.

These three functions suggest the product

$$P_{cp} = k_f k_g k_h \; f*(W) \; g(I) \; h(T) \tag{13}$$

as an initial description of multivariate dependency. The scaling
factor in (13) is formally equal to the product of the marginal
scaling factors k_f, k_g, and k_h. However, the scaling factor (k_c) in
the multivariate relationship should be estimated directly from the
multivariate data, rather than being obtained as the numerical product
of marginally estimated marginal scaling factors. We are led to the
form

$$P_{cp} = k_c \; f*(W) \; g(I) \; h(T). \tag{14}$$

The form of the function f*(W) may be made explicit by substituting
equations (2), (3) and (4) in equation (5). Doing so was of no
benefit, however, since the data available for our study would not
support estimation of the numerous parameters which then appeared in
(14).

This difficulty was resolved by using the simple relationships of
equation (5) as the postulated marginal expression for dependence of
P_{cp} on W. Writing equation (5) in functional notation as

$$P_{cp} = k_f \; f(C_{ac}) = k_f C_{ac}, \tag{15}$$

our final multivariate model for P_{cp} is

$$P_{cp} = k_c \; f(C_{ac}) \; g(I) \; h(T), \tag{16}$$

measurement vectors (C_a, W, I, T) being converted to triples (C_{ac}, I,
T) by means of the transformations derived previously.

The use of (16) in place of the more appealing (14) does somewhat
weaken the value of the multivariate model for P_{cp}. But we do receive
compensation in the form of more precision in parameter estimation.

Modelling Dark Respiration

The multivariate model for R_d is constructed from the postulated
relationships (8) and (12) in a manner analogous to that leading to
(14). There are, however, two differences: first, the marginal
dependency on W is explicit and, second, there is no dependency on I.

The consequent multivariate model for R_d as a function of W and T
is given by

$$R_d = k_d \, W \, \exp \, (-E/RT_k)/(W + W_{half}), \tag{17}$$

in which k_d is a (bivariate) scaling factor.

Multivariate Results

Having predicted C_{ac} for each of 575 data points (Link and Nash, 1984c), we proceeded to fit the multivariate models (16) and (17) simultaneously for $P_n = P_{cp} - R_d$. The parameter estimates obtained are given in Table 1, along with asymptotic standard errors for these estimates. The fit is highly significant (F = 3161, p = 0.0001, R^2 = 0.85), and yields residual plots consistent with the requirement that errors be normally and independently distributed.

This multivariate fit for P_n was then tested against 129 field data points. Firstly, in order to estimate C_{ac} for the field data, we did a multivariate full-polynomial, third-degree regression analysis of the dependence of C_{ac} on C_a, W, I, and T for the laboratory data. The regression analysis selected one of several possible predicting equations as the best. This particular equation, which did not involve I, gave a usefully strong fit (F = 532, p = 0.0001, R^2 = 0.91). Values of C_{ac} were then computed from combinations of C_a, W, and T observed in the field. Secondly, with these C_{ac} estimates, we

Table 1. Parameter Estimates and Asymptotic Standard Errors as Determined by Non-linear Regression for the Lichen Gas Exchange Model given by Equations 1, 16, and 17.

Parameter		Estimate	Asymptotic std. error
k_c		2.68×10^5	1.85×10^6
c	$m^2 s(mmol \ photon)^{-1}$	96.5	8.8
$\Delta H^{\neq}$	$J \ mol^{-1}$	5.86×10^4	1.58×10^4
ΔS	$J \ mol^{-1} {}^\circ C^{-1}$	364	88
ΔH_1	$J \ mol^{-1}$	1.08×10^5	0.28×10^5
k_d		2.16×10^3	6.09×10^3
W_{half}		0.77	0.20
E	$J \ mol^{-1}$	3.64×10^4	0.69×10^4

predicted field P_n with the multivariate fit for P_n. This yielded significant predictions ($F = 289$, $p = 0.0001$, $R^2 = 0.70$). For 9 individual experimental units, 6 were adequately predicted while 3 were under-predicted (n ranged from 5 to 44). The model does, however, appear to under-predict net photosynthesis by an average of 17 per cent for all data taken together (n = 129). We suspect that a portion of under-prediction is due to different experimental environments in the laboratory, storage of thalli under essentially unnatural conditions, and population variation.

FINAL REMARKS

We have provided univariate and multivariate gas exchange models with physiologically meaningful structure. Our formulation differs from previous ones (for example: Reed et al., 1976; Tenhunen et al., 1976a,b; Gates, 1980), but much of the inspiration for our development was drawn from these previous efforts.

The multivariate models given here, whether for net photosynthesis or its components, carboxylation rate and dark respiration, are, of course, just initial advances towards a multi-dimensional understanding of lichen gas exchange. The models need further development. In particular, it should be possible to enhance the method by which marginal relationships are used to construct a possible corresponding multivariate relationship.

ACKNOWLEDGEMENTS

We wish to acknowledge the financial assistance of NSF grant DEB 792153.

REFERENCES

Bard, Y., 1974, "Nonlinear Parameter Estimation," Academic Press, New York and London.

Bewley, J.D., 1979, Physiological aspects of desiccation tolerance, Annual Review of Plant Physiology, 30: 195-238.

Collins, C.R., and Farrar, J.F., 1978, Structural resistances to mass transfer in the lichen Xanthoria parietina, New Phytologist, 81: 71-83.

Gates, D.M., 1980, "Biophysical Ecology," Springer-Verlag, New York, Heidelberg and Berlin.

Green, T.G.A., and Snelgar, W.P., 1981a, Carbon dioxide exchange in lichens. Partition of total CO_2 resistances at different thallus water contents into transport and carboxylation components, Physiolgia Plantarum, 52: 411-416.

Green, T.G.A., and Snelgar, W.P., 1981b, Carbon dioxide exchange in lichens. Relationship between net photosynthetic rate and CO_2 concentration, Plant Physiology, 68: 199-201.

Green, T.G.A., and Snelgar, W.P., 1982, Carbon dioxide exchange in lichens. Relationship between the diffusive resistance of carbon dioxide and water vapour, The Lichenologist, 14: 255-260.

Green, T.G.A., Snelgar, W.P., and Brown, D.H., 1981, Carbon dioxide exchange in lichens. Carbon dioxide exchange through the cyphellate lower cortex of Sticta latifrons Rich, New Phytologist, 88: 421-426.

Helwig, J.T., and Council, K.A., 1981, "Statistical Analysis System," SAS Institute Inc., Cary.

Lange, O.L., 1980, Moisture content and CO_2 exchange in lichens I. Influence of temperature on moisture-dependent net photosynthesis and dark respiration in Ramalina maciformis, Oecologia, 45: 82-87.

Lange, O.L., Geiger, I.L., and Schulze, E-D., 1977, Ecophysiological investigations on lichens of the Negev Desert. V. A model to simulate net photosynthesis and respiration of Ramalina maciformis, Oecologia, 28: 247-259.

Lange, O.L., and Tenhunen, J.D., 1981, Moisture content and CO_2 exchange of lichens. II. Depression of net photosynthesis in Ramalina maciformis at high water content is caused by increased thallus carbon dioxide diffusion resistance, Oecologia, 51: 426-429.

Larson, D.W., and Kershaw, K.A., 1975, Measurement of CO_2 exchange in lichens: a new method, Canadian Journal of Botany, 53: 1535-1541.

Link, S.O., 1983, Ecophysiological studies and a mathematical model of carbon dioxide gas exchange for the lichen, Parmelia praesignis Nyl., Ph.D. Dissertation, Arizona State University, Tempe, AZ.

Link, S.O., Moser, T.J., and Nash III, T. H., 1984, Relationships among initial rate, closed chamber, and 14-CO_2 techniques with respect to lichen photosynthetic CO_2 dependencies, Photosynthetica, 18: 90-99.

Link, S.O., and Nash III, T.H., 1984a, Ecophysiological studies of the lichen, Parmelia praesignis Nyl. Population variation and the effect of storage conditions, New Phytologist, 96: 249-256.

Link, S.O., and Nash III, T.H., 1984b, A mathematical description of the effect of resaturation on net photosynthesis in the lichen Parmelia praesignis Nyl., New Phytologist, 96: 257-262.

Link, S.O., and Nash III, T.H., 1984c, The influence of water on CO_2 exchange in the lichen Parmelia praesignis Nyl., Oecologia, 64: 204-210.

Marquardt, D.W., 1970, Generalized inverses, ridge regression, biased linear estimation, and nonlinear estimation, Technometrics, 12: 591-612.

Neter, J., and Wasserman, W., 1974, "Applied Linear Statistical Models," R.D. Irwin, Inc., Homewood.

Paterson, D.R., Paterson, E.W., and Kenworthy, J.B., 1983, Physiological studies on temperate lichen species. I. A mathematical model to predict assimilation in the field, based on laboratory responses, New Phytologist, 94: 605-618.

Reed, K.L., Hamerly, E.R., Dinger, B.E., and Jarvis, P.G., 1976,

An analytical model for field measurement of photosynthesis, Journal of Applied Ecology, 13: 925-942.

Snelgar, W.P., Green, T.G.A., and Beltz, C.K., 1981a, Carbon dioxide exchange in lichens: Estimation of internal thallus CO_2 transport resistances, Physiologia Plantarum, 52: 417-422.

Snelgar, W.P., Green, T.G.A., and Wilkins, A.L., 1981b, Carbon dioxide exchange in lichens: Resistances to CO_2 uptake at different thallus water contents, New Phytologist, 88: 353-361.

Tenhunen, J.D., Yocum, C.S., and Gates, D.M., 1976a, Development of a photosynthetic model with an emphasis on ecological applications. I. Theory. Oecologia, 26: 89-100.

Tenhunen, J.D., Weber, J.A., Yocum, C.S., and Gates, D.M., 1976b, Development of a photosynthetic model with an emphasis on ecological applications. II. Analysis of a data set describing the P_m surface, Oecologia, 26: 101-119.

Tenhunen, J.D., Hesketh, J.D., and Gates, D.M., 1980, Leaf photosynthesis models, in: "Predicting Photosynthesis for Ecosystems," J.D. Hesketh, and J.W. Jones, eds, pp. 123-181, CRC press Inc., Boca Raton.

PHOTOSYNTHETIC CAPACITY CHANGES IN LICHENS AND THEIR POTENTIAL ECOLOGICAL SIGNIFICANCE

K.A. Kershaw

Department of Biology, McMaster University
Hamilton, Ontario, Canada

INTRODUCTION

For a number of reasons lichens are highly suitable terrestrial plants in which to interpret seasonal changes in photosynthesis on a mechanistic basis. Because they lack stomata it is possible to control diffusive resistances for CO_2 both throughout short-term experiments and seasonally. The expression of metabolic rates on a biomass (usually dry weight) basis is not complicated by significant growth during experiments. At least in Peltigera, photorespiration is absent (Coxson et al., 1982; but see Green et al., this volume) and cannot, therefore, be responsible for seasonal changes in net photosynthesis.

Capacity changes in lichen photosynthesis and respiration were first documented by Stålfelt (1939). He showed that photosynthesis of Evernia prunastri measured in May and December, increased in summer while the respiration rate remained constant (Fig. 1). However, both photosynthetic and respiratory rates changed in Ramalina farinacea and there was an apparent shift in the temperature optimum of net photosynthesis. Although the close agreement between E. prunastri respiration rates at different times of the year suggest that experimentally there was adequate control over inter-replicate variation, the statistical significance of these differences is still difficult to assess.

The failure of Stålfelt's work to significantly influence lichen physiological studies probably reflects a lack of any reasonable mechanistic understanding of photosynthetic events at that time. We can now appreciate that Stålfelt probably described photosynthetic events which have only recently become more fully explicable. As his

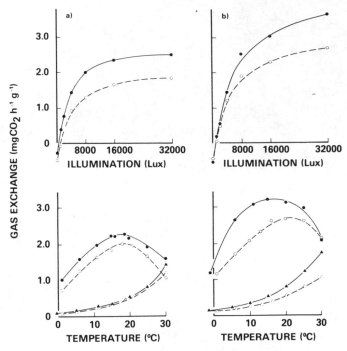

Fig. 1. Responses of gas exchange in a) _Evernia prunastri_ and b)
         Ramalina farinacea to illumination or temperature measured in
         May (closed symbols, solid line) and December (open symbols,
         dotted line). Gas exchange measured in the light (circles)
         or dark (triangles). Redrawn from Stalfelt (1939).

initial isolated observations were insufficiently replicated they
tended to be overlooked but subsequent studies indicate that he
probably did describe significant seasonal photosynthetic changes.

     A primary difficulty in assessing seasonal photosynthetic
capacity changes has been to establish what constitutes a
statistically significant difference rather than what is merely normal
experimental variation. In this context it is important to note that
Kärenlampi et al. (1975), Kallio and Kärenlampi (1975), Kershaw
(1977b) and Nash et al. (1980) have all documented the considerable
range of photosynthetic, respiratory or net photosynthetic rates which
occur between young and old parts of many lichen thalli. In _Peltigera
polydactyla_ for example (Kershaw, 1977b), mean maximum rates of net
photosynthesis in the young thallus lobes were 4.2 mg CO_2 $g^{-1}h^{-1}$
compared to 2.6 mg CO_2 $g^{-1}h^{-1}$ for older parts of the thallus and
respiration rates were 5.6 mg CO_2 $g^{-1}h^{-1}$ and 3.0 mg CO_2 $g^{-1}h^{-1}$ for the
same tissues (Fig. 2). Nash et al. (1980) demonstrated an even
greater range of photosynthetic capacity in the upper, central and
lower portions of the podetia of _Cladonia stellaris_ and _Cladonia_

The onset of winter conditions signals a variety of changes in photosynthetic and respiratory activity in members of the genus Umbilicaria. These changes have been documented in detail by Larson (1978, 1980). Larson showed that the winter induced an increase in respiration rate with or without a change in photosynthetic capacity in U. deusta and U. mammulata. Alternatively there may be a decreased respiration rate in winter, as has been detected in U. vellea and, by Tegler and Kershaw (1980), in Cladonia rangiferina. A third type of response is shown by U. papulosa (Larson, 1980), Peltigera praetextata and P. polydactyla (Kershaw, 1977 a,b) in which there is no change in respiratory rate but the photosynthetic capacity is altered at all temperatures. Collema furfuraceum (Kershaw and MacFarlane, 1982) represents a final group, in which there are no apparent changes in winter rates of CO_2-exchange.

Other patterns of photosynthetic capacity change have been detected. Kershaw and Smith (1978), working on seasonal changes in the temperature optima for net photosynthesis in Stereocaulon paschale, only detected major capacity changes at 35°C in mid-summer. The photosynthetic capacities of P. praetextata and P. scabrosa, growing under tree canopies, have been shown to be maintained at low and high light levels (Kershaw and MacFarlane, 1980).

THE MECHANISMS OF PHOTOSYNTHETIC CAPACITY CHANGES IN FREE-LIVING ALGAE

A number of possible mechanisms of photosynthetic capacity changes exist in free-living algae (Prézelin et al., 1977; Prézelin and Sweeney, 1978, 1979; Prézelin, 1981) and these can be deduced from the response of photosynthesis to light intensity, as seen in photosynthetic-illumination (PI) curves (Fig. 4). PI curves are explicable in terms of light absorption and the size and number of photosynthetic reaction centres (Herron and Mauzerall, 1972; Prézelin, 1981). The slope of the PI curve at low light intensities is a function of the amount of chlorophyll (plus accessory pigments) because of the inter-relationships between the number of reaction centres, the number of photosynthetic units (PSUs) and the cellular chlorophyll content. This part of the PI curve is controlled by the amount of chlorophyll present and is a direct measure of photosynthetic efficiency of the cell. The maximum photosynthetic rate (P_{max}) is considered to be independent of illumination level and, whilst being a function of the number of reaction centres, is dependent on the rate constants of the rate-limiting step(s) of the Calvin cycle. The temperature sensitivity of P_{max} is due to the effect on these rate limiting steps of the process leading to sugar formation.

Changes in the organization and/or activity of the photosynthetic apparatus can occur at a number of places within the photosynthetic system and these, and the implications of the form of the PI curve, are discussed below. Further detailed discussion of how the alterations induced in free-living algae by environmental changes are

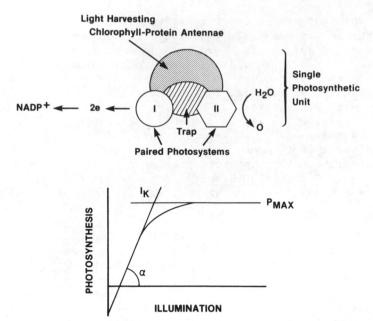

Fig. 4.   A schematic representation of the photosynthetic process and
          idealised photosynthetic-illumination (PI) curve indicating
          photosynthetic efficiency (α), light saturation point (I_k)
          and maximum rate of photosynthesis (P_{max}).

mediated by specific cellular events are given by Prézelin (1981),
Ramus (1981) and Richardson et al. (1983). It must be noted that
more than one strategy may be utilised at a time and the balance of
methods used may vary with the time of year.

Changes in Photosynthetic Unit Size

        Assuming that the chlorophyll-a (chl-a) cores of photosynthetic
reaction centres are fairly uniform, any changes in the amounts of
light harvesting pigments, present in the antenna complex, will
represent changes in PSU size (Fig. 5A). With additional light-
trapping pigments present, the slope of the PI curve will increase at
low light intensities because of the greater light trapping efficiency
of the modified system. As the size, but not the density, of the PSUs
is altered, P_{max} values will not be altered when expressed on a
cellular or dry weight basis. Modifications of this type have been
reported in response to culture at low light intensities in higher
plants (Brown et al., 1974; Alberte et al., 1976), diatoms (Perry et
al., 1981), dinoflagellates (Prézelin and Sweeney, 1978, 1979) and
probably also cyanobacteria (Jorgensen, 1969). In dinoflagellates
major increases in pigmentation occur within a generation when cultures
are transferred from high to low light intensities (Prézelin, 1981).

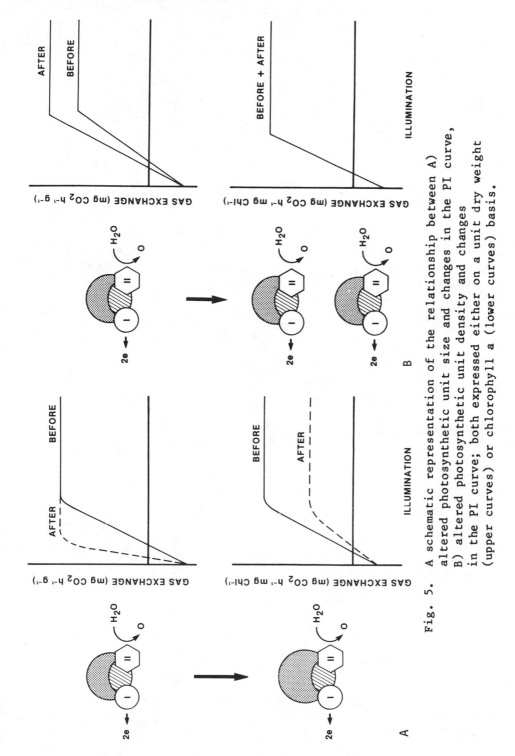

Fig. 5. A schematic representation of the relationship between A) altered photosynthetic unit size and changes in the PI curve, B) altered photosynthetic unit density and changes in the PI curve; both expressed either on a unit dry weight (upper curves) or chlorophyll a (lower curves) basis.

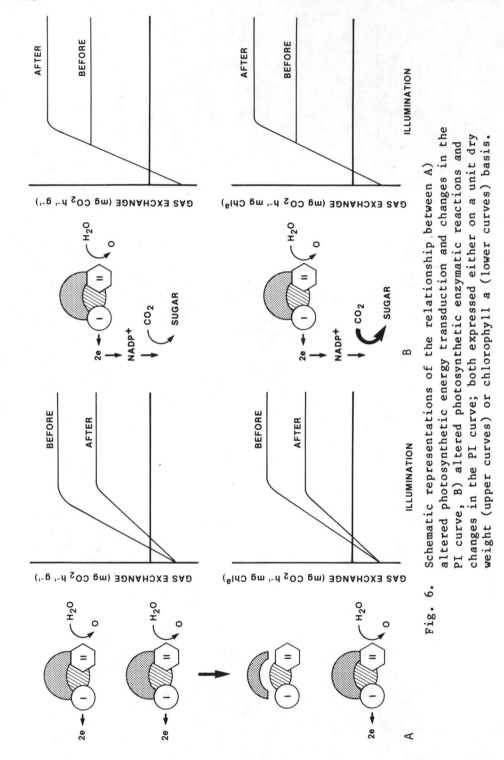

Fig. 6.   Schematic representations of the relationship between A)
altered photosynthetic energy transduction and changes in the
PI curve, B) altered photosynthetic enzymatic reactions and
changes in the PI curve; both expressed either on a unit dry
weight (upper curves) or chlorophyll a (lower curves) basis.

Changes in Photosynthetic Unit Density

When PSU density increases, there is an increase in total pigment concentration (on a cellular or dry weight basis) which will result in improved photosynthetic efficiency and an increased P_{max}. Because the PSU:chl-a ratio remains constant, the form of the PI curve will not change when expressing photosynthesis on a unit chlorophyll basis (Fig. 5B).

It is considered (Prézelin, 1981) that few reports of this strategy occur in the literature, although examples are quoted for dinoflagellates (Prézelin and Sweeney, 1979), green algae and diatoms (Jorgensen, 1969, 1970) and higher plants (Patterson et al., 1977; Terri et al., 1977; Perry et al., 1981). More complex adjustments have been reported from cyanobacteria, where, at low light intensities, a simultaneous increase in the density of Photosystem I PSUs and the antenna complex of Photosystem II occurs (Vierling and Alberte, 1980).

Changes in Energy Transduction

Figure 6A shows that when energy flow through PSUs is modified, without changes to PSU size or density, both light-limited and light-saturated rates of photosynthesis can be altered. Such changes can occur rapidly and reversibly, although the precise mechanism is still unknown. The daily photosynthetic capacity changes in dinoflagellates represent an example of this mechanism (Prézelin et al., 1977; Harding et al., 1981).

Alterations to Enzymatic Rates

Enzymatic reactions involved in the fixation of CO_2 by the Calvin cycle are not directly influenced by changes in PSU size or density. While light-limited rates of photosynthesis may remain constant, P_{max} can vary dramatically (Fig. 6B) with possible changes in amounts and activity of enzymes and isozymes as well as cofactors and allosteric effectors.

INTERPRETATION OF LICHEN PHOTOSYNTHETIC ADAPTATIONS

The models presented above for free-living algae enable the seasonal changes in lichen photosynthetic events to be provisionally re-interpreted on a more mechanistic basis, using seasonal PI curves expressed on either a mg CO_2 $h^{-1}g^{-1}$ or mg CO_2 $h^{-1}\mu g$ chl-a^{-1} basis. Final confirmation will require actual measurements of PSU size, electron transport rates and enzyme activities.

The PI-curves, at optimal thallus water content, for P. praetextata growing under trees (Fig. 7) show complex seasonal patterns (Kershaw and Webber, 1984). A comparison of winter and early

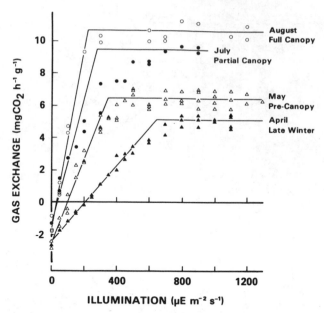

Fig. 7. The response of gas exchange rates to illumination for
Peltigera praetextata at different seasons.
Photosynthetic efficiency (α) was derived from linear
regressions calculated from 0 to 200 (April, May) and 0 to
100 μmol m^{-2}s^{-1} (July, August) and P$_{max}$ is taken as the mean
of the values above full light saturation (I$_k$).

summer characteristics show, respectively, changes in P$_{max}$, at light
saturation, from c. 5.0 to 6.5 mg CO$_2$ h^{-1}g^{-1}, light compensation from
200 to 100 μmol m^{-2}s^{-1} and I$_k$ from c. 650 to 250 μmol m^{-2}s^{-1}, which
indicates increased energy transduction resulting from formation of
additional PSUs. This trend developed further under full-canopy
conditions, with an increase in the initial slope of the PI-curve (α),
which is a measure of the quantum efficiency of the process, and P$_{max}$
to more than 10 mg CO$_2$ h^{-1}g^{-1}, while respiration declined from c. 2.75
to 1.75 mg CO$_2$ h^{-1}g^{-1}. The suggestion that additional PSUs were
synthesised is supported by the almost identical nature of pre-canopy
and full-canopy PI-curves when expressed on a μg chl-a rather than a
dry weight basis (Fig. 8); differences between the curves correspond
to distinct dark respiration rates. Under the mild and snow-free
conditions of the 1982-3 winter, P$_{max}$ and α declined while I$_k$ increased
during November and December (Fig. 9). However, even with an increase
in dark respiration rate, P$_{max}$ in February was 8.4 mg CO$_2$ h^{-1}g^{-1}, α =
0.27 and I$_k$ = 400 μmol m^{-2}s^{-1}, which is not like the previous winter
(Fig. 7). Treatment with low temperatures and short days (Kershaw,
1977a, b; Kershaw et al., 1983; MacFarlane et al., 1983) also failed
to achieve similar values to those observed during the previous,
colder, more traditional, winter. By April 1983 the values of

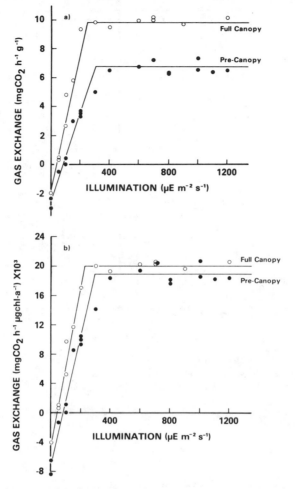

Fig. 8.   Photosynthetic-illumination curves for <u>Peltigera praetextata</u>
          sampled in early June (pre-canopy stage) and August (full
          canopy stage) expressed on a unit dry weight (a) or
          chlorophyll a (b) basis.

P_{max} = c. 5.9 mg CO_2 $h^{-1}g^{-1}$ and I_k = 340 μmol $m^{-2}s^{-1}$ were similar to
those of the previous year.   These changes are summarised in Fig. 10.

     While data from the seasonal net photosynthetic response matrices
of Kershaw (1977a,b) and Kershaw and MacFarlane (1980) showed similar
changes, they had been unable to provide a mechanistic basis for these
responses.   The above discussion has provided a firmer basis for such
interpretations.

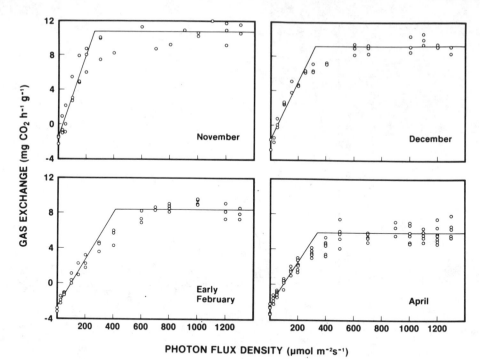

Fig. 9. A series of photosynthetic-illumination curves showing the
decline of photosynthetic efficiency (α) and maximum
photosynthetic rate (P_{max}) in <u>Peltigera praetextata</u> during
autumn and winter.
For derivation of lines see Fig. 7. (Calculated between 0
and 200 μmol $m^{-2}s^{-1}$).

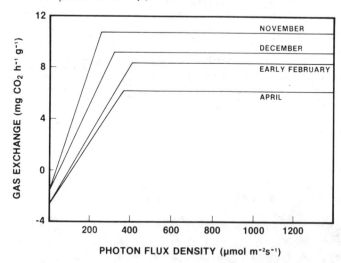

Fig. 10. Summary diagram of the sequential decline in α and P_{max} in
<u>Peltigera praetextata</u> from the data presented in Fig. 9.

DISCUSSION

A comparison of the present data on photosynthetic capacity changes in P. praetextata during the period from winter to summer, show close agreement with the values provided by Kershaw (1977a). He reported maximum photosynthetic rates of c. 2.5 mg CO_2 $h^{-1}g^{-1}$ in winter and c. 6.0 mg CO_2 $h^{-1}g^{-1}$ in spring and early summer. The data quoted in this paper were obtained with metal-halogen lamps which generate higher illuminations than the 450 μmol $m^{-2}s^{-1}$ used in these previous studies. We can now conclude that the apparent doubling in photosynthetic capacity reported earlier (Kershaw, 1977a) reflected a large change in quantum efficiency (α) but the absolute change in P_{max} is in fact closer to c. 1.5 mg CO_2 $h^{-1}g^{-1}$.

Kershaw and MacFarlane (1980) also reported that P. praetextata had very high photosynthetic rates when the tree canopy was maximal during summer. While the present work shows that high P_{max} values at this time are associated with reduced respiration rates, analysis of the data from PI-curves expressed on a mg CO_2 $h^{-1}g^{-1}$ basis indicate that the increased photosynthetic rates are due to the synthesis of extra PSUs which improve the quantum efficiency of the lichen.

Where P. praetextata grows on the forest floor, light intensities range from approximately 30 - 100 μmol $m^{-2}s^{-1}$ when maximum insolation occurs in mid-summer, with values up to 400 μmol $m^{-2}s^{-1}$ being achieved in sunflecks. Because P. praetextata avoids the deepest shade it is probably that maximal photosynthetic rates will occur under a light regime equivalent to that of spring, before the canopy is developed. The synthesis of further PSUs ensures the maintenance of quantum efficiency with increasing canopy density. The concurrent decrease in respiration rate will also assist in achieving a positive carbon balance.

The cost/benefit approach to photosynthetic acclimation has been discussed by Björkman (1981), Kershaw and MacFarlane (1982), Richardson et al. (1983) and Kershaw (1985). For seasonal acclimation to occur it is necessary to ensure that the energy gains achieved will exceed the costs involved. Richardson et al. (1983) identified two types of energy costs in acclimation. Firstly, a capital energy cost involved in the synthesis of additional pigments, enzymes and structural components and, secondly, a maintenance cost to repair and replace necessary structural components.

Because Larson (1980) has criticised the loose usage of the term "photosynthetic acclimation", we propose that, in agreement with the definitions used for free-living algae by Prézelin (1981), Ramus (1981) and Richardson et al. (1983), this term should be confined to changes in enzymic events in the Calvin cycle. Such changes are detected by alterations in photosynthetic rate above light saturation occurring without any change in the quantum efficiency of the process

<u>below</u> light saturation.

Using the strict definition of photosynthetic acclimation, temperature acclimation of photosynthetic optima has only been confirmed for lichens in <u>P. rufescens</u> (Brown and Kershaw, 1984) and <u>Caloplaca trachyphylla</u> (Coxson and Kershaw, 1984) which acclimate to <u>low</u> temperatures during winter snow-melt chinook sequences in southern Alberta. The analogous response reported by Larson (1980) with <u>U. deusta</u> probably now requires re-examination. There are a number of puzzling features about the earlier reports of summer temperature acclimations in <u>C. nivalis</u> (Larson and Kershaw, 1975a), <u>A. ochroleuca</u> (Larson and Kershaw, 1975b), <u>B. nitidula</u> (Kershaw, 1975) and <u>S. paschale</u> (Kershaw and Smith, 1978). In these reports, acclimation appeared to occur at the low, non-saturating, light intensities only available at the time, implying that capacity changes resulted from altered energy transduction. However, these summer changes could also be due to an enhanced thermal stability of the membranes associated with light harvesting with improved electron transport efficiency, as has been reported in higher plants growing in deserts (Armond et al., 1978; Mooney et al., 1978; Badger et al., 1982). The possibility of membrane adaptations in lichens is further supported by observations indicating that lichens may show seasonal changes in their response to thermal stress (Tegler and Kershaw, 1981).

Photosynthetic acclimation studies with higher plants have emphasised the fundamental importance of temperature in the ecology of terrestrial species, whereas studies with aquatic algae, in thermally buffered environments, have stressed changes in the size and density of PSUs. Lichens may be looked upon as equivalent to "terrestrial algae" inhabiting environments with large changes in both temperature and light intensity. The data discussed here show that lichens may adopt a wide range of strategies, corresponding to the particular conditions experienced in their diverse habitats.

REFERENCES

Alberte, R.S., Hesketh, J.D., and Kirby, J.A., 1976, Comparisons of photosynthetic activity and lamellar characteristics of virescent and normal green peanut leaves, <u>Zeitschrift für Pflanzen-physiologie</u>, 77: 152–159.
Armond, P.A., Scheiber, U., and Björkman, O., 1978, Photosynthetic acclimation to temperature in the desert shrub, <u>Larrea divaricata</u>. II. Light harvesting efficiency and electron transport, <u>Plant Physiology</u>, 61: 411–415.
Badger, M.R., Björkman, O., and Armond, P.A., 1982, An analysis of photosynthetic response and adaptation to temperature in higher plants: Temperature acclimation in the desert evergreen <u>Nerium oleander</u> L., <u>Plant, Cell and Environment</u>, 5: 85–99.
Björkman, O., 1981, Ecological adaptation of the photosynthetic apparatus, <u>in</u>: "Photosynthesis. VI. Photosynthesis and Produc-

tivity, Photosynthesis and Environment," G. Akoyunoglou, ed. pp.
    191-202, Balaban International Science Services, Philadelphia.
Brown, D., and Kershaw, K.A., 1984, Photosynthetic capacity changes in
    Peltigera. II. Contrasting seasonal patterns of net photo-
    synthesis in two populations Peltigera rufescens, New
    Phytologist, 96: 447-457.
Brown, J.S., Alberte, R.S., and Thornber, J.P., 1974, Comparative
    studies on the occurrence and spectral composition of chloro-
    phyll-proteins in a wide variety of plant material, in:
    "Proceedings of the Third International Congress on Photo-
    synthesis," M. Avron, ed., pp. 1951-1962, Elsevier, Amsterdam.
Coxson, D.S., and Kershaw, K.A., 1984, Low temperature acclimation of
    net photosynthesis in the crustaceous lichen Caloplaca trachy-
    phylla (Tuck.) A. Zahlbr., Canadian Journal of Botany, 62: 86-95.
Harding, L.W., Meeson, B.W., Prézelin, B.B., and Sweeney, B.M., 1981,
    Diel periodicity of photosynthesis in marine phytoplankton,
    Marine Biology, 61: 95-105.
Herron, H.H., and Mauzerall, D., 1972, The development of photo-
    synthesis in a greening mutant of Chlorella and an analysis of
    the light saturation curve, Plant Physiology, 50: 141-148.
Jorgensen, E.G., 1969, The adaptation of plankton algae. IV. Light
    adaptation in different algal species, Physiologia Plantarum, 22:
    1307-1315.
Jorgensen, E.G., 1970, The adaptation of plankton algae. V.
    Variation in the photosynthetic characteristics of Skeletonema
    costatum cells grown at low light intensity, Physiologia
    Plantarum, 23; 11-17.
Kärenlampi, L., Tammisola, J., and Hurme, H., 1975, Weight increase
    of some lichens as related to carbon dioxide exchange and thallus
    moisture, in: "Fennoscandian Tundra Ecosystems. Part I. Plants
    and Microorganisms," F.E. Wielgolaski, ed., pp. 135-137,
    Springer-Verlag.
Kallio, P., and Kärenlampi, L., 1975, Photosynthesis in mosses and
    lichens, in: "Photosynthesis and Productivity in Different
    Environments," J.P. Cooper, ed., pp. 393-423, Cambridge
    University Press, Cambridge.
Kershaw, K.A., 1975, Studies on lichen dominated systems. XIV. The
    comparative ecology of Alectoria nitidula and Cladina alpestris,
    Canadian Journal of Botany, 53: 2608-2613.
Kershaw, K.A., 1977a, Physiological-environmental interactions in
    lichens. II. The pattern of net photosynthetic acclimation in
    Peltigera canina (L.) Willd. var. praetextata (Floerke em.
    Sommerf.) Hue. and P. polydactyla (Neck.) Hoffm., New
    Phytologist, 79: 377-390.
Kershaw, K.A., 1977b, Physiological-environmental interactions in
    lichens. III. The rate of net photosynthetic acclimation in
    Peltigera canina (L.) Willd. var. praetextata (Floerke em.
    Sommerf.) Hue. and P. polydactyla (Neck.) Hoffm., New
    Phytologist, 77: 391-402.
Kershaw, K.A., 1985, "Physiological Ecology of Lichens," Cambridge

University Press, Cambridge.

Kershaw, K.A., and MacFarlane, J.D., 1980, Physiological-environmental interactions in lichens. X. Light as an ecological factor, New Phytologist, 84: 687-702.

Kershaw, K.A. and MacFarlane, J.D., 1982, Physiological-environmental interactions in lichens. XIII. Seasonal constancy of nitrogenase activity, net photosynthesis and respiration in Collema furfuraceum (Arn.) DR., New Phytologist, 90: 723-734.

Kershaw, K.A., MacFarlane, J.D., Webber, M.R., and Fovargue, A., 1983, Phenotypic differences in the seasonal pattern of net photosynthesis in Cladonia stellaris, Canadian Journal of Botany, 61: 2169-2180.

Kershaw, K.A., and Smith, M.M., 1978, Studies on lichen dominated systems. XXI. The control of seasonal rates of net photosynthesis by moisture, light and temperature in Stereocaulon paschale, Canadian Journal of Botany, 56: 2825-2830.

Kershaw, K.A., and Webber, M.R., 1984, Photosynthetic capacity changes in Peltigera. I. The synthesis of additional photosynthetic units in P. praetextata, New Phytologist, 96: 437-446.

Larson, D.W., 1978, Patterns of lichen photosynthesis and respiration following prolonged frozen storage, Canadian Journal of Botany, 56: 2119-2123.

Larson, D.W., 1980, Seasonal change in the pattern of net CO_2 exchange in Umbilicaria lichens, New Phytologist, 84: 349-369.

Larson, D.W., and Kershaw, K.A., 1975a, Acclimation of arctic lichens, Nature, 254: 421-423.

Larson, D.W., and Kershaw, K.A., 1975b, Studies in lichen dominated systems. XIII. Seasonal and geographical variation of net CO_2 exchange of Alectoria ochroleuca, Canadian Journal of Botany, 53: 2598-2607.

Larson, D.W., and Kershaw, K.A., 1975c, Studies on lichen-dominated systems. XVI. Comparative patterns of net CO_2 exchange in Cetraria nivalis and Alectoria ochroleuca collected from a raised-beach ridge, Canadian Journal of Botany, 53: 2884-2892.

MacFarlane, J.D., Kershaw, K.A., and Webber, M.R., 1983, Physiological-environmental interactions in lichens. XVII. Phenotypic differences in the seasonal pattern of net photosynthesis in Cladonia rangiferina, New Phytologist, 94: 217-233.

Mooney, H.A., Björkman, O., and Collatz, G.J., 1978, Photosynthetic acclimation to temperature in the desert shrub Larrea divaricata. I. Carbon dioxide exchange characteristics of intact leaves, Plant Physiology, 61: 406-410.

Nash, III, T.H., Moser, T.J., and Link, S.O., 1980, Nonrandom variation of gas exchange within arctic lichens, Canadian Journal of Botany, 58: 1181-1186.

Patterson, D.T., Bunce, J.A., Alberte, R.S., and Van Volkenburg, E., 1977, Photosynthesis in relation to leaf characteristics of cotton from controlled and field environments, Plant Physiology, 59: 384-387.

Perry, M.J., Larsen, M.C., and Alberte, R.S., 1981, Photoadaptation in

marine phytoplankton: Response of the photosynthetic unit,
    Marine Biology, 62: 91-101.
Prézelin, B.B., 1981, Light reactions in photosynthesis, in: "Physio-
    logical Bases of Phytoplankton Ecology," Bulletin 210, T. Platt,
    ed., Department of Fisheries and Oceans, Ottawa.
Prézelin, B.B., Meeson, B.W., and Sweeney, B.M., 1977, Characteriza-
    tion of photosynthetic rhythms in marine dinoflagellates. I.
    Pigmentation, photosynthetic capacity and respiration, Plant
    Physiology, 60: 384-387.
Prézelin, B.B., and Sweeney, B.M., 1978, Photoadaptation of photo-
    synthesis in Gonyaulax polyhedra, Marine Biology, 48: 27-35.
Prézelin, B.B., and Sweeney, B.M., 1979, Photoadaptation of photo-
    synthesis in two bloom-forming dinoflagellates, in: "Proceedings
    of the 2nd International Conference on Toxic Dinoflagellate
    Blooms," C. Ventsch, ed., pp. 101-106.
Ramus, J., 1981, The capture and transduction of light energy, in:
    "The Biology of Seaweeds," Botanical Monographs Vol. 17,
    C.S. Lobban, and M.J. Wynna, eds, pp. 458-492, Blackwell
    Scientific Publications, Oxford.
Richardson, K., Beardall, J., and Raven, J.A., 1983, Adaptation of
    unicellular algae to irradiance: an analysis of strategies, New
    Phytologist, 93: 157-191.
Stålfelt, M.G., 1939, Der Gasaustausch der Flechten, Planta, 29:
    11-31.
Tegler, B., and Kershaw, K.A., 1980, Studies on lichen-dominated
    systems. XXIII. The control of seasonal rates of net photo-
    synthesis by moisture, light, and temperature in Cladonia
    rangiferina, Canadian Journal of Botany, 58: 1851-1858.
Tegler, B., and Kershaw, K.A., 1981, Physiological-environmental
    interactions in lichens. XII. The seasonal variation of the heat
    stress response of Cladonia rangiferina, New Phytologist, 87:
    395-401.
Terri, J.A., Patterson, D.T., Alberte, R.S., and Castelberry, R.M.,
    1977, Changes in the photosynthetic apparatus of maize in
    response to simulated natural temperature fluctuations, Plant
    Physiology, 60: 370-373.
Vierling, E., and Alberte, R.S., 1980, Functional organization and
    plasticity of the photosynthetic unit of the cyanobacterium
    Anacystis nidulans, Plant Physiology, 50: 93-98.

ELECTROPHORETIC AND GAS EXCHANGE PATTERNS OF TWO POPULATIONS OF

PELTIGERA RUFESCENS

D. Brown and K.A. Kershaw

Department of Biology, McMaster University
Hamilton, Ontario, Canada

INTRODUCTION

Measurements of net photosynthesis have been used extensively to determine the degree of ecotypic variation and photosynthetic plasticity in a variety of plants. Gas exchange patterns have been compared for species found along gradients of altitude (Mooney and Shropshire, 1967; Chapin and Oechel, 1983), light (Kershaw et al., 1983; MacFarlane et al., 1983), and microtopography (Kershaw, 1975; Larson and Kershaw, 1975) as well as contrasting maritime and continental (Pearcy, 1977; Mooney, 1980) or temperate and arctic (Brown and Kershaw, 1984) conditions. In all of these papers a similar logic is followed. Populations are chosen because of distinct differences in morphology, microclimatic or macroclimatic conditions. Measurements are then made of the response of net photosynthesis to light, temperature and season and any variation observed is used to support the initial premise that the populations were distinct. A circular argument is therefore generated, without any attempt to establish the genetic basis of these differences.

The genotypic variation in populations has been frequently investigated by the technique of protein electrophoresis (Shecter and De Wet, 1975; Liu et al., 1978; Burdon et al., 1980; Shumaker and Babble, 1980; Soudek and Robinson, 1983). The same technique has been successfully used to demonstrate environmentally-induced changes in enzyme isozyme patterns (McGown et al., 1969; De Jong, 1973; Shannon et al., 1973; Kelly and Adams, 1977; Krasnuk et al., 1978 a,b; McMillan, 1980) and as a taxonomic aid in lichen studies (Fahselt and Jancey, 1977; Fahselt and Hageman, 1983).

111

Kershaw et al. (1983) and MacFarlane et al. (1983) combined an electrophoretic study of populations of <u>Cladonia rangiferina</u> and <u>C. stellaris</u> with an examination of the net photosynthetic response at different times of the year. While some variation in gas exchange patterns were detected, little variation in electrophoretic patterns was detected. Differences between populations appeared to be maintained only by environmentally-induced modifications to the phenotype and thus the sun and shade morphotypes detected were not true ecotypes.

This paper deals with the enzymic variability of temperate and arctic populations of <u>Peltigera rufescens</u>. These populations have been shown to differ markedly in their seasonal gas exchange patterns and ability to acclimate to high temperatures (Brown and Kershaw, 1984).

MATERIALS AND METHODS

Collection and Storage of Material

<u>Peltigera rufescens</u>, collected from rock outcrops in the Muskoka Lakes region of Central Ontario and from a raised beach system near Churchill, Manitoba, was stored in an air dry state in growth chambers maintained at the light intensity, temperature and photoperiod corresponding to those of the original habitat (Brown and Kershaw, 1984). The material was equilibrated to these conditions for one week before experimentation.

Photosynthetic Measurements

Infrared gas analysis was used to construct seasonal net photosynthetic response matrices, with a factorial combination of five light levels (0, 300, 500, 800 and 1000 μmol m^{-2}s^{-1}) and five temperatures (5, 15, 25, 35 and 45°C) by the method of Brown and Kershaw (1984). The experimental induction of photosynthetic temperature acclimation was studied with the Muskoka population. The results are presented as mg CO_2 g^{-1}h^{-1} for the net photosynthetic rates at optimum thallus moisture content.

Enzyme Extraction and Detection

Enzyme extracts for electrophoresis were obtained from samples of <u>P. rufescens</u>, collected in June 1983, from Muskoka and Churchill and also from Hawley Lake, Ontario. Several lobes were dissected from the mat, soaked overnight and allowed to dry under the existing storage regime. Protein was extracted from 6-8 g of air dry thalli in 0.2 M potassium phosphate buffer containing 0.02 M 2-mercaptoethanol, 0.2 M sodium metabisulphite and 0.1% (w/v) bovine serum albumin (BSA), adjusted to pH 6.8 following the procedure of MacFarlane et al. (1983).

Electrofocusing was carried out on a horizontal flat bed cell (BioRad model 1415). Polyacrylamide gels (100 x 125 x 1.5 mm), of 5% acrylamide, 0.15% bisacrylamide and 2% BioRad ampholine, were photo-polymerized for 1 h using the BioRad CTL casting system. Protein was loaded into preformed pockets in the gel, focused under a constant power of 1.5 watts for 1 h, and then 6 watts for an additional 3 h, using a catholyte of 1 M NaOH and anolyte of 0.33 M H_3PO_4. Following electrofocusing, gels were washed in the staining buffer, stained, photographed and dried. Representative drawings were made from each gel. Bands were numbered sequentially from the origin.

Phosphoglucose isomerase (PGI), NADP-dependent glyceraldehyde-3-PO_4-dehydrogenase (G3PDH), malate dehydrogenase (MDH), esterase, glucose-6-PO_4-dehydrogenase (G6PDH) and phosphoglucomutase (PGM) were assayed by the methods of Shaw and Prasad (1970). Peroxidase was stained according to Brewbaker et al. (1968) and general protein was detected by the silver stain of Switzer et al. (1979).

RESULTS

Gas Exchange Studies

The Churchill population of P. rufescens showed no seasonal differences in net photosynthetic capacity. Responses to light and temperature in November, September and July (Fig. 1) showed similar patterns (Brown and Kershaw, 1984). In July, net photosynthesis, measured at 1000 μmol $m^{-2}s^{-1}$ and optimum water content, was little different at temperatures of 25, 35 and 45°C; rates being 3.9, 4.4 and 3.2 mg CO_2 $g^{-1}h^{-1}$, respectively. Gas exchange was slightly reduced at 15°C (2.4 mg CO_2 $g^{-1}h^{-1}$) and dropped to one third of the 35°C rate at 5°C. No temperature effect on gas exchange was observed at lower light intensities, except that at 45°C stress responses probably accounted for the depressed rates under 300 and 500 μmol $m^{-2}s^{-1}$ illumination. Light saturation occurred at or above 500 μmol $m^{-2}s^{-1}$.

Pronounced seasonal differences were observed in the Muskoka populations (Fig. 2). In December the highest net photosynthetic rate was observed at 25°C and 1000 μmol $m^{-2}s^{-1}$ (7.7 mg CO_2 $g^{-1}h^{-1}$), with slightly depressed rates at 35 and 45°C (5.8 and 4.8 mg CO_2 $g^{-1}h^{-1}$) and substantially lower rates at 15 and 5°C (3.3 and 2.1 mg CO_2 $g^{-1}h^{-1}$). Material tested in January had a similar pattern of response with maximum values at 5, 15, 25 and 35°C (under 800 μmol $m^{-2}s^{-1}$) of 2.2, 5.4, 5.9 and 3.9 mg CO_2 $g^{-1}h^{-1}$, respectively (Brown and Kershaw, 1984). By mid-summer a different pattern was detected (Fig. 2), with the temperature optimum raised to 35°C and photosynthetic capacity increasing substantially with higher levels of illumination. Fewer rate changes were observed at lower temperatures and light intensities.

Neither population showed any seasonal differences in dark

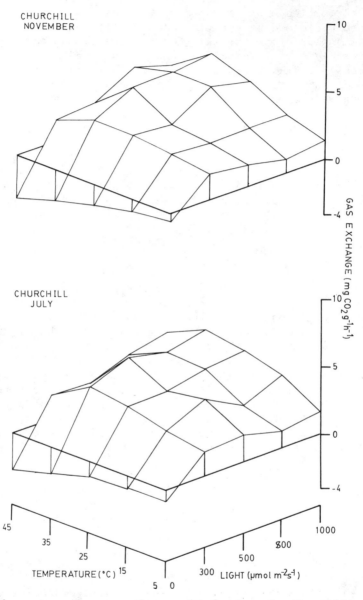

Fig. 1.   The seasonal gas exchange matrix for the Churchill population
of <u>Peltigera rufescens</u> at optimum thallus moisture content.
Maximum standard error = 0.8 mg CO_2 $g^{-1}h^{-1}$.

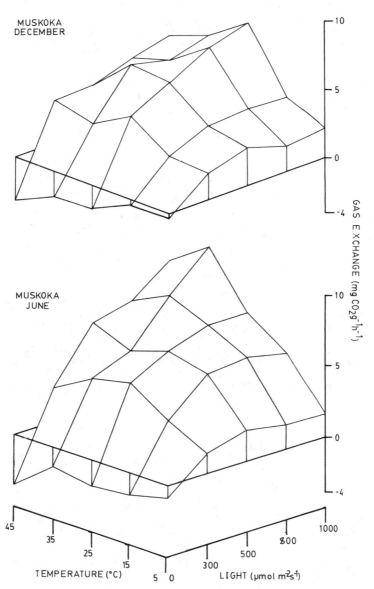

Fig. 2. The seasonal gas exchange matrix for the Muskoka population of <u>Peltigera rufescens</u> at optimum thallus moisture content. Maximum standard error = 0.8 mg CO_2 $g^{-1}h^{-1}$.

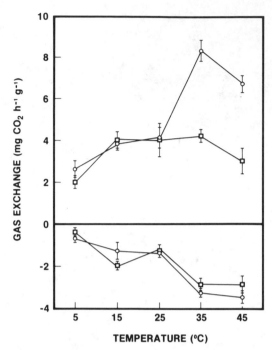

Fig. 3.   The response of net photosynthesis (800 μmol m^{-2} s^{-1}) and
          respiration, measured at optimum thallus moisture content, to
          storage temperature in the Muskoka <u>Peltigera rufescens</u>
          population (April collection).
          Storage conditions: 12h photoperiod either O = warm (15/25°C)
          or □ = cold (-1/5°C) night/day temperature regime.   Vertical
          bars give the standard error.

respiration.   The Churchill population appeared to have lower
respiration rates at all temperatures tested (Figs 1 & 2).

     Because the Muskoka population showed seasonal changes in net
photosynthesis, experiments were conducted to establish if these
alterations could be duplicated in the laboratory (Fig. 3).   When
material was tested in April, after storage at 2°C (winter conditions)
net photosynthetic rates were c. 4.0 and 3.0 mg CO_2 g^{-1}h^{-1} at 35 and
45°C.   After incubation for a further 9 days at 25°C these rates had
increased to 8.5 and 7.0 mg CO_2 g^{-1}h^{-1}, respectively.   Net photosynthetic rates at 5, 15 and 25°C were unchanged after storage and
respiration rates remained unchanged at all temperatures.   In July,
low temperature or short-day treatments applied singly produced only
transient photosynthetic responses (Fig. 4).   A combined low
temperature and short day treatment was required in order to induce a
lower, stable photosynthetic rate.   This induced winter response was
rapidly reversed by increasing the temperature or day length (Fig. 4).

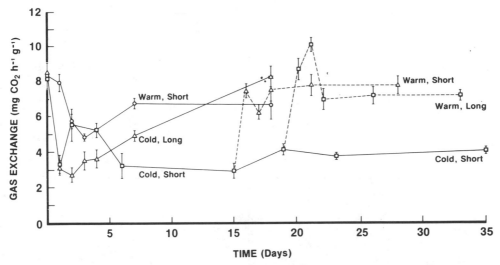

Fig. 4. The time course of temperature acclimation in the Muskoka
population of <u>Peltigera rufescens</u> collected in July.
Photoperiod regimes are long = 15 h and short = 9 h;
temperature regimes are warm = 17/35°C and cold −1/5°C
night/day temperatures. Combined conditions and transfers
between conditions are shown on the diagram. Gas exchange
values measured at optimum thallus moisture content, 35°C and
800 μmol m^{-2}s^{-1}. Vertical bars give the standard error.

Observations made in August on the effect of temperature and
photoperiod on the Muskoka <u>P. rufescens</u> population were analysed in
terms of net photosynthesis at different light intensities expressed
on a unit dry weight or chlorophyll-a basis (Fig. 5) (see Kershaw,
this volume, for a discussion of this approach). After 7 days low
temperature and short day length storage, the maximum photosynthetic
rate at 35°C declined significantly from c. 7.5 to 5.0 mg CO_2 g^{-1}h^{-1}.
This decline was evident when either expressed on a biomass or
chlorophyll basis and occurred without any change in the quantum
efficiency at light intensities below saturation. Material from both
storage treatments showed no change in either photosynthetic or
respiratory rates when tested at 25°C (Fig. 5). An inverse induction
experiment was attempted in late December. This failed to elicit any
acclimation response to an increased temperature and/or day length
(Brown and Kershaw, 1984).

Enzyme Electrophoresis Studies

Isozyme patterns from the Muskoka, Churchill and Hawley Lake
populations of <u>P. rufescens</u> are presented in Figs 6 to 9. Bands are
numbered from the origin (top).

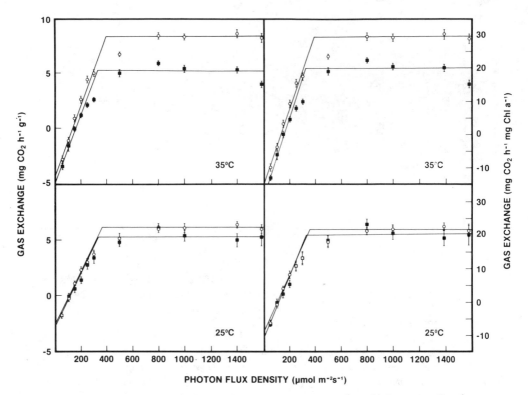

Fig. 5.   The response of <u>Peltigera rufescens</u> (Muskoka population,
           August collection) gas exchange rates to photon flux density
           and incubation temperature.
           Samples were stored under night/day temperatures and
           photoperiods equivalent to O = summer (25/35°C, 16 h) or
           ■ = winter (-1/5°C, 8 h) regimes. Gas exchange, at optimum
           thallus moisture content, is expressed on a unit thallus dry
           weight or unit chlorophyll-a basis. Vertical bars give the
           standard error.

        Phosphoglucomutase (Fig. 6A) showed little isoenzymatic
variability. The Muskoka population possessed two similarly stained
bands running close together. The Churchill and Hawley Lake
populations had two other bands in common, although one of these (3)
stained lightly in the latter population; band 4 was the most active.

        The Muskoka and Churchill populations both contained single bands
of glyceraldehyde-3-PO$_4$-dehydrogenase (Fig. 6B), the former being the
more acidic of the two isoenzymes. No activity was detected in the
Hawley Lake population.

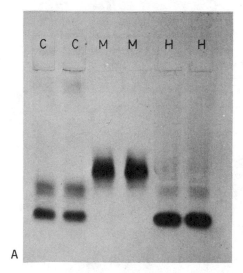
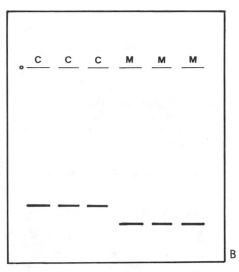

Fig. 6.   The electrophoretic patterns of <u>Peltigera rufescens</u> for A)
          phosphoglucomutase isozyme activity (pH gradient 4.0-6.0),
          B) (diagrammatic) NADP-dependent glyceraldehyde-3-PO$_4$-
          dehydrogenase isozyme activity (pH gradients 3.0-10.0).
          (C) = Churchill, (M) = Muskoka and (H) = Hawley Lake
          populations. Origin = o.

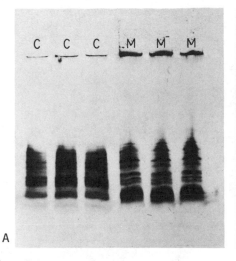
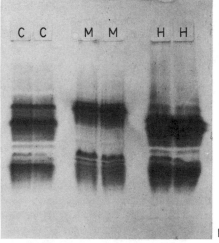

Fig. 7.   The electrophoretic patterns of <u>Peltigera rufescens</u> for A)
          glucose-6-PO$_4$-dehydrogenase isozyme activity, B) phospho-
          glucose isomerase isozyme activity (pH gradients 4.0-6.0).
          Symbols as in Fig. 6.

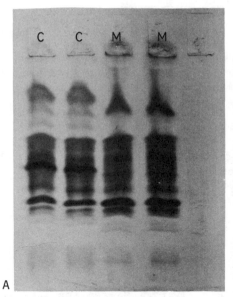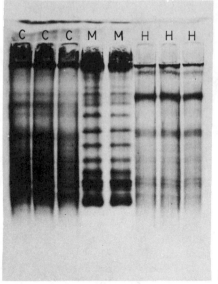

Fig. 8.   The electrophoretic patterns of <u>Peltigera rufescens</u> for A)
          malate dehydrogenase isozyme activity (pH gradient 3.0-10.0),
          B) esterase isozyme activity (pH gradient 4.0-6.0).
          Symbols as in Fig. 6

Glucose-6-PO$_4$-dehydrogenase (Fig. 7A) is a moderately
polymorphic, but highly conservative, enzyme with 13 of the 15 bands
identified as being common to the Muskoka and Churchill populations.
The Muskoka population had higher activity in the more acidic bands
while bands 10 and 11 showed the greatest activity in the Churchill
material; no activity was observed in the Hawley Lake population.

Phosphoglucose isomerase (Fig. 7B) is like G6PDH in being
polymorphic but is less conservative with only four bands being common
to the three populations. Of the 15 bands detected, two were unique
to the Muskoka, two to the Churchill and three to the Hawley Lake
populations. The three populations also showed distinct differences
in the activities of the different isoenzymes.

Malate dehydrogenase was not detected in the Hawley Lake samples
but for the other two populations there were 13 common isozymes as
well as four unique to Churchill and one to Muskoka (Fig. 8A). There
were also differences in staining intensity for specific bands between
these two populations.

A total of 21 bands with esterase activity were detected (Fig.
8B). The Muskoka population showed the greatest differences with 6
unique bands out of a total of 16. The Churchill and Hawley Lake

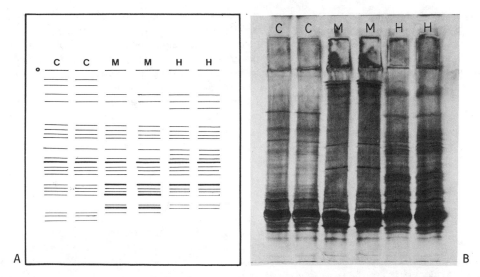

Fig. 9.   The electrophoretic patterns of <u>Peltigera rufescens</u> for A)
          (diagrammatic) peroxidase isozyme activity, B) general
          protein stain (pH gradients 3.0-10.0).
          Symbols as in Fig. 6.

material had no unique isozymes and of the 11 bands scored in the
Hawley Lake population, ten were present in the Churchill sample.

    Peroxidase was the most polymorphic of the specific enzymes
tested with a total of 25 bands being observed (Fig. 9A). However, it
was also the most conservative with 17 bands common to all three
populations. Four bands were found only in the Churchill populations,
while Hawley Lake had two unique bands. There were no unique
peroxidase bands in the Muskoka population although bands 14, 17 and
19 were highly stained.

    The silver stain for protein (Fig. 9B) was sensitive enough to
detect a total of 40 bands; 25 in Muskoka, 29 in Churchill and 32 in
Hawley Lake material. Seventeen bands were common to all populations
while a further 5 bands were common between Churchill-Hawley Lake and
Churchill-Muskoka. Hawley Lake had the greatest number of unique
bands (8), while Churchill and Muskoka samples had only 2 and 3
respectively.

DISCUSSION

    Several important points can be deduced from the seasonal gas
exchange matrices (Figs. 1 & 2). During winter the Churchill and
Muskoka populations have similar net photosynthetic rates at 5, 35 and
45°C. However at 15 and 25°C the Muskoka material had higher

photosynthetic rates. As MacFarlane and Kershaw (1980) have recorded
temperatures of dry thalli in the Muskoka P. rufescens population in
excess of 50°C, the doubling in summer of net photosynthetic rates at
35 and 45°C suggests that this population is well adapted to the
warmer climate. During the summer at Muskoka, rainfall is restricted
to thunderstorms and substantial carbon gains would be possible under
the high light intensities and high boundary layer temperatures that
follow such events. In contrast, frontal activity is the major source
of summer rainfall at Churchill. Even with alternating sun and shower
events, thallus surface temperatures rarely exceed 25°C (Kershaw and
Watson, 1983) due to the low irradiation and low to moderate
temperatures which occur. It is, therefore, unlikely that an ability
to increase photosynthetic capacity at high temperatures would ever
actually be expressed in an increased carbon gain.

     In higher plants, Mooney et al. (1978) and Badger et al. (1982)
reported seasonal changes in temperature optima for photosynthesis
(corrected for photorespiration) which are similar to those seen in
the Muskoka P. rufescens population.

     Other seasonal photosynthetic adaptations have been detected in
different lichen species. Caloplaca trachyphylla (Coxson and Kershaw,
1984) and Umbilicaria deusta (Larson, 1980) show increased net
photosynthetic capacities at high light intensities and low
temperatures during winter snow-melt. In the summer, photosynthetic
capacity increases at 35°C in Stereocaulon paschale but, as this
occurs at all light intensities (Kershaw and Smith, 1978), it may
reflect a change in temperature stress resistance at that time of year
(Brown and Kershaw, 1984).

     The demonstration of different seasonal net photosynthetic
responses in the two P. rufescens populations emphasizes that
metabolic patterns defined in one habitat do not necessarily represent
the situation throughout the ecological range of that species. While
net photosynthetic variations have been reported in different ecotypes
of a wide range of plants (Mooney and Shropshire, 1967; Smith and
Hadley, 1974; Kershaw, 1975; Larson and Kershaw, 1975; Tegler and
Kershaw, 1980; Chapin and Oechel, 1983; MacFarlane et al., 1983), it
is important to re-emphasize the risks involved in making over-
generalised statements based on data from only one population of a
given species.

     There can also be considerable year-to-year variation in photo-
synthetic patterns. For example, the Muskoka populations showed
characteristically high photosynthetic rates during the summers of
1981, 1982 and 1984, whereas the rates determined at 35°C remained at
the winter levels during 1983 (Table 1). The year-to-year variations
reported for P. praetextata (Kershaw and Webber, 1984) further
emphasize our rather tenuous understanding of the phenomenon.

Table 1. Net Photosynthetic Capacity During the Summer Months[1]

Date		Net Photosynthesis $mg\ CO_2\ g^{-1}h^{-1}$		Date		Net Photosynthesis $mg\ CO_2\ g^{-1}h^{-1}$	
		Mean	SE			Mean	SE
September	1980	6.6	0.9	August	1982	8.4	0.2
April	1981	8.2	0.8	October	1982	8.1	0.3
June	1981	7.9	0.6	April	1983	3.2	0.3
July	1981	8.3	0.7	June	1983	4.0	0.5
September	1981	7.2	0.5	October	1983	5.3	0.5
June	1982	11.1	0.7	May	1984	8.7	0.6

[1] Measured at 35°C, 800 $\mu mol\ m^{-2}s^{-1}$ and optimum moisture content.

The induction of acclimation to new temperature regimes is biphasic (Fig. 4). Any change in the storage conditions causes a temporary decline in the gas exchange rate, which we interpret as a "shock" response. By day 10, the non-inductive treatments (cold/long day and warm/short day) have recovered from this shock while the cold/short day material has stabilised at a net photosynthetic capacity similar to the winter rates. A similar gradual acclimation has been seen in higher plants which has been interpreted in terms of a synthesis of the appropriate isoenzymes (Berry and Björkman, 1980; Badger et al., 1982), which may be fructose-1,6-bisphosphatase or ribulose-1,5-bisphosphate carboxylase.

Prézelin (1981) and Richardson et al. (1983) have discussed the possible mechanisms, in free-living algae, for net photosynthetic capacity changes on the basis of alterations in photosynthetic units (detected below light saturation) or enzymes of the Calvin Cycle (detected above light saturation). Sawada and Miyachi (1974 a,b) have reported similar patterns in higher plants. The application of this approach to lichen acclimation studies is dealt with by Kershaw (this volume). Using such arguments it is very probable that, because capacity changes in the Muskoka populations are only detectable at high, saturating light intensities, they represent enzymatic adaptations.

The electrophoretic data presented here indicate several enzymatic differences between the P. rufescens populations from Churchill and Muskoka. Thus there are no common isozyme bands for PGM and G3PDH and significant differences were detected in PGI and esterase patterns. This is in contrast to the greater similarity in the peroxidase and protein patterns, both of which have been reported elsewhere to show large inter-population and inter-species variations

(Sorenson et al., 1971; Shecter and De Wet, 1975; Shumaker and Babble, 1980).

The Hawley Lake population showed a general similarity to the Churchill population in its PGM and esterase pattern but was more like the Muskoka material for PGI and peroxidase. However, all of the populations were distinct from each other. The Hawley Lake material is conspicuously different in that it apparently lacked G6PDH, MDH and G3PDH. The lack of these essential enzymes is unlikely to be a genuine genetical distinction but, probably reflects their instability during extraction and storage. The lichens' physiological state prior to enzyme extraction may have an important bearing on enzyme stability as has been reported for higher plants by Brulfert et al. (1979).

The large differences in gas exchange and electrophoretic banding patterns observed in the morphologically similar populations of P. rufescens from Churchill and Muskoka are in marked contrast to earlier studies. For example, MacFarlane et al. (1983) compared the gas exchange and electrophoretic patterns from sun and shade populations of Cladonia rangiferina from Muskoka and Hawley Lake and found that, despite large morphological differences, the electrophoretic patterns for esterase, acid phosphatase and leucine aminopeptidase were almost identical. Only quantitative differences in staining intensity were observed in MDH patterns but PGI did have an unique band in the sun population. The Hawley Lake population possessed very similar electrophoretic patterns despite being morphologically and physiologically distinct from the Muskoka material. Comparable results were obtained with a parallel study of C. stellaris (Kershaw et al., 1983) and very little between- or within-population variation in electrophoretic patterns has been detected in Parmelia spp. (Fahselt and Jancey 1977) or Cetraria arenaria (Fahselt and Hageman, 1983).

The results from the P. rufescens populations suggest that the high variability in isozyme patterns (especially for enzymes associated with carbohydrate metabolism) detected by electrophoresis, combined with seasonal differences in gas exchange patterns (Figs 1 & 2), do reflect true metabolic adaptations to distinct environments. We conclude that two kinds of adaptation are possible. Firstly, considerable morphological differences can develop, without any metabolic alterations, which are solely phenotypic responses to the environment. Secondly, enzymic differences may occur without being correlated with identifiable morphological differences. We are currently studying the kinetic background to the enzyme variability which we have established in P. rufescens.

ACKNOWLEDGEMENTS

We wish to thank Dr. I. Brodo of the National Museum of Canada for confirming the identification of the populations of Peltigera,

Mr. B.S. Smith for his technical assistance, the Churchill Northern Studies Center for their field support and the National Science and Engineering Research Council of Canada and the Department of Indian and Northern Affairs for their financial support during this research.

REFERENCES

Badger, M.R., Björkman, O., and Armond, P.A., 1982, An analysis of photosynthetic response and adaptation to temperature in higher plants: Temperature acclimation in the desert evergreen Nerium oleander L., Plant, Cell and Environment, 5: 85-99.

Berry, J., and Björkman, O., 1980, Photosynthetic response and adaptation to temperature in higher plants, Annual Review of Plant Physiology, 31: 491-543.

Brewbaker, J.L., Upadhya, M.D., Makinen, Y., and MacDonald, T., 1968, Isoenzyme polymorphism in flowering plants. III. Gel electrophoretic methods and applications, Physiologia Plantarum, 21: 930-940.

Brown, D., and Kershaw, K.A., 1984, Photosynthetic capacity changes in Peltigera. II. Contrasting season patterns of net photosynthesis in two populations of Peltigera rufescens, New Phytologist, 96: 447-457.

Brulfert, J., Arrabaca, M.C., Guerrier, D., and Queiroz, O., 1979, Changes in the isozymic pattern of phosphoenolpyruvate. An early step in photoperiodic control of Crassulacean Acid Metabolism level, Planta, 146: 129-133.

Burdon, J.J., Marshall, D.R., and Groves, R.H., 1980, Isozyme variation in Chondrilla juncea L. in Australia, Australian Journal of Botany, 28: 193-198.

Chapin III, F.S., and Oechel, W.C., 1983, Photosynthesis, respiration, and phosphate absorption by Carex aquatilis ecotypes along latitudinal and local environmental gradients, Ecology, 64: 743-751.

Coxson, D.C., and Kershaw, K.A., 1984, Low-temperature acclimation of net photosynthesis in the crustaceous lichen Caloplaca trachyphylla (Tuck.) A. Zahlbr., Canadian Journal of Botany, 62: 86-95.

De Jong, D.W., 1973, Effect of temperature and daylength on peroxidase and malate (NAD) dehydrogenase isozyme composition in tobacco leaf extracts, American Journal of Botany, 60: 846-852.

Fahselt, D., and Hageman, C., 1983, Isozyme banding patterns in two stands of Cetraria arenaria Karnef., The Bryologist, 86: 129-134.

Fahselt, D., and Jancey, R.C., 1977, Polyacrylamide gel electrophoresis of protein extracts from members of the Parmelia perforata complex, The Bryologist, 80: 429-438.

Kelly, W.A., and Adams, R.P., 1977, Seasonal variation of isozymes in Juniperus scopulorum: Systematic significance, American Journal of Botany, 64: 1092-1096.

Kershaw, K.A., 1975, Studies on lichen dominated systems. XIV. The comparative ecology of Alectoria nitidula and Cladina alpestris, Canadian Journal of Botany, 53: 2608-2613.

Kershaw, K.A., MacFarlane, J.D., Webber, M.R., and Fovargue, A., 1983, Phenotypic differences in the seasonal pattern of net photosynthesis in Cladonia stellaris, Canadian Journal of Botany, 61: 2169–2180.

Kershaw, K.A., and Smith, M.M., 1978, Studies on lichen dominated systems. XXI. The control of seasonal rates of net photosynthesis by moisture, light and temperature in Stereocaulon paschale, Canadian Journal of Botany, 56: 2825–2830.

Kershaw, K.A., and Watson, S., 1983, The control of seasonal rates of net photosynthesis by moisture, light and temperature in Parmelia disjuncta Erichs., The Bryologist, 86: 31–43.

Kershaw, K.A., and Webber, M.R., 1984, Photosynthetic capacity changes in Peltigera I. The synthesis of additional photosynthetic units in P. praetextata, New Phytologist, 96: 437–446.

Krasnuk, M., Witham, F.H., and Jung, G.A., 1978a, Hydrolytic enzyme differences in cold-tolerant and cold-sensitive alfalfa, Agronomy Journal, 70: 597–605.

Krasnuk, M., Jung, G.A., and Witham, F.H., 1978b, Dehydrogenase levels in cold-tolerant and cold-sensitive alfalfa, Agronomy Journal, 70: 605–613.

Larson, D.W., 1980, Seasonal change in the pattern of net CO_2 exchange in Umbilicaria lichens, New Phytologist, 84: 349–369.

Larson, D.W., and Kershaw, K.A., 1975, Studies on lichen dominated systems. XIII. Seasonal and geographical variation of net CO_2 exchange of Alectoria ochroleuca, Canadian Journal of Botany, 53: 2598–2607.

Liu, E.H., Sharitz, R.R., and Smith, M.H., 1978, Thermal sensitivities of malate dehydrogenase isozymes in Typha, American Journal of Botany, 65: 214–220.

MacFarlane, J.D., and Kershaw, K.A., 1980, Physiological-environmental interactions in lichens. IX. Thermal stress and lichen ecology, New Phytologist, 84: 669–685.

MacFarlane, J.D., Kershaw, K.A., and Webber, M.R., 1983, Physiological-environmental interactions in lichens. XVII. Phenotypic differences in the seasonal pattern of net photosynthesis in Cladonia rangiferina, New Phytologist, 94: 217–233.

McGown, B.H., Beck, G.E., and Hall, T.C., 1969, The hardening response of three clones of Dianthus and the corresponding complement of peroxidase isoenzymes, Journal of the American Society for Horticultural Science, 94: 691–693.

McMillan, C., 1980, Isozymes of tropical seagrasses from the Indo-Pacific and the Gulf of Mexico-Caribbean, Aquatic Botany, 8: 163–172.

Mooney, H.A., 1980, Photosynthetic plasticity of populations of Heliotropium curassavicum L. originating from differing thermal regimes, Oecologia, 45: 372–376.

Mooney, H.A., Björkman, O., and Collatz, G.J., 1978, Photosynthetic acclimation to temperature in the desert shrub, Larrea divaricata I. Carbon dioxide exchange characteristics of intact leaves, Plant Physiology, 61: 406–410.

Mooney, H.A., and Shropshire, F., 1967, Population variability in
    temperature related photosynthetic acclimation, Oecologia
    Plantarum, 11: 1-13.
Pearcy, R.W., 1977, Acclimation of photosynthetic and respiratory
    carbon dioxide exchange to growth temperature in Atriplex
    lentiformis (Torr.) Wats., Plant Physiology, 59: 795-799.
Prézelin, B.B., 1981, Light interactions in photosynthesis, in:
    "Physiological Bases of Phytoplankton Ecology," Bulletin 210,
    T. Platt, ed., pp. 1-43, Department of Fisheries and Oceans,
    Ottawa.
Richardson, K., Beardall, J., and Raven, J.A., 1983, Adaptation of
    unicellular algae to irradiance: an analysis of strategies, New
    Phytologist, 93: 157-191.
Sawada, S., and Miyachi, S., 1974a, Effects of growth temperature on
    photosynthetic carbon metabolism in green plants I. Photo-
    synthetic activities of various plants acclimatized to varied
    temperatures, Plant and Cell Physiology, 15: 111-120.
Sawada, S., and Miyachi, S., 1974b, Effects of growth temperature on
    photosynthetic carbon in green plants II. Photosynthetic $^{14}CO_2$-
    incorporation in plants acclimatized to varied temperatures,
    Plant and Cell Physiology, 15: 225-238.
Shannon, M.C., Ballal, S.K., and Harris, J.W., 1973, Starch gel
    electrophoresis of enzymes from nine species of Polyporus,
    American Journal of Botany, 60: 96-100.
Shaw, C.R., and Prasad, R., 1970, Starch gel electrophoresis of
    enzymes - A compilation of recipes, Biochemical Genetics, 4:
    297-320.
Shecter, Y., and De Wet, J.M.J., 1975, Comparative electrophoresis
    and isozyme analysis of seed protein from cultivated Sorghum,
    American Journal of Botany, 62: 254-261.
Shumacker, K.M., and Babble, G.R., 1980, Patterns of allozymic
    similarity in ecologically central and marginal populations of
    Hordeum jubatum, Evolution, 34: 110-116.
Smith, E.M., and Hadley, E.B., 1974, Photosynthetic and respiratory
    acclimation to temperature in Ledum groenlandicum populations,
    Arctic and Alpine Research, 6: 13-27.
Snelgar, W.P., Green, T.G.A., and Beltz., C.K., 1981, Carbon dioxide
    exchange in lichens: estimation of internal thallus CO_2 transport
    resistance, Plant Physiology, 52: 417-422.
Sorenson, W.G., Larsh, H.W. Larsh, and Hamp, S., 1971, Acrylamide gel
    electrophoresis of proteins from Aspergillus species, American
    Journal of Botany, 58: 588-593.
Soudek, Jr. D., and Robinson, G.G.C., 1983, Electrophoretic analysis
    of the species and population structure of the diatom
    Asterionella formosa, Canadian Journal of Botany, 61: 418-433.
Switzer III, R.C., Merril, C.R., and Shifrin, S., 1979, A highly
    sensitive silver stain for detecting proteins and peptides in
    polyacrylamide gels, Analytical Biochemistry, 98: 231-237.
Tegler, B., and Kershaw, K.A., 1980, Studies on lichen-dominated
    systems. XXIII. The control of seasonal rates of net photo-

synthesis by moisture, light and temperature in _Cladonia_
rangiferina, _Canadian Journal of Botany_, 58: 1851–1858.

MULTIPLE ENZYME FORMS IN LICHENS

D. Fahselt

Department of Plant Sciences
University of Western Ontario
London, Ontario, Canada N6A 5B7

INTRODUCTION

Some aspects of lichen biology are poorly understood. For example, reproductive strategies are essentially unknown. It has been commonly assumed that lichens mainly reproduce asexually and on this basis a relatively high degree of genetic uniformity might be expected. Nevertheless, there is little available evidence pertaining to the genetic structure of lichen populations.

Isozyme studies in lichens are far behind those of other organisms. This has been in part related to problems caused by extracellular phenolics and the difficulty of disrupting cells while maintaining a reasonable concentration of active protein (Martin, 1972; Fahselt, 1980). The existence of at least two symbiotic partners is another complication. In our laboratory we have investigated isozymes with a view to their potential usefulness in evolutionary/populational studies of lichens; isoelectricfocusing (IEF) has been used as a routine means of separating enzyme forms as this has proved to be the most satisfactory method.

COMPARISON OF MYCOBIONT, PHYCOBIONT AND LICHEN PROTEINS

Cladonia cristatella

As a first step in attempting to understand the respective contributions of symbionts to the isozymes in crude lichen extracts a comparative study of separated mycobiont and phycobiont and intact lichen enzymes was undertaken.

Since density gradient centrifugation of disrupted thalli of
Cladonia cristatella did not yield perfectly clean preparations of
either symbiont, isolates of Trebouxia erici Ahmadjian and the C.
cristatella mycobiont were obtained from V. Ahmadjian and grown in
culture. Though certain aspects of symbiont physiology are different in culture than in the lichenized form, some enzymes are
known to be active in both states (Stewart and Rowell, 1977).
While it would be unproductive to attempt an analysis of enzymes in
pathways such as those involved in the formation of secondary
products, it may be useful to compare enzymes of primary metabolic
pathways which are likely to be operative regardless of whether or
not the cultures are lichenized. There is a possibility that differences between the growing conditions of symbionts and intact
thalli could affect isozyme production but, according to Gottlieb
(1977), one of the advantages of electrophoretic data is that,
aside from developmental effects, it is generally little influenced
by environmental factors.

In our experience with lichens there is a limited range of
procedures which provide extracts of active protein suitable for
IEF (Fahselt, 1980; Fahselt and Hageman, 1983; Hageman and Fahselt,
1984). Our standard method for extracting thalli usually involves
vigorous grinding in liquid nitrogen (to disrupt cells), the use of
either polyvinyl pyrolidone (PVPP) or acetone pre-extractions (to
minimize interference by extracellular phenolics), ammonium sulphate precipitations (to concentrate protein) as well as periods of
centrifugation and dialysis. This protocol is appropriate for many
lichens but does not disclose a given enzyme with equal facility in
all species tested, probably as a consequence of basic physiological differences among species.

In order to access a comparable subset of proteins in the
lichen and cultures of its pure symbionts the same extraction
procedure should be used consistently. However, as shown by
serious tailing of some bands during IEF, the standard extraction
method developed for use with lichens is not satisfactory for the
cultured algal partner of C. cristatella. Unlike extracts of
lichens, those of pure T. erici tend to be viscous and mucid, a condition which may be related to alterations in surface chemistry of
the phycobiont due to culturing. Differences in the cell wall or
extracellular matrix of the phycobiont in relation to the
lichenized compared to the nonsymbiotic condition have been described elsewhere by Ahmadjian (1973a), Ahmadjian and Jacobs (1983)
and Bubrick et al. (1982). A simple buffer extraction of fresh
algal cultures which involves no acetone pre-extraction, no
ammonium sulphate precipitation or dialysis is most effective with
T. erici; it results in the same number of electromorphs with the
same isoelectric points (pI's) as the standard extraction method.
However, bands are straighter and sharper, thus facilitating comparisons among samples.

With <u>C. cristatella</u> mycobionts the two alternative extraction procedures yield similar zymograms for some enzymes, but for others there are obvious differences. Since the standard lichen extraction method, which partitions proteins at various stages, would be expected to finally yield fewer electromorphs than the simple buffer extract it is suspected that the extra bands which appear following use of this more involved standard method could be artefacts. For those enzymes where mycobiont banding patterns vary according to the type of extraction a comparison of the two methods is made and only the reproducible electromorphs are considered to be native isozymes.

The lichen, <u>C. cristatella</u> from Princeton, Mass., USA, was compared with symbionts isolated by V. Ahmadjian from this and other sites within the same general geographic area. Mycobionts were grown in liquid malt-yeast extract (Ahmadjian, 1973b), and the phycobionts were cultured on Tris-buffered mineral medium (Smith and Wiedeman, 1964) in agar slants. Whole lichens were first tested for 11 enzyme systems, several of which were shown to be polymorphic within the collection site (Table 1). Symbionts were compared to intact lichens on the basis of only those enzymes which displayed essentially no thallus-to-thallus variability within the

Table 1. Enzyme Systems Assayed in <u>Cladonia cristatella</u> Collection Site near Princeton, Massachusetts

Enzyme	EC Code	Assay method
Isocitrate dehydrogenase[2]	1.1.1.42	Adams & Joly (1980)
6-Phosphogluconate dehydrogenase	1.1.1.44	Adams & Joly (1980)[3]
Glucose-6-phosphate dehydrogenase	1.1.1.49	Brewer & Sing (1970)[4]
Glutamate dehydrogenase	1.4.1.4	Adams & Joly (1980)[5]
Catalase[1]	1.11.1.6	Harris & Hopkinson (1976)
Peroxidase[2]	1.11.1.7	Brewer & Sing (1970)[6]
Superoxide dismutase	1.15.1.1	Harris & Hopkinson (1976)
Esterase	3.1.1.1	Brewer & Sing (1970)
Alkaline phosphatase[2]	3.1.3.1	Brewer & Sing (1970)
Acid phosphatase[2]	3.1.3.2	Brewer & Sing (1970)
Carbonic anhydrase	4.2.1.1	Brewer & Sing (1970)

[1]No activity observed.
[2]Varied within the collection site.
[3]pH of 8.5 substituted for pH 8.0
[4]pH of 8.5 used instead of pH 7.2
[5]NADP$^+$ used in place of NAD$^+$
[6]3,3',5,5'-tetramethylbenzidine substituted for benzidine

population; these were esterase (EST), carbonic anhydrase (CAN),
superoxide dismutase (SOD), glutamate dehydrogenase (GDH), glucose-
6-phosphate dehydrogenase (GPD) and 6-phosphogluconic acid dehydrogenase (6PG). Each of the enzyme extractions from C.cristatella
was duplicated on at least two different occasions. As long as
good resolution was achieved, the banding patterns were the same
each time. Work in our laboratory indicates that replicate extractions nearly always give the same result in other lichen species as
well. Storage of pre-extracted thallus materials at -20°C for as
long as 10-12 months has previously been shown to have no
appreciable effect on isozymes (Fahselt and Hageman, 1983) and
storage at -10°C did not alter Coumassie Blue banding patterns from
those in fresh thalli (Fahselt and Jancey, 1977). The C.
cristatella study samples were stored at -20°C for less than 10
months.

The concentration of protein in lichen extracts was approximately 2-3 mg ml^{-1} and in extracts from symbionts about 0.8-1.0 mg
ml^{-1}. In order to firmly establish the presence of protein in the
extracts, electrofocused gels were stained with Coumassie Blue.
While a one-dimensional separation may not have been sufficient to
resolve all of the major proteins which reacted with the dye, myco-

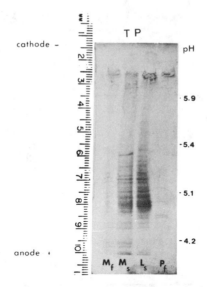

Fig. 1. Zymogram showing Coumassie Blue staining pattern of crude
          protein extracts of the Cladonia cristatella lichen and
          cultured symbionts. M_f = mycobiont extracts prepared fresh
          prior to IEF; M_s = mycobiont extracts made using the standard
          procedure developed for lichens; L_s = lichen extracts
          prepared using the standard method; P_f = phycobiont extracts
          made fresh prior to IEF.

biont zymograms appeared electrophoretically similar to each other
regardless of the method of extraction. However, when the simpler
method (M_f) was used, the bands were weaker and fewer were detectable since in this case no steps were taken to concentrate the
protein. The basic similarity between fungus and lichen zymograms
would support the idea that most of the detectable lichen protein
could well have been produced by the mycobiont (Fig. 1). Since the
isoelectric points of some lichen bands corresponded to those of
algal proteins it must be accepted that detectable lichen proteins
may also have originated from the algal partner. That one or two
lichen electromorphs had no apparent counterpart in either the mycobiont or the phycobiont could have been a reflection of quantitative differences between the lichen and its symbionts or an indication of novel protein in the intact thallus.

The dehydrogenases of C. cristatella which were tested were
NADP[+]- rather than NAD[+]-dependent and they were more active at pH
8.5 than at 8.0 or 9.0. The effect of extraction technique on
these enzymes was marked. The standard extraction method resulted
in considerable activity in the more cathodal parts of the M_s zymogram which did not appear when a simpler extraction procedure (M_f)
was used (Fig. 2a,b,c). This phenomenon, while observed in C.
cristatella, is by no means typical of all species; in many lichens
just a few well-defined dehydrogenase bands are observed. When the
standard extraction method was used for both lichen and mycobiont,

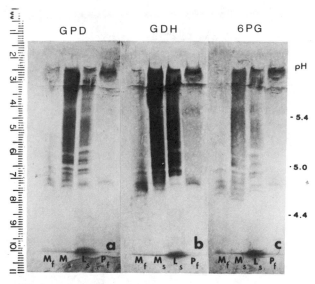

Fig. 2.   Zymograms of some dehydrogenase enzymes of the Cladonia
          cristatella lichen and symbionts. a) glucose-6-phosphate
          dehydrogenase, b) glutamate dehydrogenase, c)
          6-phosphogluconic acid dehydrogenase. Symbols as Fig. 1.

their banding patterns were similar, showing densely-staining
regions near the cathode. Since there is doubt about the authen-
ticity of the more cathodal electromorphs of the mycobiont
extracted this way, electrophoretically-similar lichen bands are
also suspect. The more anodal dehydrogenase electromorphs of the
mycobiont were observed quite independently of the extraction
method and in many cases bands with the same pI's were also noted
in the lichen. Activity in cultured Trebouxia was limited to one
main band in each of the three enzyme systems and this band was
similar in mobility to one of the enzyme forms of the lichen. On
the basis of IEF separations, detectable lichen dehydrogenase
isozymes were probably produced by the fungal partner with some
contribution from the phycobiont.

Since a limited range of extraction protocols yield detectable
proteins from the lichen thallus, the direct testing of different
extraction techniques on the lichen is precluded. However, it is
clear that the effects of extraction can be evaluated to some
extent by experimentation with the mycobiont.

An overall similarity was observed among the zymograms of GDH,
GPD, and 6PG. There could concievably be an evolutionary link
between these three functionally-related enzymes.

Esterase and carbonic anhydrase are two more enzymes similar
in function and with zymograms resembling each other (Fig. 3a). In

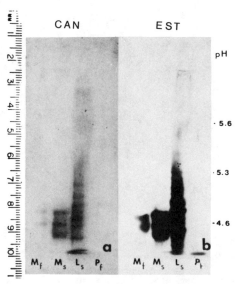

Fig. 3.   Zymograms of the Cladonia cristatella lichen and pure
          symbionts. a) carbonic anhydrase, b) esterase.
          Symbols as Fig. 1.

plants carbonic anhydrases never exhibit esterase activity, while
in mammalian cells some CAN isozymes do (Tashian, 1977). In view
of the electrophoretic similarities between the two enzyme systems
in C. cristatella as well as in other species, it seems possible
that CAN in lichens may also have esterase activity. Mycobiont
zymograms for CAN and EST were similar regardless of which extraction method was used, although, without the procedural steps to
concentrate the M_f extract, the intensity of the bands again was
much less. Activity of CAN and EST was rarely detectable in
extracts of the phycobiont even if gels were deliberately over-
developed to intensify weak bands (Fig. 3b). Mycobiont isozymes
varied somewhat from one collection site to another (Fig. 4). For
both CAN and EST the lichens collected from Princeton were most
similar to a mycobiont isolated from the same population (e.g.
Ahmadjian culture #29). Mycobiont cultures from other collection
sites resembled the Princeton lichen to a lesser degree.

Superoxide dismutase is widely distributed among living organisms (McCord, 1979) and though the mycobiont showed some evidence of it activity was not detected here as reliably as in the
phycobiont. The phycobiont and the lichen each had one major band
of SOD activity when the Harris and Hopkinson (1976) assay for SOD
was used (Fig. 5), and evidence of other possible minor SOD bands
was seen primarily in gels stained for dehydrogenases. Lichen SOD
was electrophoretically most similar to that of the phycobiont and
may well have been produced by the lichenized T. erici.

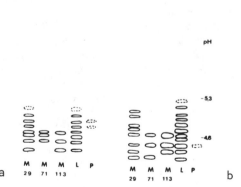

Fig. 4.   Sketches of Cladonia cristatella gels. Mycobionts and lichens
           were extracted using the standard method for lichens and the
           phycobiont was extracted fresh just prior to IEF. Bands
           shown with dotted lines were not detectable in all trials.
           M_{29} = mycobiont culture from single spore originating from
           Princeton, Mass. M_{73} = single spore culture of the mycobiont
           from the Smoky Mountains; M_{113} = single spore isolate from
           Provincetown, Mass. a) carbonic anhydrase, b) esterase.

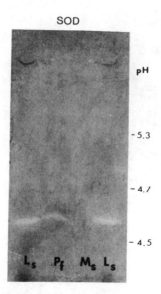

Fig. 5. Zymogram of <u>Cladonia cristatella</u> superoxide dismutase.
         Symbols as Fig. 1.

        Lichen zymograms exhibit a relatively large number of bands
for some enzyme systems. For example, plants have two CAN isozymes
and mammals have three (Tashian, 1977) but more electromorphs than
this are evident in lichen extracts. Esterase assays, for example,
could be detecting a range of enzymes with specificities for a
number of reactions and this could be one explanation for the many
bands. A large number of enzyme forms might be due to post-translational alteration of proteins (Womack, 1983) and there is the
likelihood that different isozymes are made in the cytoplasm and in
organelles. However, it is not known if any of these possibilities
are more likely to occur in lichens than in other groups of organisms. At present the most obvious difference is that in lichens
at least two symbiotic partners are involved and, since there is no
close phylogenetic relationship between them, each may have dis-
tinctive isoenzymes.

Parmelia caperata

        The possibility that the intact thallus has enzyme forms which
are different from those in either of the symbionts should not be
overlooked. An important but little known study is the one by E.
Martin (1973) on the invertases of <u>Parmelia caperata</u>. This thesis
contains critical data concerning the interaction between lichen
symbionts at the molecular level. Invertase is a glycoprotein and
in the <u>P. caperata</u> lichen the single enzyme form is different from
that of either of the symbionts. The enzyme in the intact lichen

has a higher molecular weight, a higher K_m (Table 2), a higher pI and is more heat stable than the invertases of either of the cultured symbionts. In terms of substrate specificity the lichen enzyme resembles an algal one, but its optimal pH range is more reminiscent of a mycobiont enzyme. With respect to its carbohydrate content, lichen invertase is intermediate. Martin (1973) concluded that invertase extracted from the lichen is probably formed of units produced by both of the symbionts.

While the quarternary structure of lichen invertase is not known, the yeast (Saccharomyces cerevisiae) enzyme is regarded as an oligomer, variously a dimer (e.g. Frevert and Ballou, 1974) or an octamer (e.g. Chu et al., 1983). Martin's own data (1973) provides some evidence that there are two forms of extracellular invertase produced by Trebouxia, one is possibly a multimeric form and the other an active subunit of it. It is therefore quite likely that lichen invertase is also a oligomer of some type. The simplest explanation, based on molecular weights and proportions of carbohydrate, as determined by Martin, would be that lichen invertase is composed of the entire mycobiont enzyme plus one half of the protein component from the extracellular phycobiont enzyme (Fig. 6).

Table 2.   Physical/Chemical Characteristics of Parmelia caperata
           Invertase

Source	MW (x 10^3 Da.)	pH Optimum	Carbohydrate Content (%)	Km (mM)	Substrate specificity Relative Activity[1]		
					S	R	M
Thallus	490 ± 10	3.2–4.8	43	3.9	100	29	2
Trebouxia (medium)	355 ± 10	3.0–3.2	30	0.31	100	40	2
Trebouxia (cell)	320	3.4	–	0.21	100	38	–
Mycobiont A	370 ± 20	5.0–5.5	56	1.48	100	76	83
Mycobiont B	375 ± 10	4.5–5.5	56	1.48	100	89	87

[1]S = sucrose, R = raffinose, M = β-methylfructoside.
From Martin (1973)

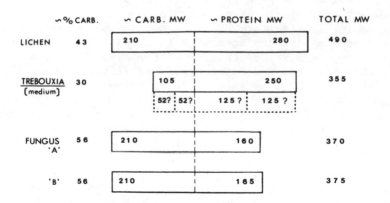

Fig. 6. Schematic drawing to represent approximate molecular weights
of the carbohydrate and protein components of intact Parmelia
caperata and symbiont invertases.

There is evidence that catalytic activity can be maintained in
enzyme forms which differ from each other by the degree of association of subunits. Both associated and dissociated subunits of
glutamate dehydrogenase have been reported to be capable of full
catalytic activity (Fisher et al., 1982), and in Neurospora both
light and heavy forms of invertase showed activity (Metzenberg,
1964). Chu et al. (1983) stated that the smallest enzymatically
active unit of yeast invertase is a dimer aggregating externally to
form various large oligomers which are also active. Alkaline phosphatases are usually dimers but, in some higher animals, active
aggregates and complexes of this enzyme have been observed
(Stigbrand et al., 1982). Association of enzyme subunits in different ways is not then inconceivable and association of mycobiont-
and phycobiont-derived subunits in lichens may thus explain any
unique enzyme forms found in the symbiotic state.

Since novel protein sometimes does occur in the dual organism
(Martin, 1973), one cannot expect to attribute all lichen electromorphs to one symbiont or the other, and certainly interpretations
based solely on IEF could not resolve the question of the symbiont
responsible. Thus, it will be not only necessary but appropriate
to regard lichen zymograms as products of the symbiosis itself.

GENETIC VARIABILITY WITHIN LICHEN POPULATIONS

Evidence suggests that individuals of one lichen species, even thalli of sterile species growing together in the same population, can be unexpectedly heterogeneous.

Secondary products of lichens provide indirect evidence of genetic variability. Lichen products have been quantified using paired-ion chromatography (PIC) (Fahselt, 1984), a variant of high performance liquid chromatography which is highly reproducible and especially suited for ionic compounds. Ionic chemical species which might otherwise be eluted too quickly are combined with a suitable counter-ion to make an ion pair which has improved retention. With PIC a methanolic mobile phase may be used in place of highly acidic aqueous solvents or the more disagreeable organic solvents, and it provides better resolution of compounds such as usnic acid.

In stands of the sterile Cladonia stellaris, paired-ion chroma-tography has demonstrated that there are highly significant dif-ferences in the levels of usnic and perlatolic acids among thalli;

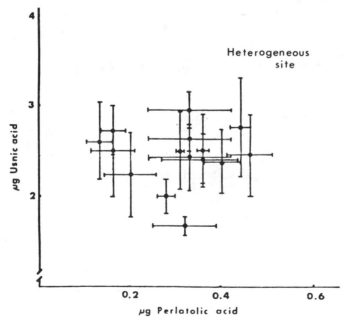

Fig. 7.   Variability of usnic and perlatolic acid levels in 15 thalli of Cladonia stellaris collected from one stand where substrate, exposure and overstorey varied internally. Units are µg per 4mg dry lichen. (Bars indicate one standard deviation).

actual concentrations showed as much as two to four fold differences from one thallus to the next (Fahselt, 1984). Interthalline differences in levels of secondary products were no
greater in apparently heterogeneous sites (Fig. 7) deliberately
selected to include a range of internal conditions of light availability, cover, substrate and aspect than they were in sites homo-
geneous with respect to substrate, exposure and dominant vegetation
(Fig. 8). Judging by what is known about the stability of lichen
products and the kinds of factors which cause them to vary under
controlled conditions (Culberson et al., 1983), a genetic and
probably an enzymatic basis for the observed thallus-to-thallus
differences is likely.

Further evidence of genetic variability within lichen populations is provided by isozyme analysis; considerable isozyme polymorphism is noted among thalli collected at one time from an
apparently uniform habitat. To date, enzyme polymorphism has been
demonstrated in both of the non-apotheciate species, Cetraria
arenaria (Fahselt and Hageman, 1983) and Umbilicaria mammulata
(Hageman and Fahselt, 1984). A number of epiphytic lichens
currently under study, including asexual species such as Usnea

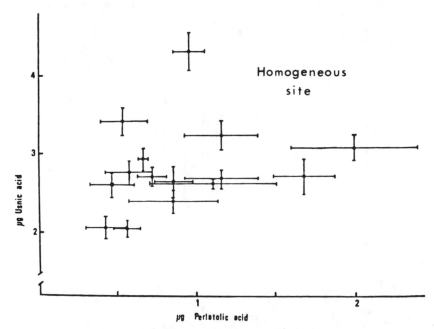

Fig. 8. Variability of usnic and perlatolic acid level in 15 thalli
of Cladonia stellaris collected from one stand which was
relatively uniform with respect to soils, light availability
and dominant vegetations. Units are µg per 4mg dry lichen.
(Bars indicate one standard deviation).

EST

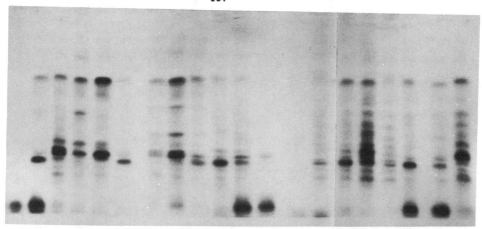

Fig. 9.   Zymogram showing esterase isozymes in 20 thalli of Usnea
          subfloridana collected from within the same apparently
          homogeneous stand of moribund conifers in May 1983.

subfloridana (Fig. 9), all demonstrate considerable thallus-to-
thallus variation in banding patterns within a stand. Enzyme polymorphism in lichens occurs in esterases, phosphatases or carbonic
anhydrases but is not restricted to these enzymes.

    The observed intraspecific differences in isozymes are smaller
than interspecific differences. Those enzyme systems which are
invariant within one taxonomic species tend to be peculiar to that
species and set it apart from others. While effects of environmental factors on lichen isozymes have not been exhaustively
tested, the results of MacFarlane et al. (1983) and Kershaw et al.
(1983) have shown that sun and shade forms of C. stellaris and C.
rangiferina did not differ with respect to the enzyme systems
tested. It would seem unlikely, when stands are deliberately
chosen to be as uniform as possible, that an appreciable portion of
the intrastand variation of isozymes could be attributed to environmental causes.

    It is inadvisable to assume that all thalli within one collec-
tion site, even sterile or non-apotheciate forms, are genetically
identical to one another. This is an important consideration to
bear in mind when sampling either for systematic studies or physiological analyses.

REFERENCES

Ahmadjian, V., 1973a, Resynthesis of lichens, in: "The Lichens,"

V. Ahmadjian, and M.E. Hale, eds, pp. 565–579. Academic
    Press, New York.

Ahmadjian, V., 1973b, Methods of isolating and culturing lichen
    symbionts and thalli, in: "The Lichens," V. Ahamdjian, and
    M.E. Hale, eds, pp. 653–659. Academic Press, New York.

Ahmadjian, V., and Jacobs, J.B. 1983, Algal-fungal relationships
    in lichens: recognition, synthesis and development, in "Algal
    Symbiosis," L.J. Goff, ed., pp. 147–172. Cambridge University
    Press, Cambridge.

Adams, W.T., and Joly, R.J., 1980, Genetics of allozyme variants in
    loblolly pine, Journal of Heredity, 71: 33–40.

Brewer, G., and Sing, C.F., 1970, "An introduction to isozyme
    techniques," Academic Press, New York.

Bubrick, P., and Galun, M., 1980, Proteins from the lichen
    Xanthoria parietina which bind to phycobiont cell walls.
    Correlation between banding patterns and cell wall chemistry,
    Protoplasma, 104: 167–173.

Bubrick, P., Galun, M., Ben-Yaacov, M., and Frensdorff, A., 1982,
    Antigenic similarities and differences between symbiotic and
    cultured phycobionts from the lichen, Xanthoria parietina,
    FEMS Microbiology Letters, 13: 435–438.

Chu, F.K., Watorek, W., and Maley, F., 1983, Factors affecting
    oligomeric structure of yeast external invertase, Archives
    Biochemistry and Biophysics, 223: 543–555.

Culberson, C.F., Culberson, W.L., and Johnson, A., 1983, Genetic
    and environmental effects on growth and production of
    secondary compounds in Cladonia cristatella, Biochemical
    Systematics and Ecology, 11: 77–84.

Fahselt, D., 1980, Alternative method for analyzing protein
    characters in lichens, The Bryologist, 82: 340–343.

Fahselt, D., 1984, Interthalline variability in levels of lichen
    products within stands of Cladina stellaris, The Bryologist,
    87: 50–56.

Fahselt, D., and Hageman, C., 1983, Isozyme banding patterns in two
    stands of Cetraria arenaria Karnef, The Bryologist, 86:
    129–134.

Fahselt, D., and Jancey, R.C., 1977, Polyacrylamide gel electro-
    phoresis of protein extracts from members of the Parmelia
    perforata complex, The Bryologist, 80: 429–438.

Fisher, H.F., Cross, D.G., and McGregor, L.L., 1962, Catalytic
    activity of subunits of glutamic dehydrogenase, Nature, 196:
    895–896.

Frevert, J., and Ballou, C.E., 1982, Yeast (Saccharomyces
    cerevisiae) invertase (EC 3.2.1.26) polymorphism correlated
    with variable states of oligosaccharide chain phosphorylation,
    Proceedings of National Academy of Sciences (USA), 79:
    6147–6150.

Gottlieb, L.D., 1977, Electrophoretic evidence and plant
    systematics, Annals of the Missouri Botanical Garden, 64:
    161–180.

Hageman, C., and Fahselt, D., 1984, Interspecific variability of isozymes of the lichen Umbilicaria mammulata, Canadian Journal of Botany, 62: 617-623.

Harris, H., and Hopkinson, D.A., 1976, "Handbook of enzyme electrophoresis in human genetics," Elsevier, New York.

Kershaw, K.A., MacFarlane, J.D., Webber, M.R., and Fovargue, A., 1983, Phenotypic differences in the seasonal pattern of net photosynthesis in Cladonia stellaris, Canadian Journal of Botany, 61: 2169-2180.

MacFarlane, J.D., Kershaw, K.A., and Webber, M.R., 1983, Physiological-environmental interactions in lichens. XVII Phenotypic differences in seasonal pattern of net photosynthesis in Cladonia rangiferina, New Phytologist, 94: 217-233.

Martin, E.J., 1972, Some problems in the isolation and characterization of enzymes from lichens, Virginia Journal of Sciences, 23: 113.

Martin, E.J., 1973, Lichen physiology: the invertases of the lichen Parmelia caperata (L.) Ach. and its isolated symbionts. Ph.D. Thesis. University of Michigan, Ann Arbor, Mich., U.S.A.

McCord, J.M., 1979, Superoxide dismutases. Occurrence, structure, function and evolution, in: "Isozymes: Current Topics in Biological and Medical Research," Vol. 3, M.C. Rattazzi, J.G. Scandolios, and G.S. Whitt, eds, pp. 1-21.

Metzenberg, R.L., 1964, Enzymatically active subunits of Neurospora invertase, Biochimica et Biophysica Acta, 89: 291-302.

Smith, R.L., and Wiedeman, V.E., 1964, A new alkaline growth medium for algae, Canadian Journal of Botany, 42: 1582-1586.

Stewart, W.D.P., and Rowell, P., 1972, Modifications of nitrogen fixing algae on lichen symbioses, Nature, 265: 371-372.

Stigbrand, T., Millan, J.L., and Fishman, W.H., 1982, The genetic basis of alkaline phosphatase variation, Current Topics in Biological and Medical Research, 6: 93-117.

Tashian, R.E., 1977, Evolution and regulation of carbonic anhydrase isozymes, Current Topics in Biological and Medical Research, 2: 21-62.

Womack, J.E., 1983, Post-translational modification of enzymes: processing genes, Current Topics in Biological and Medical Research, 7: 175-186.

STUDIES ON THE NITROGEN METABOLISM OF THE LICHENS PELTIGERA APHTHOSA AND PELTIGERA CANINA

P. Rowell[a], A.N. Rai[a,b] and W.D.P. Stewart[a]

[a] Department of Biological Sciences and A.F.R.C. Research
Group on Cyanobacteria, University of Dundee
Dundee DD1 4HN, Scotland, UK
[b] Department of Biochemistry, School of Life Sciences
North-Eastern University, Shillong – 793014, India

INTRODUCTION

Cyanobacteria are O_2-evolving photosynthetic prokaryotes, many species of which fix N_2. Most cyanobacteria are exclusively free-living forms but a few occur in symbiotic association with a variety of eukaryotes. Those which occur in symbiotic association are, almost invariably, N_2-fixing heterocystous forms (see Stewart et al., 1980; 1983 for recent reviews). In this paper we will consider aspects of the fixation, transfer and assimilation of nitrogen in the lichens Peltigera aphthosa and Peltigera canina.

NITROGEN FIXATION

The main thallus of P. aphthosa consists of a mycobiont (an asco-mycetous fungus) and a phycobiont (a green alga, Coccomyxa). The cyanobiont (a Nostoc), which is the secondary photosynthetic partner, occurs in superficial cephalodia on the upper surface of the thallus. Cephalodia represent about 2.6% (dry weight) of the thallus biomass and the cyanobiont accounts for most (about 82%) of the cephalodial protein (see Sampaio et al., 1979; Rai et al., 1981a). The cyanobiont is the N_2-fixing partner (Millbank and Kershaw, 1969; Englund, 1977; Rai et al., 1980) and it has a high average heterocyst frequency (22%) and correspondingly high nitrogenase activity when compared with free-living Nostoc spp. (Hitch and Millbank, 1975). The cephalodia are smaller and more numerous at the growing apex than in the rest of the thallus and the cyanobiont within these cephalodia has a lower (14%) heterocyst frequency (Englund, 1977). Nitrogenase activity is higher in the central region of the thallus than at its base or apex (Englund, 1977). Nitrogen fixed by the cyanobiont is transferred to

145

the remainder of the thallus (Millbank and Kershaw, 1969; Kershaw and
Millbank, 1970; Rai et al., 1981b). Using $^{15}N_2$ as a tracer, N_2 incorporation into thallus discs and into excised cephalodia has been
studied (Rai et al., 1981b). ^{15}N incorporation into attached and
excis d cephalodia continued for 3 days but labelling did not increase
further, whereas the labelling of discs with attached cephalodia
increased for 6 days. Most the ^{15}N fixed in the cephalodia was transferred to the mycobiont and the Coccomyxa of the main thallus over the
6 day period, confirming previous findings (Kershaw and Millbank,
1970), although the labelling of the Coccomyxa was greater than that
noted in the previous study.

     P. canina consists of a mycobiont (an ascomycetous fungus) and a
cyanobiont (a Nostoc). Nitrogen fixed by the cyanobiont is transferred to the mycobiont (Millbank and Kershaw, 1969; Kershaw and
Millbank, 1970; Stewart and Rowell, 1977; Rai et al., 1983a) and the
heterocyst frequency and nitrogenase activity are similar to those of
free-living Nostoc CAN (Hitch and Millbank, 1975; Stewart and Rowell,
1977) but lower than those of the cyanobiont of P. aphthosa, probably
because the cyanobiont of P. canina also provides fixed carbon for the
whole thallus (Smith et al., 1969; Stewart et al., 1980). Different
parts of thalli of P. canina show markedly different nitrogenase
activities (Millbank, 1971; Hitch and Stewart, 1973). The location of
nitrogenase activity in thalli of P. canina is considered below,
together with the location of enzymes of ammonium assimilation.

     The cyanobionts of P. aphthosa and P. canina release fixed carbon
in the form of glucose and, in P. aphthosa, the Coccomyxa releases
ribitol (Richardson et al., 1968; Smith et al., 1969). In addition to
this, the provision of carbon skeletons for ammonium assimilation may
be dependent on dark CO_2 fixation. In P. aphthosa (Rai et al., 1981c;
1983b), sustained dark nitrogenase activity was found to be supported
by the catabolism of polyglucose, accumulated in the light and by dark
CO_2 fixation via phosphoenolpyruvate carboxylase activity of the mycobiont. In the absence of such dark CO_2 fixation, the accumulation of
ammonium, due to a lack of carbon skeletons for its assimilation,
resulted, directly or indirectly, in inhibition of nitrogenase
activity (Rai et al., 1983b).

REGULATION OF NITROGEN FIXATION

     Nitrogenase activity of P. canina is markedly less sensitive to
inhibition by addition of exogenous ammonium than that of most free-
living cyanobacteria (Stewart and Rowell, 1977). Since inhibition of
nitrogenase activity of free-living cyanobacteria by ammonium did not
occur when glutamine synthetase (GS) had been inactivated (Stewart and
Rowell, 1975) or in mutant strains with reduced GS (Sakhurieva et al.,
1982; N.W. Kerby, P. Rowell and W.D.P. Stewart, unpublished data), it
is possible that the lack of inhibition by ammonium relates, in part
at least, to the low GS activity of the cyanobiont.

Addition of ammonium to excised cephalodia of P. aphthosa had no
effect on nitrogenase activity, whereas its addition to thallus discs
resulted in complete inhibition of nitrogenase activity within 24h
(Rai et al., 1980). This inhibition, when cephalodia were attached to
the main thallus, could be overcome by treatment with L-methionine-
D,L-sulphoximine (MSX) to inactivate GS of the main thallus (mainly GS
of the Coccomyxa). In contrast to ammonium, exogenous glutamine
inhibited nitrogenase activity of excised cephalodia and when ammonium
was added to thallus discs, glutamine accumulated in the cephalodia.
Thus, glutamine synthesised in the main thallus, probably via GS of
the phycobiont, may be involved in regulation of nitrogenase activity
of the cyanobiont (Rai et al., 1980).

ENZYMES OF AMMONIUM ASSIMILATION

We have studied the activities of enzymes of ammonium assimilation in P. aphthosa and P. canina extensively (Stewart and Rowell,
1977; Sampaio et al., 1979; Rai et al., 1981a,b; 1983a; Stewart et
al., 1983). When compared with the free-living isolates, the cyanobionts of these lichens had very low activities of GS and glutamate
synthase (GOGAT) (Table 1). In contrast the Coccomyxa in the main
thallus of P. aphthosa had substantial GS/GOGAT activities. We have
found no evidence for GS/GOGAT activities in the mycobionts of either
lichen, but both had high activities of NADP$^+$-dependent glutamate dehydrogenase (GDH). This GDH activity was highly localised in both
lichens; in cytochemical studies (Stewart and Rowell, 1977) it was
found that GDH activity was mainly associated with fungal hyphae close
to the cyanobionts (see also Lallement and Savoye, this volume) and
was much higher in cephalodia of P. aphthosa than in the main thallus.
The freshly isolated cyanobionts, like the free-living isolates, had
little GDH activity (Stewart and Rowell, 1977; Rai et al., 1980).
Thus, the cyanobionts have low activities of GS and GOGAT, which are
responsible for ammonium assimilation in free-living cyanobacteria
(Dharmawardene et al., 1973; Stewart and Rowell, 1975; Thomas et al.,
1975) and the mycobionts have high activities of GDH, which is
involved in ammonium assimilation in free-living fungi (Sims and
Folkes, 1964; Lara et al., 1982).

Figure 1 shows data on the activities of nitrogenase, GS and GDH
in extracts of a series of sections from thalli of P. canina cut from
the growing apex to the base of the thallus. GDH activity was
generally high throughout but decreased towards the apex. The GDH
activities obtained here are higher than those previously obtained
(Stewart and Rowell, 1977), although the reason for this is uncertain.
In contrast there was substantial GS activity in extracts of the
apical sections but activity was very low in the other extracts. The
highest GS activites obtained here are, however, considerably lower
than the values obtained for free-living Nostoc CAN (Stewart and
Rowell, 1977), even taking into account the fact that only about 36%
of the protein in thallus extracts is cyanobacterial protein (Sampaio

Table 1.   GS and GOGAT Activities in Cell-free Extracts of Symbiotic
           Cyanobacteria from Peltigera aphthosa and P. canina.

Material	GS activity	GS protein	GOGAT activity
P. aphthosa/ Nostoc APH	5.5	6	.2.3
P. canina/ Nostoc CAN	6	<5	2.5

All values are expressed as a percentage of the values obtained for
the respective free-living isolates. GS (Mg^{2+}-dependent biosynthetic)
activities were 65 and 72 nmol product formed min^{-1} mg protein^{-1} and
GOGAT activites were 31 and 34 nmol product formed min^{-1} mg protein^{-1}
for cell-free extracts of free-living Nostoc APH and Nostoc CAN,
respectively. For further details see Stewart and Rowell (1977), Rai
et al. (1980; 1981a,b; 1983a), Stewart et al. (1983). GS antigen
levels were determined by rocket immunoelectrophoresis, using antisera
against GS purified from Nostoc CAN and Anabaena cylindica (for
details of methods see Sampaio et al., 1979; Stewart et al., 1983; Ip
et al., 1984). For calculation of activities and antigen levels it
was assumed that the cyanobionts of P. aphthosa and P. canina
constituted 82% and 36% of the total protein in cell-free extracts of
excised cephalodia and whole thalli, respectively, (Sampaio et al.,
1979). GS constituted 1% of the total protein in cell-free extracts
of free-living Nostoc CAN.

et al., 1979). Nitrogenase activity was lowest in sections close to
the base of the thallus and there was a broad band of activity from
the apex to the centre of the thallus.

     The locations of nitrogenase, GS and GDH, relative to each other,
are important when considering the pathways of ammonium assimilation.
It is clear that, close to the growing apex, ammonium fixed by nitrogenase could be assimilated by either GS or GDH whereas, towards the
centre of the thallus, where there is little or no GS activity, it is
likely that ammonium assimilation is via GDH alone.

     At present, little is known about the factors which affect the
synthesis and/or activity of key enzymes such as GS and GOGAT of the
cyanobionts of P. aphthosa and P. canina. When N_2-fixing heterocystous cyanobacteria are treated with amino acid analogues such as
MSX, to inactivate GS (Stewart et al., 1982), ammonium is liberated
into the medium (Stewart and Rowell, 1975). Similarly, mutant strains
deficient in the synthesis of GS, or which synthesise an inactive

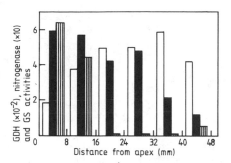

Fig. 1.   Localisation of nitrogenase, GS and NADP+-dependent GDH
activities in thalli of <u>Peltigera canina</u>.
A series of 8 mm x 8mm sections, cut from each of several
individual thalli from the apex to the base, were used either
for estimation of the rate of acetylene reduction
(nitrogenase activity) or for the preparation of cell-free
extracts for assay of GS and GDH activities. Cell-free
extracts were prepared by homogenising each thallus section
in a small volume of 50 mM Tris/HCl buffer (pH 7.5)
containing 10mM 2-mercaptoethanol, 1mM ethylenediamine
tetraacetic acid and 5mM MgCl$_2$, centrifuging at 12,000 x g
for 10 min, then dialysing against the same buffer for 18h at
4°C. Enzyme activities were assayed as previously described:
Stewart and Rowell (1977) for GDH (□), Sampaio et al. (1979)
for Mg^{2+}-dependent biosynthetic activity of GS (▥) and
Stewart et al. (1967) for nitrogenase (■). All enzyme
activites are expressed as nmol product formed min^{-1} mg
protein^{-1} and are the means of 3-6 determinations.

enzyme, liberate ammonium extracellularly (Polukhina et al., 1982;
Sakhurieva et al., 1982; Stewart et al., 1985; Kerby, Rowell and
Stewart, unpublished data). The low GS activities in the cyanobionts
of <u>P. aphthosa</u> and <u>P. canina</u> are probably due to a lack of GS protein
(Stewart et al., 1982) rather than to an inhibition or inactivation of
the enzyme. Our recent findings, using antisera against cyanobacterial GS (Stewart et al., 1983), indicate that the low GS activity
in the cyanobionts is accompanied by a corresponding reduction in the
level of GS protein, as determined by rocket immunoelectrophoresis
(Table 1). Thus, it is likely that the synthesis of GS does not
occur, and this may be because transcription of <u>gln A</u>, the structural
gene for GS (Tumer et al., 1983), has been inhibited. It remains to
be determined whether such inhibition is caused, either directly or
indirectly, by specific metabolic regulators produced by the mycobiont. There may, however, be a more general modification of the
regulation of nitrogen metabolism of the cyanobiont. This may involve
the <u>ntr</u> regulatory system (De Bruijn et al., 1984; Dixon et al., 1984)
which has been shown to function in some nitrogen-fixing prokaryotes
and may also occur in cyanobacteria (Machray et al., 1984).

Recently, sarcosine was identified as a major component of the free amino acid pools of Peltigera praetextata and was suggested to be a possible regulator of GS and nitrogenase in this lichen (Hallbom, 1984). This was based on the fact that, on treatment of the free-living Nostoc from P. canina with sarcosine, nitrogenase activity increased, GS activity decreased and ammonium was liberated. The effect of sarcosine was, thus, similar to the effect of MSX (Stewart and Rowell, 1975).

We have examined the free amino acid pools of P. canina but we have been unable to confirm the presence of a significant level of sarcosine. Figure 2 shows the free amino acid pools of P. canina and, although there is a peak (X) which elutes close to sarcosine, it does not co-chromatograph with it. In other respects the amino acid pool composition is similar to that of P. praetextata (Hallbom, 1984). The major free pool amino acids are similar in different regions of the thallus. Alanine, glutamate and glutamine are the major components which are always present at the highest concentrations and which may function as nitrogen storage compounds (Jager and Weigel, 1978). Although the identity of X is not known, its occurrence correlates with the presence of GS activity; it is most abundant in the amino acid pools of the apical region of the thallus, where GS activity is highest.

The component X (Fig. 2), in fact, co-elutes with γ-methylglutamine, which is a product of metabolism of methylammonium by GS (see Rai et al., 1984). The synthesis of methylammonium has been demonstrated in some lichens (Bernard and Goas, 1968; Bernard and Larher, 1971) and its metabolism via GS could occur. Exogenous ^{14}C-labelled methyl-

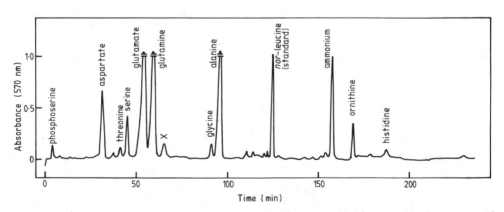

Fig. 2. Free amino acid pool composition of Peltigera canina thalli. Amino acids were extracted and analysed as previously described (Rai et al., 1984) using a LKB 4400 amino acid analyser with a lithium citrate buffer system. X represents an unidentified component (see text).

ammonium was taken up by P. canina and, in short-term uptake experiments, most (>95%) of the ^{14}C taken up remained in trichloroacetic
acid-soluble material (our unpublished data). When sections of thalli
of P. canina, cut from the apex where GS activity is highest or from
the region showing lowest GS activity (see Fig. 1), were incubated
with methylammonium, peak X (Fig. 2) increased to a much greater
extent in the apical sections (Table 2), probably due to the synthesis
of γ-methylglutamine. This finding is consistent with the presence of

Table 2. Effects of Methylammonium on the Free Amino Acid Pools of
Peltigera canina and Free-living Nostoc CAN.

| Amino acid | P. canina | | | | Nostoc CAN | |
| | Apex | | Centre | | | |
	−MA	+MA	−MA	+MA	−MA	+MA
Aspartate	35.4	32.7	28.5	28.4	0.64	0.86
Threonine	3.7	4.3	4.7	3.7	0.09	0.09
Serine	12.9	8.6	11.5	7.2	0.34	0.44
Glutamate	34.2	25.6	39.3	34.5	2.30	1.85
Glutamine	52.5	89.9	83.8	63.2	0.28	0.11
X/γ-methyl-glutamine	6.0	12.1	1.2	2.0	ND	0.64
Glycine	5.2	4.3	4.4	4.5	0.43	0.41
Alanine	45.3	41.0	54.5	63.2	0.14	0.13
Methionine	3.6	4.3	8.1	7.5	ND	ND
Ammonium	2.9	4.6	3.6	3.9	0.46	0.32
Ornithine	3.9	9.9	7.3	4.5	ND	ND

Sections (8mm x 8mm) were cut from thalli of P. canina, either from
the apex or from the part (centre) showing lowest GS activity (see
Fig. 1). The sections were incubated in nitrogen-free BG-11 medium
(pH 7.5) at 25°C, at a photon flux density of 50 μmol m^{-2} s^{-1} for 2h
in the absence (−MA) or presence (+MA) of 3mM methylammonium chloride.
Exponentially-growing cultures of Nostoc CAN in nitrogen-free BG-11
medium (Rippka et al., 1979) were harvested, washed and resuspended in
10 mM Hepes/NaOH buffer (pH 7.0), then incubated in the same buffer at
25°C, at a photon flux density of 50 μmol m^{-2} s^{-1} for 30 min. in the
absence (−MA) or presence (+MA) of 30 μM methylammonium chloride. For
details of extraction and analysis of free amino acids see legend to
Fig. 2. Data are expressed as nmol amino acid mg fresh weight^{-1} for
P. canina and as nmol amino acid μg chl^{-1} for free-living Nostoc CAN.
For calculation of quantities of X/γ-methylglutamine the colour
recovery was assumed to be identical to that of glutamine. The lichen
material used contained 0.31 μg chl. mg fresh weight^{-1}. ND = not
detectable.

metabolically active GS in the apical sections. There was also an increase in the glutamine pool, although the reason for this is not clear. Although the data in Table 2 indicate a lower glutamine pool in the apical sections, it should be noted that the pool sizes of the major free amino acids generally show variations of this magnitude and that these variations do not follow a clearly identifiable pattern.

AMMONIUM TRANSPORT

At present little is known of the ways in which ammonium and other nitrogenous compounds are transferred between the partners in lichen symbioses. The liberation of ammonium by free-living cyanobacteria, when ammonium assimilation via GS/GOGAT is inhibited, is probably due to passive diffusion of ammonia into the medium.

Intracellular accumulation of ammonium could result from passive diffusion of ammonia into the cell and ion trapping (protonation of ammonia) when the external pH is high relative to the internal pH (pH 7 to 7.5 in free-living cyanobacteria) (Hawkesford et al., 1983). However, many microorganisms have been shown to have active transport systems for ammonium (see Cook and Anthony, 1978; Kleiner et al., 1981; Rai et al., 1984). Using ^{14}C-labelled methylammonium, the free-living cyanobacterium Anabaena variabilis (Rai et al., 1984) and Nostoc CAN, the free-living isolate of the cyanobiont of P. canina, (Fig. 3; Kerby, Rowell and Stewart, unpublished data) have been shown to have ammonium (methylammonium) transport systems. Methylammonium uptake was biphasic. There was a rapid (<60s) first phase which resulted in the accumulation (about 40 fold) of an intracellular methylammonium pool which could be displaced by addition of ammonium; this also prevented further methylammonium uptake. A second, slower phase was dependent on metabolism of intracellular methylammonium via GS to form γ-methylglutamine which accumulated intracellularly (Table 2). Such metabolism was prevented when GS was inactivated by treatment with MSX, but methylammonium transport itself was not inhibited.

Such a transport system may be important in free-living cyanobacteria for the uptake of external ammonium and also for preventing the loss, by passive diffusion, of ammonia from N_2-fixing cells. There is evidence (Laane et al., 1980; Wiegel and Kleiner, 1982) that some rhizobia may lack an ammonium transport system and it has been suggested that this may be important in relation to the liberation of ammonium in symbiosis.

The data in Fig. 3 indicate that free-living Nostoc CAN, like the symbiotic Anabaena azollae of the water fern Azolla caroliniana (Rai et al., 1984), has an ammonium transport system. However, it is possible that in P. canina the ammonium transport system of the cyanobiont is absent or inhibited, and this may effectively increase the "leakiness" of the plasma membrane to ammonia. Also, an effective scavenging by the mycobiont of ammonium, liberated by the cyanobiont,

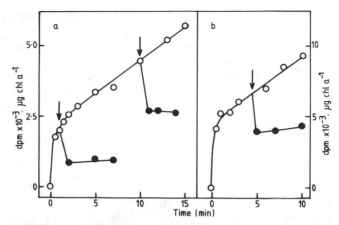

Fig. 3.   Methylammonium uptake by free-living Nostoc CAN.
Exponentially grown cultures of Nostoc CAN in nitrogen-free
BG-11 medium at 25°C and at a photon flux density of 50 μmol
m^{-2} s^{-1} were centrifuged, washed and resuspended in (a) 10mM
Hepes/NaOH buffer (pH 7.0) or (b) 10mM Tricine/NaOH buffer
(pH 9.0). $^{14}CH_3NH_3^+$ (30 μm final concentration; 10 kBq ml^{-1})
was added, at t = 0 (O).   NH$_4$Cl (3mM final concentration)
was added, at the times arrowed, to duplicate samples (●)
and incorporation of radioactivity into the cells was
determined at timed intervals (for further details see Rai et
al., 1984). dpm = disintegrations per minute.

may be important in maintaining a low external ammonium concentration
and thus ensuring continued liberation of ammonium by the cyanobiont.

PATHWAYS OF AMMONIUM ASSIMILATION

We have studied the pathways of ammonium assimilation in
P. aphthosa and P. canina using $^{15}N_2$ as a tracer (Rai et al., 1981b;
1983).

$^{15}N_2$-Incorporation in P. canina

The rate of $^{15}N_2$ incorporation into thalli of P. canina was
estimated as 0.85 nmol N$_2$ h^{-1} mg^{-1} dry weight (Rai et al., 1983a).
After $^{15}N_2$ incorporation for 3h, 75% of the total label was in the
ethanol-soluble fraction and most of this was in the five major free
pool nitrogenous compounds: ammonium, glutamate, glutamine, alanine
and aspartate.

When thalli, which had been pre-treated with digitonin to disrupt
fungal membranes (Chambers et al., 1976), were allowed to incorporate
$^{15}N_2$ for 6h, 54% of the ^{15}N fixed was liberated as extracellular
ammonium whereas less than 2% was liberated as amino acids. The

remainder of the ^{15}N was retained in the digitonin-treated discs
having, presumably, been assimilated by the cyanobiont. This
confirmed previous findings (Stewart and Rowell, 1977) that newly-
fixed nitrogen was liberated, at least partly, as ammonium by
digitonin-treated thalli of P. canina and indicates that ammonium is
the main form in which nitrogen is transferred from the cyanobiont to
the mycobiont. Such results contrast with earlier findings (Millbank,
1974).

Determination of the kinetics of ^{15}N-labelling of ammonium and
the major free pool amino acids over a short (30 min) period (Fig. 4;
Rai et al., 1983a) revealed that the highest initial labelling was in
ammonium, followed by glutamate and the amide-nitrogen of glutamine,
although the rate of enrichment was low (0.7 atom % excess ^{15}N in
ammonium after 30 min). When the above 30 min labelling period was
followed by a 30 min "chase" period, in which the ^{15}N$_2$-containing gas
phase was replaced by air (Fig. 4), the labelling in ammonium
decreased rapidly, followed by glutamate and the amide-nitrogen of
glutamine. In contrast, the labelling of alanine continued to
increase at the beginning of the chase period, then remained constant.

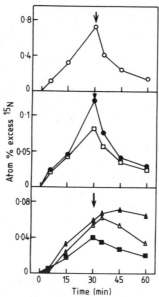

Fig. 4. ^{15}N-labelling of ammonium and the major free amino acids
after exposure of thalli of Peltigera canina to ^{15}N$_2$ (75 atom
% excess ^{15}N in N$_2$/O$_2$/CO$_2$ (79.96/20/0.04, by volume) for 30
min and, after returning to air at the time arrowed, for a
further 30 min.
O = ammonium; ● = glutamate; □ = amide-N of glutamine; ■ =
amino-N of glutamine; Δ = aspartate; ▲ = alanine. (From Rai
et al., 1983a).

These data are consistent with the synthesis of glutamate and
glutamine by primary amination reactions, e.g. GDH and GS respectively. Pretreatment of thalli with MSX and azaserine inhibited
[15]N-incorporation into glutamine, but did not inhibit labelling of
glutamate, alanine and aspartate, confirming that this was the case.

   In the above experiments, intact thalli or batches of randomly
sampled discs were used and it was recognised that compartmentali-
sation of the thalli, with respect to enzyme activites and amino acid
pools, must be taken into consideration. The existence of metabolic
amino acid pools, which may be small and show a rapid turnover, and
separate storage pools must be taken into account. The rapid incorporation into, and loss of label from, some components, particularly
ammonium and, to a lesser extent, glutamate and the amide-nitrogen of
glutamine, is consistent with the existence of small pools of these
components which show a rapid turnover. In addition, the activities
of GS and nitrogenase are not uniformly distributed throughout the
thallus (see above and Lallement and Savoye, this volume). Thus,
glutamine is probably synthesised via GS of the cyanobiont in the
apical region of the thallus, where growth of the cyanobacterium is
necessary to maintain growth of the thallus, and is subsequently transferred to the rest of the thallus and incorporated into storage pools.
The very low activity of GS, coupled with high nitrogenase activity
and high fungal GDH activity, in the central part of the thallus
suggests that it is mainly in this part of the thallus that ammonium
fixed in the cyanobiont is liberated, transferred to the mycobiont and

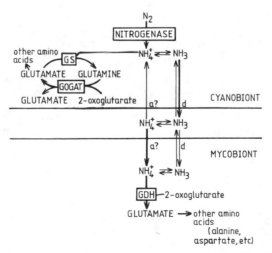

Fig. 5. Possible routes of transport and assimilation of fixed
         nitrogen in Peltigera canina.
         (For details see text and Rai et al., 1983a). a, active
         transport process; d, diffusion.

assimilated via GDH (Fig. 5). Since GS activity is also low in the
apical region (when compared with free-living Nostoc CAN) it is likely
that some of the ammonium fixed by the cyanobiont in this region is
also liberated and assimilated via GDH of the mycobiont.

$^{15}N_2$-Incorporation in P. aphthosa

The kinetics of $^{15}N_2$ incorporation into ammonium and amino acid
pools were also determined for isolated cephalodia of P. aphthosa (Rai
et al., 1981b). The highest initial labelling was into ammonium, but
labelling levelled off after about 12 min at a very low enrichment
(about 0.8 atom % excess ^{15}N). When the 15 min period of labelling
was followed by a 30 min chase period, the label was lost rapidly and
almost completely from ammonium. This is consistent with the existence of a small pool of ammonium which turns over rapidly and into
which newly-fixed nitrogen is incorporated. ^{15}N was also incorporated
rapidly into the amide-nitrogen of glutamine and into glutamate. The
amino-nitrogen of glutamine, aspartate and alanine were labelled more
slowly and with a distinct lag. The data indicate that glutamine and
glutamate are primary products of ammonium assimilation and that
aspartate and alanine are formed secondarily. In the 30 min chase
period the labelling of ammonium, glutamine (amide-N and amino-N),
glutamate and aspartate decreased whereas the labelling of alanine
increased. The increase in the total ^{15}N label in alanine over the 30
min chase period was sufficient to account for much of the ^{15}N
displaced from the other pool amino acids, mainly glutamate. Alanine
is probably formed from other amino acids, such as glutamate, via
aminotransferases.

As with P. canina, pretreatment of cephalodia with MSX and
azaserine, to inhibit ammonium assimilation via GS/GOGAT, prevented
labelling of glutamine but did not inhibit labelling of glutamate.
Thus synthesis of glutamate is, apparently, mainly independent of the
GS/GOGAT pathway.

Digitonin-treated cephalodia released virtually all of the
nitrogen fixed as extracellular ammonium (Rai et al., 1980) although,
in contrast with previous findings (Englund, 1977), no ammonium was
liberated in the absence of digitonin. The data (Rai et al., 1980;
1981b) indicate that the main form in which nitrogen is transferred to
the mycobiont in cephalodia is ammonium, with GDH of the mycobiont
being mainly responsible for its assimilation.

The accumulation of ^{15}N label in alanine in excised cephalodia
(Rai et al., 1981b) was presumably due to the lack of a sink for
products of ammonium assimilation which would normally be transferred
to the main thallus. It is uncertain whether alanine is the normal
transport compound, or whether a lack of transfer of other nitrogenous
compounds results in accumulation of newly fixed nitrogen in the
alanine pool of cephalodia.

ACKNOWLEDGEMENTS

   The work was supported by the AFRC, SERC, EEC and the Royal
Society of London. We thank Dr N.W. Kerby for helpful discussions.

REFERENCES

Bernard, T., and Goas, G., 1968, Contribution à l'étude du métabolisme
    azoté des lichens. Caractérisation et dosages des méthylamines
    de quelques espèces de la famille des Stictaceés, Comptes Rendus,
    267: 622-624.
Bernard, T., and Larher, F., 1971, Contribution à l'étude du
    métabolisme azoté des lichens. Rôle de la glycine ^{14}C-2 dans la
    formation des méthylamines chez Lobaria laetevirens, Comptes
    Rendus, 272: 568-571.
Chambers, S., Morris, M., and Smith, D.C., 1976, Lichen physiology.
    XV. The effect of digitonin and other treatments on biotrophic
    transport of glucose from alga to fungus in Peltigera
    polydactyla, New Phytologist, 76: 485-500.
Cook, R.J., and Anthony, C., 1978, The ammonia and methylamine active
    transport system of Aspergillus nidulans, Journal of General
    Microbiology, 109: 265-274.
De Bruijn, F.J., Sundaresan, V., Szeto, W.W., Ow, D.W. and Ausubel,
    F.M., 1984, Regulation of the nitrogen fixation (nif) genes of
    Klebsiella pneumoniae and Rhizobium meliloti. Role of nitrogen
    regulation (ntr) genes, in: "Advances in Nitrogen Fixation
    Research," C. Veeger, and W.E. Newton, eds, pp. 627-633, Nijhoff,
    Junk, Pudoc, The Hague.
Dharmawardene, M.W.N., Haystead, A., and Stewart, W.D.P., 1973,
    Glutamine synthetase of the nitrogen-fixing alga Anabaena
    cylindrica, Archiv für Mikrobiologie, 90: 281-295.
Dixon, R.A., Alvares-Morales, A., Clements, J., Drummond, M., Merrick,
    M., and Postgate, J.R., 1984, Transcriptional control of the nif
    regulon in Klebsiella pneumoniae, in: "Advances in Nitrogen
    Fixation Research," C. Veeger, and W.E. Newton, eds, pp. 635-642,
    Nijhoff, Junk, Pudoc, The Hague.
Englund, B., 1977, The physiology of the lichen Peltigera aphthosa,
    with special reference to the blue-green phycobiont (Nostoc sp.),
    Physiologia Plantarum, 41: 298-304.
Hallbom, L., 1984, Sarcosine: a possible regulatory compound in the
    Peltigera praetextata - Nostoc symbiosis, FEMS Microbiology
    Letters, 22: 119-121.
Hawkesford, M.J., Rowell, P., and Stewart, W.D.P., 1983, Energy
    transduction in cyanobacteria, in: "Photosynthetic Prokaryotes:
    Cell Differentiation and Function," G.C. Papageorgiou, and
    L. Packer, eds, pp. 199-218, Elsevier, New York.
Hitch, C.J.B., and Millbank, J.W., 1975, Nitrogen metabolism in
    lichens. VI. The blue-green phycobiont content, heterocyst
    frequency and nitrogenase activity in Peltigera species, New
    Phytologist, 74: 473-476.

Hitch, C.J.B., and Stewart, W.D.P., 1973, Nitrogen fixation by lichens in Scotland, New Phytologist, 72: 509-524.

Ip, S-M., Rowell, P., Aitken, A., and Stewart, W.D.P., 1984, Purification and characterisation of thioredoxin from the N_2-fixing cyanobacterium Anabaena cylindrica, European Journal of Biochemistry, 141: 497-504.

Jager, H.J., and Weigel, H.J., 1978, Amino acid metabolism in lichens. The Bryologist, 81: 107-113.

Kershaw, K.A., and Millbank, J.W., 1970, Nitrogen metabolism in lichens. II. The partition of cephalodial fixed nitrogen between the mycobiont and phycobionts of Peltigera aphthosa, New Phytologist, 69: 75-79.

Kleiner, D., Phillips, S., and Fitzke, E., 1981, Pathways and regulatory aspects of N_2 and NH_4^+ assimilation in N_2-fixing bacteria, in: "Biology of Inorganic Nitrogen and Sulphur," H. Bothe, and A. Trebst, eds, pp. 131-140, Springer, Berlin.

Laane, C., Krone, W., Konings, W., Haaker, H., and Veeger, C., 1980, Short-term effect of ammonium chloride on nitrogen fixation by Azotobacter vinlandii and by bacteriods of Rhizobium leguminosarum, European Journal of Biochemistry, 103: 39-46.

Lara, M., Blanco, L., Campomanes, M., Calva, E., Palacios, R., and Mora, J., 1982, Physiology of ammonium assimilation in Neurospora crassa, Journal of Bacteriology, 150: 105-112.

Machray, G.C., Hagan, C.E., Boxer, M., and Stewart, W.D.P., 1984, Nitrogen-regulatory genes in nitrogen fixing cyanobacteria, in: "Advances in Nitrogen Fixation Research," C. Veeger, and W.E. Newton, eds, p. 740, Nijhoff, Junk, Pudoc, The Hague.

Millbank, J.W., 1971, Nitrogen metabolism in lichens. IV. The nitrogenase activity of the Nostoc phycobiont in Peltigera canina, New Phytologist, 71: 1-10.

Millbank, J.W., 1974, Nitrogen metabolism in lichens. V. The forms of nitrogen released by the blue-green phycobiont in Peltigera spp., New Phytologist, 73: 1171-1181.

Millbank, J.W., and Kershaw, K.A., 1969, Nitrogen metabolism in lichens. I. Nitrogen fixation in the cephalodia of Peltigera aphthosa, New Phytologist, 68: 721-729.

Polukhina, L.E., Sakhurieva, G.N., and Shestakov, S.V., 1982, Ethylenediamine resistant Anabaena variabilis mutants with derepressed nitrogen-fixing system, Microbiologiya, 51: 90-95.

Rai, A.N., Rowell, P., and Stewart, W.D.P., 1980, NH_4^+ assimilation and nitrogenase regulation in the lichen Peltigera aphthosa Willd., New Phytologist, 85: 545-555.

Rai, A.N., Rowell, P., and Stewart, W.D.P., 1981a, Glutamate synthase activity in symbiotic cyanobacteria, Journal of General Microbiology, 126: 515-518.

Rai, A.N., Rowell, P., and Stewart, W.D.P., 1981b, $^{15}N_2$ incorporation and metabolism in the lichen Peltigera aphthosa Willd., Planta, 152: 544-552.

Rai, A.N., Rowell, P., and Stewart, W.D.P., 1981c, Nitrogenase activity and dark CO_2 fixation in the lichen Peltigera aphthosa

Willd., Planta 151: 256–264.

Rai, A.N., Rowell, P., and Stewart, W.D.P., 1983a, Interactions between cyanobacterium and fungus during $^{15}N_2$-incorporation and metabolism in the lichen Peltigera canina, Archives of Microbiology, 134: 136–142.

Rai, A.N., Rowell, P., and Stewart, W.D.P., 1983b, Mycobiont-cyanobiont interactions during dark nitrogen fixation by the lichen Peltigera aphthosa, Physiologia Plantarum, 57: 285–290.

Rai, A.N., Rowell, P., and Stewart, W.D.P., 1984, Evidence for an ammonium transport system in free-living and symbiotic cyanobacteria, Archives of Microbiology, 137: 241–246.

Richardson, D.H.S., Hill, D.J., and Smith, D.C., 1968, Lichen physiology. XI. The role of the alga in determining the pattern of carbohydrate movement between lichen symbionts, New Phytologist, 97: 469–486.

Rippka, R., Deruelles, J., Waterbury, J.B., Herdman, M., and Stanier, R.Y., 1979, Generic assignments, strain histories and properties of pure cultures of cyanobacteria, Journal of General Microbiology, 111: 1–61.

Sakhurieva, G.N., Polukhina, L.E., and Shestakov, S.V., 1982, Glutamine synthetase in Anabaena variabilis mutants with derepressed nitrogenase, Mikrobiologiya, 51: 308–312.

Sampaio, M.J.A.M., Rai, A.N., Rowell, P., and Stewart, W.D.P., 1979, Occurrence, synthesis and activity of glutamine synthetase in N_2-fixing lichens, FEMS Microbiology Letters, 6: 107–110.

Sims, A.P., and Folkes, B.F., 1964, A kinetic study of the assimilation of ^{15}N-ammonia and the synthesis of amino acids in an exponentially growing culture of Candida utilis, Proceedings of The Royal Society of London, Series B, 159: 479–502.

Smith, D.C., Muscatine, L., and Lewis, D.H., 1969, Carbohydrate movement from autotrophs to heterotrophs in parasitic and mutualistic symbiosis, Biological Reviews, 44: 17–90.

Stewart, W.D.P., Fitzgerald, G.P., and Burris, R.H., 1967, In situ studies on N_2 fixation using the acetylene reduction technique, Proceedings of The National Academy of Sciences, USA, 58: 2071–2078.

Stewart, W.D.P., and Rowell, P., 1975, Effects of L-methionine-D,L-sulphoximine on the assimilation of newly fixed NH_3, acetylene reduction and heterocyst production in Anabaena cylindrica, Biochemical and Biophysical Research Communications, 65: 846–856.

Stewart, W.D.P., and Rowell, P., 1977, Modifications of nitrogen-fixing algae in lichen symbioses, Nature, 265: 371–372.

Stewart, W.D.P., Rowell, P., Cossar, J.D., and Kerby, N.W., 1985, Physiological studies on N_2-fixing cyanobacteria, in: "Nitrogen Fixation and CO_2 Metabolism" P.W. Ludden, and J.E. Burris, eds, pp. 269–279, Elsevier, New York.

Stewart, W.D.P., Rowell, P., Hawkesford, M.J., Sampaio, M.J.A.M., and Ernst, A., 1982, Nitrogenase and aspects of its regulation in cyanobacteria, Israel Journal of Botany, 31: 168–189.

Stewart, W.D.P., Rowell, P., and Rai, A.N., 1980, Symbiotic
    nitrogen-fixing cyanobacteria, in: "Nitrogen Fixation,"
    W.D.P. Stewart, and J.R. Gallon, eds, pp. 239-277, Academic
    Press, London.
Stewart, W.D.P., Rowell, P., and Rai, A.N., 1983, Cyanobacteria-
    eukaryotic plant symbioses, Annales de Microbiologie (Institut
    Pasteur), 134B: 205-228.
Thomas, J., Wolk, C.P., Shaffer, P.W., Austin, S.M., and Galonsky, A.,
    1975, The initial organic products of fixation of ^{13}N-labelled
    nitrogen gas by the blue-green alga Anabaena cylindrica,
    Biochemical and Biophysical Research Communications, 67: 501-507.
Tumer, N.E., Robinson, S.J., and Haselkorn, R., 1983, Different
    promoters for the Anabaena glutamine synthetase gene during
    growth using molecular or fixed nigrogen, Nature, 306: 337-342.
Wiegel, J., and Kleiner, D., 1982, Survey of ammonium (methylammonium)
    transport by aerobic N_2-fixing bacteria – the special case of
    Rhizobium, FEMS Microbiology Letters, 15: 61-63.

NITROGEN LOSSES FROM DIAZOTROPHIC LICHENS

J.W. Millbank

Department of Pure & Applied Biology
Imperial College
London, SW7 2AZ, U.K.

INTRODUCTION

Nitrogen loss from lichens has received little attention in
the past and has only relatively recently become the subject of
study by a few groups. Early investigations were concerned with
the total input of nitrogen into an ecosystem and naturally follow
the realisation that since a number of lichens contain a member of
the Cyanophyceae they might therefore be capable of fixing
atmospheric nitrogen. This valuable feature was first definitely
demonstrated by Bond and Scott (1955) using ^{15}N. Later, when more
studies of lichen physiology had been reported and thallus moisture
content was recognised as being of great importance, it was
appreciated that release of nitrogenous compounds from thalli could
be a normal feature, unassociated with death and decay, and so
additional to the more obvious contribution provided by the latter.

STUDIES OF THE CONTRIBUTION BY THE WHOLE THALLUS

Data on the contribution of whole thalli due to normal growth,
death and decay, are few, being handicapped by a number of factors.
Thus lichens are among the slowest growing plants, grow poorly, if
at all, in artificial or "transplanted" situations and are very
prone to the effects of common atmospheric pollutants especially
sulphur dioxide, oxides of nitrogen, ozone and fluorides. They
also evidently thrive under conditions of regular environmental
change and stress and hence normal incubation techniques are often
the reverse of optimal. Field studies, or analyses of material
gathered from the field are therefore almost obligatory. For all
these reasons investigations are inevitably rather time-consuming.

Resulting from their slow growth and delicate and exacting microclimatic and environmental requirements, lichens are rarely the dominant members of an ecosystem and so their contribution to the nutritional status of an area is usually assumed to be small or even insignificant. A few areas exist, however, where lichens are accepted as having an important role. Such areas are the drier heathland and woodlands of the Arctic and Subarctic, the tundra at similar latitudes in North America and Europe, and the old-growth Douglas fir forests of North Western U.S.A. The last are on the slopes of the Western Cascade mountains of Oregon, altitude 500-600m, latitude 45°N, which receive an annual rainfall of about 2250 mm, mostly as heavy falls numbering about 40 per year. Nitrogen-fixing lichens are abundant and important in these special habitats.

The need for field study and the other technical handicaps mentioned have resulted in a great preponderance of estimates involving inferential conclusions from short-term sampling data, often augmented by extrapolation of metabolic activities derived from laboratory experiments. The acetylene reduction technique for estimating nitrogenase activity figures prominently in these studies. Using it, Granhall and Selander (1973) reported high rates of nitrogenase activity in Stereocaulon paschale, Nephroma arcticum and Peltigera scabrosa, as did Crittenden (1975) in P. aphthosa and P. canina with corroborative data from Horne (1972) and Schell and Alexander (1973).

A few attempts have been made to quantify the nitrogen input in arctic and subarctic ecosystems, over a long time period. Huss-Danell (1977) studied the activity of S. paschale in a pine forest area of North Sweden (lat. 64°N) having a very low rainfall. She concluded that the nitrogen fixed amounted to about 1 kg hectare^{-1} during the 5 months May-September. The estimates were derived from long (4 hour) in situ incubations under acetylene, using small closed vessels. Although replication and sampling were thorough, the technique suffered from the drawbacks that the thallus temperature could become very high and carbon dioxide supply depleted or even exhausted when using such lengthy incubations. Nitrogenase activity is also stimulated by long exposure to acetylene. Kallio and Kallio (1975) studied N. arcticum and S. paschale in the sub-arctic at Kevo, Finland. Again, a comprehensive series of readings were taken but the same criticisms apply. The results of these studies are shown in Table 1. Comprehensive criticisms of these results appear in Crittenden and Kershaw (1978, 1979), largely based on Kershaw and his collaborators' detailed studies of environmental effects on nitrogenase activity; for details of these see Kershaw (1985). In summary, Crittenden and Kershaw concluded that, due to the unquantifiable effects of 'pretreatment' on lichen metabolism, predictions of the initiation and subsequent rates of nitrogen

Table 1.   Estimates of the Contribution by Lichens to the Nitrogen
           Economy of Various Ecosystems

Location	Species	
	Stereocaulon paschale	Nephroma arcticum
	$(kg\ N\ ha^{-1}\ annum^{-1})$	
(a) Pine forest, North Sweden	1	–
(b) Pine forest, Kevo, Finland	38	–
(c) Birch forest, Kevo, Finland	5	10
(d) Sub-alpine heath, Kevo, Finland	17	0.5
(e) Arctic tundra, Barrow, Alaska	1	–

(a)   Huss-Danell (1977).   (b, c, d) Kallio & Kallio (1975).
(e)   Alexander (1981).

fixation are not possible with any worthwhile degree of accuracy.
They therefore advocate frequent direct in situ measurements of
nitrogenase activity using "proven techniques of reliable accuracy"
as being the only acceptable, if indirect, method of estimating
nitrogen input.

The tundra and teiga ecosystems of Alaska have been
extensively studied by Alexander and her collaborators particularly
resulting from the I.B.P. support. An account of this work has
been given in some detail (Alexander et al., 1978) and a later
summary, taking into account the subsequent doubts and reservations
concerning the quantitative reliability of the acetylene reduction
technique, appeared in the proceedings of the 4th symposium on
nitrogen fixation, Canberra, Australia (Alexander, 1981). In these
studies, a considered estimate of the lichen contribution as being
between 0.5 and 1.5 kg hectare^{-1} annum^{-1}, and representing about
50% of the total nitrogen input to the ecosystem was reported.

In an entirely different context are the reports of Carroll,
Denison and their collaborators of the contribution by lichens to
the nitrogen economy of the Redwood forests of north western
U.S.A., notably in Oregon. These have been described in a
considerable number of reports and articles (Denison, 1973, 1979;
Pike, 1978; Pike et al., 1972, 1975, 1977; Bernstein and Carroll,
1979; Caldwell et al., 1979; Carroll, 1979, 1980; Carroll et al.,
1980; Sollins et al., 1980; Silvester, 1983). There is abundant
evidence of a very significant contribution by the epiphytic lichen

population, especially Lobaria oregana, though L. pulmonaria also
contributes. Detailed surveys of the abundance of all the arboreal
epiphytes have been made as well as estimates of the content and
turnover rates of N, P, K, Ca and Mg in the lichen population. The
studies have also been extended to a number of other forest
ecosystems. Four are mentioned in Pike (1978); others are from
Forman (1975), Forman and Dowden (1977), Todd et al. (1978), Becker
(1980) and Rhoades (1981).

In the studies from the University of Oregon, the growth rate
of L. oregana, its biomass, nitrogen content and rate of litterfall
have all been utilised to derive figures for the lichen's
contribution to the nitrogen input of the ecosystem. Clearly,
assumptions and approximations have had to be made but the derived
estimates are much the same whatever basis is taken. Pike et al.
(1977) reported the total biomass of L. oregana on one 400 year old
Douglas fir specimen as being 13.1 kg; 10-15 kg Lobaria per tree
can be assumed, and this amounts to 500-900 kg hectare^{-1}. If a
growth rate of 0.3 g g^{-1} annum^{-1} is taken and a nitrogen content of
2.1%, then a rate of nitrogen fixation of 3.2 kg N hectare^{-1}
annum^{-1} results from 500 kg Lobaria hectare^{-1}; 5.7 kg N from 900 kg
Lobaria (Denison, 1979). Equivalent data from Pike (1978) gives a
rate of nitrogen fixation of 4.5 kg hectare^{-1} annum^{-1} by Lobaria.

Complications in calculations result when account is taken of
the fact that Lobaria and other epiphytes decay in situ and give
rise to a form of "epiphytic soil". Attempts to include this
effect have therefore been made. Such "soil" will result in lower
figures of "litterfall" nitrogen but will, however, provide
nitrogen for lichen growth, which will be reflected in increased
biomass. Nitrogen is also known to be leached from the lichen
thalli and this augments the nitrogen present in "throughfall"
water that has been derived from the non-fixing tissues over which
it has flowed. For these reasons some authors have taken more or
less arbitrary fractions of "throughfall" nitrogen and "litter"
nitrogen when deriving estimates of nitrogen fixation. For
example, "total Lobaria litter N + ½ throughfall N; or ½ litterfall
N + ½ throughfall N" (Sollin et al., 1980). Figures of about 3 kg
N hectare^{-1} annum^{-1} result from such assumptions. Clearly any
number of assumptions and approaches are possible but, as pointed
out by Silvester (1983), a surprisingly consistent result of about
4 kg N hectare^{-1} annum^{-1} emerges each time.

LEACHING OF FIXED NITROGEN DURING GROWTH

The appreciation that nitrogenous compounds are lost from the
thalli of nitrogen-fixing lichens by leaching, as a normal feature
of their existence, has prompted studies to be made of the
contribution that this makes to the nitrogen economy of their
habitat. Losses of carbon compounds when thalli become moistened

by rain after a period of desiccation have been recognised since
Farrar and Smith (1976) and nitrogen compounds are also involved.
Hitch and Stewart (1973) suggested that a similar process to that
occurring in free-living cyanobacteria may take place and Pike et
al. (1972) and Denison (1973) shared the view. More recent
investigations (Millbank, 1978, 1982; Pike, 1978) have clearly
shown that nitrogenous compounds are leached from healthy lichen
thalli, especially when dry thalli are rewetted. Changes in the
membrane conformation of the thalli with hydration have been
advanced to account for the phenomenon, but definitive evidence is
lacking. Laboratory simulations of "rewetting" episodes, however,
suffer from the shortcoming of a failure to mimic adequately the
natural variations in relative humidity and rainfall intensity, and
so may very well overestimate the losses. Some examples of the
losses from flooding are given in Table 2.

To study nitrogen leakage quantitatively under realistic
field conditions Crittenden (1983) investigated Stereocaulon and
Cladonia, arranged in specially constructed stainless steel and
polyester collecting vessels, in a lichen-rich birch woodland at
Kevo, Finland. As an alternative approach I have further developed
the controlled environment chamber previously described (Millbank
and Olsen, 1981) to simulate more closely the conditions
obtaining in the wet autumn, winter and spring seasons on the South
West coast of Scotland. The chamber permits the use of $^{15}N_2$
labelled atmospheric nitrogen for periods of a month and enables

Table 2. Effects of Irrigation on Lichens

Treatment		Total nitrogen loss, $\mu g/g^{-1}$	
1. Immersion of air-dry thalli for 3 min.		300	
	Fife	Inverness	N. Wales
2. Air-dry thalli moistened.	268	214	112
After 15 min. immersed for 3 mins.	168	131	58
Cycle repeated three times.	103	83	22
3. Continuous mist of sterile filtered rainwater	8 in first 100 ml.		
	10 in second 100 ml.		
	2 in third and subsequent 100 ml.		

1, 2: Millbank (1978, 1982) Peltigera membranacea.
   3: Pike (1978) Lobaria oregana.

any migration of the nitrogen fixed by the lichen to be monitored.
Using P. membranacea (a species closely related to P. canina)
transfer from the lichen to the associated mosses and run-off water
can now be partly quantified.

Field Studies

In studies with Stereocaulon and Cladonia Crittenden (1983)
showed that both species absorbed ammonium-nitrogen from rain and
released organic nitrogen. Losses of organic nitrogen per unit
area of mats of the nitrogen-fixer Stereocaulon were up to 6.5
times greater than that from the non-nitrogen-fixing Cladonia. The
nature of the organic nitrogen is not known. Some appeared to be
polypeptide, and betaines may well form an important part. Losses
of organic nitrogen from Stereocaulon were greatest following the
resumption of rain after a period during which the lichen remained
moist. Potassium ions were not released under such circumstances,
and Crittenden considers that the nitrogen released is probably not
reabsorbed by the lichen. During seven consecutive rainfall events
extending over 49 days the total net exchange of nitrogen was + 2.4
mg m^{-2} for Cladonia (an uptake) and – 9.4 mg m^{-2} for Stereocaulon.
The rainfall contained 8.6 mg m^{-2} of total nitrogen. Hence the
contribution of Stereocaulon was about the same as that of incident
rain. It is thought that both species studied may be resistant to
rehydration stress and thus the amounts of solutes lost on sudden
rewetting was small. On the other hand Peltigera and Nephroma,
which are often found associated with quite luxuriant ground
vegetation, are subject to few sudden extremes of hydration/
dehydration, are less stress-resistant and so are particularly
"leaky" when suddenly rehydrated (see later, and Buck and Brown,
1979). Crittenden was prudently non-commital regarding estimating
the total fixation of nitrogen per annum per square metre of
monospecific Stereocaulon in view of the quantitative uncertainty
of the acetylene reduction technique. He ventured to estimate that
wetting losses could represent up to 12% of the total nitrogen
fixed in the period of suitable conditions, initiated by the onset
of rainfall. Studies to estimate the degree of utilisation of the
leached nitrogen by the associated plants have not yet been
attempted, but it is proposed to do so (Crittenden, personal
communication). It must be admitted, though, that even almost
continuous Stereocaulon mats can only contribute a small fraction
of the total nitrogen demands of the plant community in systems
such as birch forest, lichen heathland, spruce/feathermoss forest
and silver fir forest, which have total nitrogen requirements
ranging between 12-35 kg hectare^{-1} annum^{-1}.

Laboratory Studies

Studies of Peltigera spp. (Millbank, 1981) using ^{15}N-enriched
atmospheres in a laboratory-simulated controlled environment gave

estimates of total nitrogen fixation of 2.4 to 5.8 kg hectare[-1] annum[-1], making, <u>inter alia</u>, the assumption that lichen coverage of the terrain was 10%. The low figure referred to an open dune habitat in South Wales, where thallus moisture in the summer was often extremely low, whilst the others referred to more consistently moist habitats in North East Scotland. Experiments in which lichen thalli enriched with ^{15}N were subjected to simulated rainfall episodes (Millbank, 1982) indicated that amounts of organic nitrogen up to three times the amount fixed in 24 hours, could be lost in one rainfall event if the thallus had previously become dry. However, it was realised that these treatments were much more drastic than natural conditions.

In order to obtain data on nitrogen losses from lichens under moist autumn, winter and spring conditions of South West Scotland (Dunstaffnage, near Oban) using ^{15}N labelling for maximum sensitivity, the controlled environment chamber mentioned above has undergone significant development. This now allows much more than just the long term rate of nitrogen fixation to be established. A reservoir beneath the perforated support for the lichen/moss association enables throughflow water to be collected and analysed. Automatic pumps remove the accumulated run-off water at regular intervals, with safeguards to prevent loss of the N[15] enriched atmosphere. The water distribution system for rainfall simulation is much improved and fully programmable valves enable the appropriate seasonal rainfall to be applied to the thalli in any number of discrete events per day. The temperature control permits a more gradual transition from day maximum to night minimum and the lighting arrangements are more realistic with a higher red content at dawn and dusk. The CO_2 concentration is held at c. 0.03% by volume. A partition permits separate moss specimens entirely devoid of lichen to be maintained and exposed to [15]N_2 simultaneously. Due to the requirement for a completely closed vessel, the humidity is very high and while this is fortuitously advantageous for many circumstances it means that simulations of summertime growth are not possible. Measures to control the atmospheric humidity were not economically possible although technically feasible. Access to the plant material is also not possible during a trial, for the same reasons. The normal period of simulation is a month.

The results of a winter trial are presented in Table 3. Run-off was analysed daily for nitrogen content and ^{15}N label. The level of ^{15}N labelling in the run-off water exhibited a rapid rise for the first 4 days and then became approximately constant for the rest of the 28 day trial. This showed that newly-fixed nitrogen was being eluted and lost from the lichens but formed only a very small proportion, c. 0.2% of the total newly-fixed nitrogen. However, overall losses of soluble nitrogen by leaching were appreciable, representing about 9.4% of the total amount fixed, and

Table 3. Nitrogen Fixed and Eluted from Peltigera membranacea under Conditions of Simulated Winter in South West Scotland.

	LICHEN				MOSS			
Chamber	Dry Weight	Total N	N^{15} excess	N gain	Dry Weight	Total N	N^{15} excess	Newly fixed N
1	4.68 g	82 mg	0.235%	14.1 mg	5.11 g	69 mg	0.018%	0.41 mg
2	6.20 g	239 mg	0.201%	16.6 mg	-	-	-	-

RUNOFF SOLUTION

	Total Nitrogen		Newly fixed Nitrogen
Chamber	Weight mg	N^{15} excess	g
1	1.87	0.033%	20.2
2	1.55	0.063%	33.7

Duration of trial 28 days. Day length 8 hr. Max. temp. 9°. Min. temp. 4°.
Light intensity 80 - 100 E cm^{-2} sec^{-1}. 1 g dry weight thallus covers c. 70 sq.cm.
Chamber 1 : complete system.
Chamber 2 : Moss removed; lichen only.

0.65% of the total thallus nitrogen. Thus, it would seem that the immediate products of fixation are incorporated into a mobile pool of organic nitrogen, the label becoming, therefore, diluted. Relatively immobile proteins and other large molecules are synthesised from this pool. Losses of these by death and decay of the thallus were small during these trials as care was taken to choose specimens showing an absolute minimum of senescent features (e.g. discoloured, thin thallus or soft areas). Nevertheless, 0.65% of the total thallus nitrogen was lost over four weeks; during this period fixation provided a net increase in thallus nitrogen of approximately 7.2%.

Of the nitrogen lost, comparatively little appeared to be directly absorbed and incorporated into the moss fronds. The amount of combined nitrogen in the throughfall beneath the lichen/moss association was similar in amount to that beneath the lichen alone; however, the level of ^{15}N enrichment was significantly less after permeating through the moss material, implying absorption by the mosses. This absorption represented about 13 µg N, but the labelling of the moss indicated the presence of over 400 µg of newly-fixed nitrogen. Fixation by epiphytic cyanobacteria was suspected as being responsible, and this was reinforced by the finding that in this trial moss material incubated entirely separately from lichen thalli also became labelled to a similar extent. Other trials have resulted in totally unlabelled moss material and it seems evident that there is much variation in the distribution of epiphytic cyanobacteria.

It is concluded, therefore, that a small quantity of newly-fixed and eluted combined nitrogen is directly absorbed by mosses associated with P. membranacea; epiphytic cyanobacteria can fix substantially more than this in the short term. Since the combined nitrogen released by the lichen is largely organic, immediate absorption by mosses would not normally be expected, and the normal processes of microbial activity are presumed to be responsible for the assimilation and ultimate conversion to assimilable combined nitrogen in the substratum beneath the lichen.

Assuming a 10% area coverage by the lichen, the supply of combined nitrogen by leaching from the thalli during the wet months (October to May) amounted to approximately 50 g ha^{-1} month^{-1}; an annual contribution of at least 0.4 kg ha^{-1} can thus be expected. To this of course must be added the nitrogen fixed and retained by the lichen, amounting to about 9 times this, making a total contribution to the ecosystem of 4 kg ha^{-1} annum^{-1}. Thus, in these somewhat special habitats lichens can make a worthwhile contribution to the nitrogen nutrition of their ecosystem.

ACKNOWLEDGEMENTS

It is a pleasure to record the assistance of Mr J.D. Olsen in the author's laboratory. Some of the studies were funded in part by the British Natural Environment Research Council and the Science and Engineering Research Council.

REFERENCES

Alexander, V., 1981, Nitrogen fixing lichens in tundra and teiga ecosystems, in: "Current Perspectives in Nitrogen Fixation," A.H. Gibson and W.E. Newton, eds, p. 256, Canberra, Australian Academy of Science.

Alexander, V., Billington, M., and Schell, D.M., 1978, Nitrogen fixation in arctic and alpine tundra, in: "Vegetation and Production Ecology of an Alaskan Arctic Tundra," Ecological Studies, 29, L.L. Tieszen, ed., pp. 539-558, Springer Verlag, Stuttgart.

Becker, V.E., 1980, Nitrogen-fixing lichens in the forests of the Southern Appalachian Mountains of North Carolina, The Bryologist 83: 29-39.

Bernstein, M.E., and Carroll, G.C., 1977, Microbial populations on Douglas fir needle surfaces, Microbial Ecology, 4: 41-52.

Bond, G., and Scott, G.D., 1955, An examination of some symbiotic systems for fixation of nitrogen, Annals of Botany, 19: 67-77.

Buck, G.W., and Brown, D.H., 1979, The effect of desiccation on cation location in lichens, Annals of Botany, 44: 265-277.

Caldwell, B.A., Hagedorn, C., and Denison, W.C., 1979, Bacterial ecology of an old-growth Douglas fir canopy, Microbial Ecology, 5: 91-103.

Carroll, G.C., 1979, Needle microepiphytes in a Douglas fir canopy: biomass and distribution patterns, Canadian Journal of Botany, 57: 1000-1007.

Carroll, G.C., 1980, Forest canopies; complex and independent subsystems, in: "Forests; Fresh Perspectives from Ecosystems Analysis," R.H. Waring, ed., pp. 87-107, Oregon State University Press, Corvallis.

Carroll, G.C., Pike, L.H., Perkins, J.R., and Sherwood, M.A., 1980, Biomass and distribution patterns of conifer twig microepiphytes in a Douglas fir forest, Canadian Journal of Botany, 58: 624-630.

Crittenden, P.D., 1975, Nitrogen fixation by lichens on glacial drift in Iceland, New Phytologist, 74: 41-49.

Crittenden, P.D., 1983, The role of lichens in the nitrogen economy of subarctic woodlands: Nitrogen loss from the nitrogen-fixing lichen Stereocaulon paschale during rainfall, in: "Nitrogen as an Ecological Factor," J.A. Lee, S. McNeill and I.H. Rorison, eds, pp. 43-68, Blackwell, Oxford.

Crittenden, P.D., and Kershaw, K.A., 1978, Discovering the role of
    lichens in the nitrogen cycle in boreal-arctic ecosystems,
    The Bryologist, 81: 258-267.
Crittenden, P.D., and Kershaw, K.A., 1979, Studies on
    lichen-dominated systems, 22. The environmental control of
    nitrogenase activity in Stereocaulon paschale in spruce lichen
    woodland, Canadian Journal of Botany, 57: 236-254.
Denison, W.C., 1973, Life in tall trees, Scientific American, 228:
    75-80.
Denison, W.C., 1979, Lobaria oregana, a nitrogen-fixing lichen in
    old growth Douglas fir forests, in: "Symbiotic Nitrogen
    Fixation in the Management of Temperate Forests," J.C. Gordon,
    C.T. Wheeler and D.A. Perry, eds, pp. 266-275, Oregon State
    University Press, Corvallis.
Farrar, J.F., and Smith, D.C., 1976, Ecological physiology of the
    lichen Hypogymnia physodes, III. The importance of the
    rewetting phase, New Phytologist, 77: 115-125.
Forman, R.T.T., 1975, Canopy lichens with blue-green algae; a
    nitrogen source in a Colombian rain forest, Ecology, 56:
    1176-1184.
Forman, R.T.T., and Dowden, D.L., 1977, Nitrogen fixing lichen
    roles, from desert to alpine, in the Sangre de Cristo
    mountains, New Mexico, The Bryologist, 80: 561-570.
Granhall, U., and Selander, H., 1973, Nitrogen fixation in a
    subarctic mire, Oikos, 20: 175-178.
Hitch, C.J.B., and Stewart, W.D.P., 1973, Nitrogen fixation by
    lichens in Scotland, New Phytologist, 72: 509-524.
Horne, A.J., 1972, The ecology of nitrogen fixation on Signy
    Island, South Orkney Islands, British Antarctic Survey
    Bulletin, 27: 1-18.
Huss-Danell, K., 1977, Nitrogen fixation by Stereocaulon paschale
    under field conditions, Canadian Journal of Botany, 55:
    585-592.
Kallio, S., and Kallio, P., 1975, Nitrogen fixation in lichens at
    Kevo, North Finland, in: "Fennoscandian Tundra Ecosystems,
    Part I. Plants and Microorganisms," F.E. Wielgolaski, ed.,
    pp. 292-304, Springer, New York.
Kershaw, K.A., 1985, "Physiological Ecology of Lichens," Cambridge
    University Press, Cambridge.
Millbank, J.W., 1978, The contribution of nitrogen fixing lichens
    to the nitrogen status of their environment, Ecological
    Bulletins (Stockholm), 26: 260-265.
Millbank, J.W., 1981, The assessment of nitrogen fixation and
    throughput by lichens, 1. The use of a controlled environment
    chamber to relate acetylene reduction estimates to nitrogen
    fixation, New Phytologist, 89: 647-655.
Millbank, J.W., 1982, The assessment of nitrogen fixation and
    throughput by lichens, 3. Losses of nitrogenous compounds by
    Peltigera membranacea, P. polydactyla and Lobaria pulmonaria
    in simulated rainfall episodes, New Phytologist, 92: 229-234.

Millbank, J.W., and Olsen, J.D., 1981, The assessment of nitrogen fixation and throughput by lichens, 2. Construction of an enclosed growth chamber for the use of $^{15}N_2$, _New Phytologist_, 89: 657-665.

Pike, L.H., 1978, The importance of epiphytic lichens in mineral cycling, _The Bryologist_, 81: 247-257.

Pike, L.H., Tracy, D.M., Sherwood, M.A., and Nielsen, D., 1972, Estimates of biomass and fixed nitrogen of epiphytes from old-growth Douglas fir, _in_: "Research on Coniferous Forest Ecosystems: First Year Progress in the Coniferous Forest Biome," _US/IBP_, J.F. Franklin, L.J. Dempster and R.H. Waring, eds, pp. 177-187, Pacific Northwest Forest and Range Experimental Station, Forest Service, U.S. Department of Agriculture, Portland, Oregon.

Pike, L.H., Denison, W.C., Tracy, D.M., Sherwood, M.A., and Rhoades, F.M., 1975, Floristic survey of epiphytic lichens and bryophytes growing on old-growth conifers in western Oregon, _The Bryologist_, 78: 389-402.

Pike, L.H., Rydell, R.A., and Denison, W.C., 1977, A 400-year-old Douglas fir tree and its epiphytes: biomass, surface area and their distributions, _Canadian Journal Forest Research_, 7: 680-699.

Rhoades, F.M., 1981, Biomass of epiphytic lichens and bryophytes on _Abies lasiocarpa_ on a Mt. Baker lava flow, Washington, _The Bryologist_, 84: 39-47.

Schell, D.M., and Alexander, V., 1973, Nitrogen fixation in arctic coastal tundra in relation to vegetation and micro relief, _Arctic_, 26: 130-137.

Sollins, P., Grier, C.C., McCorison, F.M., Cromack, K., Fogel, R., and Fredericksen, R.L., 1980, The internal element cycle of an old-growth Douglas-fir ecosystem in Western Oregon, _Ecological Monographs_, 50: 261-285.

Silvester, W.B., 1983, Analysis of Nitrogen Fixation, _in_: "Biological Nitrogen Fixation in Forest Ecosystems; Foundations and Applications," pp. 173-212, Martinus Nijhoff/ W. Junk, The Hague.

Todd, R.L., Meyer, R.D., and Waide, J.B., 1978, Nitrogen fixation in a deciduous forest in the South eastern United States, _Ecological Bulletin (Stockholm)_, 26: 172-177.

INFLUENCE OF AUTOMOBILE EXHAUST AND LEAD ON THE OXYGEN EXCHANGE OF TWO LICHENS MEASURED BY A NEW OXYGEN ELECTRODE METHOD

V. Lemaistre

Laboratoire de Cryptogamie
Université Pierre et Marie Curie
9 quai Saint-Bernard
F-75230 Paris Cedex, France

INTRODUCTION

The dangerous consequences of pollution resulting from human activities on all living organisms is well known. Scientists and political leaders have only recently become aware of its disastrous effects on the biosphere. One of the best examples of man-made pollution endangering the biosphere is Pb dispersion resulting from automobile emissions. Lead, added as an anti-knock agent in car fuel, has become the object of priority research for scientists and a controversial issue for European political leaders. An estimated 100,000 tons of atmospheric Pb fall on to the Northern Hemisphere each year.

Lichens have become the preferred material for studies of environmental and atmospheric pollution. Many studies have examined the influence of heavy metals on lichens. However, little is known about the impact of Pb, both as a tracer and as an active element in the evaluation of the physiological effects of automobile pollution. Déruelle (1983), using lichens, found that the impact of Pb dispersion was traceable up to 500 metres on either side of a highway. Preliminary studies (Déruelle and Petit, 1983) showed that lichens subjected to automobile emissions suffered an alteration to their photosynthesis. The purpose of the present study was to investigate and confirm if Pb, acquired under field and laboratory conditions, could change the gas exchange patterns of lichens.

MATERIALS AND METHODS

Lichen samples were collected from the Forest of Fontainebleau (Seine et Marne, France) which was selected as our study site because pollution, apart from automobile exhaust, was minimal (Lemaistre,

1981).   Two common species, <u>Parmelia caperata</u> (Pb sensitive) and
<u>Cladonia portentosa</u> (semi-sensitive), were chosen for their avail-
ability and for their known affinity for Pb (Déruelle, 1983).

Thalli were oven-dried at 105°C and digested in boiling H_2O_2 (at
80°C) and the Pb concentration was then measured with a Perkin Elmer
monobeam 107 atomic absorption spectrophotometer (sensitivity 0.6 µg
ml^{-1}).   Artificial Pb contamination was carried out by soaking thalli
in 0.001M lead nitrate solution for one to eighteen hours.

Photosynthetic measurements were obtained with an original O_2
potentiometric electrode in a closed chamber.   This system was
originally designed and constructed by Leclerc (1983) with whom we
have collaborated to develop and automate the system.

The apparatus consists of two Clark-type electrodes (Clark,
1956), in this case platinum-silver, which when polarised and immersed
in an electrolyte (KCl) show a current variation that is proportional
to the variation in the partial pressure of oxygen in the measuring
chamber (De Kouchkowsky, 1963).   The measuring system is equipped with
two electrodes (Fig. 1) with Teflon membranes (E_1, E_2), which effect
the measurements in chambers of variable size (15 - 25 cm^3) according
to the volume of the sample.   The size of the chamber is determined by
the height of the carbonate-bicarbonate buffer (B) located at the base

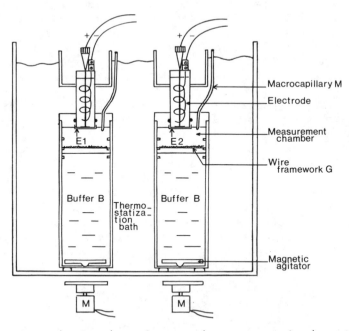

Fig. 1.   Schematic drawing of measuring system showing the two
electrodes and their measuring chambers.   For explanation see
text.

of the cylinder. The lichen thalli are placed on a small pastic wire
frame (G) which is fitted just above the surface of the buffer. A
macrocapillary, plugged through the lid of the cylinder, permits
pressure equilibration in the chamber. A copper-constantan thermo-
couple fitted through the macrocapillary monitors the temperature,
while the CO_2 level is maintained in the chamber by agitating the
buffer with a magnetic stirrer controlled by a precision motor (M).

The two devices are submerged in a running water bath to avoid
overheating. Overhead neon lamps illuminate the system with light
intensities between 100 and 850 μE m^{-2} s^{-1}. The current flowing from
the electrodes and thermocouples is measured by an Enertec-
Schlumberger 7066 microvoltmeter coupled to an Enertec-Schlumberger
7010 scanner. The system is automated by a Commodore micro-computer
which also collects and processes data. After eliminating variations
in the currents due to temperature alterations, data processing allows
for various correction factors according to the formula (Leclerc,
personal communication):

$$QO_2 = Cs\ (Vt - Vg) + Ct\ (Vg - Vl) \times S \times A \times 1.25 \times M^{-1}$$
$$= ml\ O_2\ h^{-2}\ g^{-1}\ dry\ weight$$

where QO_2 = rate of oxygen change
       Cs = O_2 solubility coefficient in buffer
       Vt = total volume (gas + buffer); Vg = gas volume; Vl = lichen
           volume
       Ct = thermal coefficient
          = $1.05 \times P_a \times 273/T_k$
where P_a = atmospheric pressure, and T_k = temperature in
           degrees Kelvin
       S = slope of electrode current versus time
       A = proportion of O_2 in air
       M = lichen dry weight
  1.25 = correction coefficient of losses due to pressure
           equilibration macrocapillaries.

Experiments were carried out for 6 hours (one 2 hour period of
light between two 2 hour periods of darkness). Temperatures varied
between 13 - 15°C. Lichen water contents were close to saturation and
never less than 85% at the end of the experimental period. Light was
saturating, the level being calculated from preliminary studies on
control samples (Fig. 2). These were, respectively 200 and 140 μE m^{-2}
s^{-1} for P. caperata and C. portentosa, with the corresponding compensation points at 50 and 25 μE m^{-2} s^{-1}.

Three series of gas exchange measurements have been made with 1)
control material collected far from car pollution, 2) naturally
polluted material from highway borders and 3) control material treated
with lead nitrate (artificial pollution). As individual experiments
showed some variability (Tables 1 and 2), the curves in Figs 3 and 4

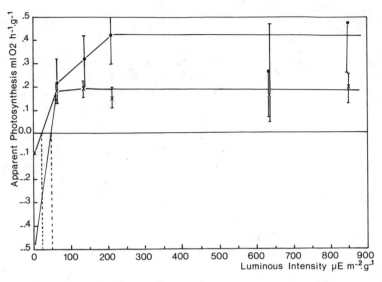

Fig. 2.   Influence of light intensity on apparent photosynthesis.
● = <u>Parmelia caperata</u>, X = <u>Cladonia portentosa</u>.

were constructed from average values derived from large numbers of
samples, the total number in each class being shown in Table 1.   In
constructing Figs 6 and 7, all of the data was combined and expressed
in five or six classes based on broad Pb concentration ranges.

A brief study of the microclimatic conditions at the sites where
control and polluted samples were collected was carried out which
indicated that temperature, light intensity and relative humidity were
similar at all sites (Lemaistre, 1983).   The results suggest that the
differences observed in response patterns do not result from
adaptations to the specific conditions of the collecting sites.

RESULTS

Lead concentration data in control and naturally- and
artificially-polluted samples are presented in Table 1.   Values for
artificially-polluted samples did not increase significantly after 1
hour and are presented as extreme values.   <u>C. portentosa</u> accumulated
less Pb than did <u>P. caperata</u> in both laboratory experiments, when Pb
was supplied under identical conditions, and in the field.

Gas exchanges of <u>C. portentosa</u> were substantially lower than
those of <u>P. caperata</u> (e.g. Table 2) with gross photosynthetic rates of
0.29 ± 0.08 and 0.76 ± 0.30 ml O_2 h^{-1} g^{-1} respectively.

Table 1. Measurement of Lead Content of Field or Laboratory Treated
          Samples of Parmelia caperata and Cladonia portentosa.

Species	Number of Measurements	Physiological Conditions	Lead content $\mu g\ g^{-1}$ dry weight
P. caperata	45	Non polluted	34± 23
C. portentosa	51		9± 6
P. caperata	37	Natural pollution	207.5±136
C. portentosa	40		55± 31
P. caperata	14	Laboratory treated	1000–5000*
C. portentosa	15	1 to 18 h soaking	250–900*

*Minimum and maximum values observed

        Figures 3 and 4 show the mean values for O_2 exchange throughout
the 6 hour experimental period; dashed lines represent electrode
stabilisation periods and are, like the first part of the initial dark
period, excluded from calculations. Naturally-polluted P. caperata
(Fig. 3) showed increased net photosynthesis and decreased
respiration, compared to the control. Artificially-polluted samples
showed a decrease in both net photosynthesis and respiration, the
latter being very small. For C. portentosa (Fig. 4) both natural and
artificial pollution enhanced net photosynthesis and respiration,
although the increases with artificial pollution were much lower. The
mean values for these gas exchanges are presented with standard errors
in Table 2. Frequency histograms of the distribution of photosynthetic rates (Fig. 5) show that the results were far more dispersed

Table 2. Mean values of Photosynthetic and Respiratory Gas Exchanges
          of Field or Laboratory Treated Samples of Parmelia caperata
          and Cladonia portentosa. (ml $O_2\ h^{-1}s^{-1}$)

Species	Respiration Dark Period 1	Net photosynthesis Light	Respiration Dark period 2
P. caperata			
Non polluted	-0.39 + 0.15	0.43 + 0.07	-0.40 + 0.12
Natural pollution	-0.33 + 0.14	0.69 + 0.20	-0.43 + 0.07
Laboratory treated	-0.25 + 0.11	0.24 + 0.14	-0.25 + 0.03
C. portentosa			
Non polluted	-0.13 + 0.02	0.15 + 0.02	-0.10 + 0.02
Natural pollution	-0.16 + 0.02	0.25 + 0.05	-0.23 + 0.06
Laboratory treated	-0.19 + 0.09	0.22 + 0.04	-0.14 + 0.03

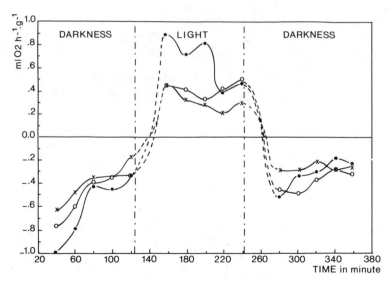

Fig. 3. Gas exchange of <u>Parmelia caperata</u> during dark and light
         exposure periods for three different degrees of pollution.
         O = control samples (mean of 42 samples)
         ● = polluted samples from the field (mean of 33 samples)
         X = laboratory treated samples (mean of 14 samples)

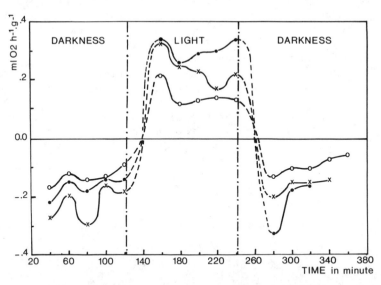

Fig. 4. Gas exchanges of <u>Cladonia portentosa</u> during dark and light
         exposure periods for three different degrees of pollution.
         O = control samples (mean of 48 samples)
         ● = polluted samples from the field (mean of 38 samples)
         X = laboratory treated samples (mean of 15 samples)

and variable for the polluted samples than for the controls.

The relationship between gross photosynthesis and thallus Pb content, using data from naturally- and artificially-polluted material (Figs 6 &7), shows a substantial increase in gross photosynthesis at relatively low Pb concentrations, being roughly 180% for both species. At high Pb concentrations, gross photosynthesis is depressed by c. 30% in P. caperata but is not reduced below the control level in C. portentosa. With both species, only one value of gross photo-synthesis at high thallus Pb concentration was obtained with naturally-polluted material.

DISCUSSION

On a dry weight basis, the gas exchanges of C. portentosa control samples were substantially lower than those of P. caperata. This may reflect different volumes of dead or inactive tissue in the two species. The size of the P. caperata individuals examined was limited by the diameter of the experimental chamber but contained no decaying material. This was probably not the case for tufts of C. portentosa, which were examined whole. Furthermore, the possible influence of

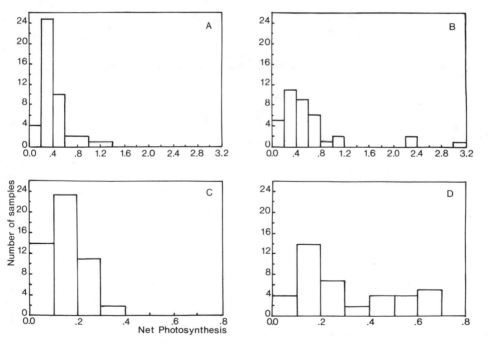

Fig. 5. Distribution of results of net photosynthesis in control and polluted samples.
A and B = Parmelia caperata, C and D = Cladonia portentosa.
A and C = control samples, B and D = Field polluted samples.

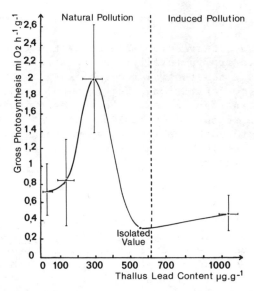

Fig. 6. Gross photosynthetic rates related to lead content of
         Parmelia caperata.

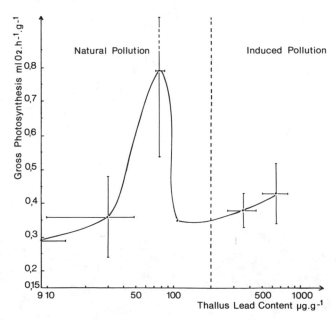

Fig. 7. Gross photosynthetic rates related to lead content of
         Cladonia portentosa.

shading effects in the tufts of Cladonia and physiological differences between the species must not be overlooked.

Standard errors reported in Fig. 2 and Table 2 represent individual variability which is roughly 40% for P. caperata and 30% for C. portentosa. This variability is also seen with Pb concentration (Table 1). The histograms in Fig. 5 demonstrate a significant increase in the variability of gas exchanges of naturally-polluted samples which seems to result from difference in Pb accumulation by individuals.

The experiments presented here demonstrate that a threshold of Pb toxicity exists in P. caperata (Fig. 6) but no such threshold has been shown in C. portentosa under the conditions used (Fig. 7). However, further work is required to determine this threshold precisely. In earlier studies, Puckett (1976), working with Umbilicaria muhlenbergii noted an increase in ^{14}C-fixation during short exposures to lead and a decrease during longer exposures, implying some threshold value was exceeded.

The depression of gas exchanges with high thallus Pb concentrations has been reported by Baddeley et al. (James, 1973), Puckett (1976), Punz (1979a) and Déruelle and Petit (1983). Interactions between a reduction in water content (Punz, 1979b), a slight K loss at low Pb concentrations (Puckett, 1976) and an alteration in the CO_2 diffusion properties of membranes (Adam, 1930) may be involved in the decreased gas exchanges. However, the observation that naturally-polluted C. portentosa seems less sensitive than P. caperata at high Pb concentrations, could be explained by the probability of the Pb being concentrated in the dead basal parts of the thallus (Goyal and Seaward, 1981a, b; Pakarinen, 1981) where it would not influence photosynthetic activity. We were only able to find one field sample for each species with a high Pb content, which suggests that when there is very high Pb contamination, individuals tend to die and disappear. However, it must be noted that the most contaminated, naturally-polluted sample of C. portentosa did not show any photosynthetic inhibition compared to control samples and this species withstood even higher thallus Pb concentrations when artificially polluted.

The enhanced rates of gas exchange with intermediate thallus Pb concentrations, around 300 μg g^{-1} for P. caperata and 80 μg g^{-1} for C. portentosa, is harder to interpret. Stimulation of gas exchange has already been noted by Baddeley et al. (James, 1973) and Puckett (1976) but no satisfactory explanation has been given. It is interesting that these authors worked with thalli immersed in solution and, therefore, showed enhancement with high thallus water contents. Further investigation of this phenomenon should help to establish the relationship with the drying of the lichen thallus since Punz (1979b) has demonstrated that Pb has an effect on the acquisition and retention of water.

It has been shown with other systems that pollutants that are toxic at high concentrations can be beneficial at low concentrations. Thus pollution by crude petroleum can stimulate photosynthesis in phytoplanktonic algae that absorb small amounts of the pollutant (Gordon et al., 1973; Parson and Waters, 1975; Karydis, 1979). Lichen respiration is stimulated by low concentrations of SO_2 but is reduced at higher levels (Pearson and Skye, 1965; Baddeley et al., 1971, 1972).

In the complex phenomenon of automobile pollution, Pb is only a single agent. We have used it as a tracer and as a potentially active agent of this pollution. Nevertheless, other components may act in conjunction with Pb and the effects of some, such as nitrogenous components which may become accessible sources of nitrogen and can induce an increase in gas exchanges, must not be neglected.

ACKNOWLEDGEMENTS

We acknowledge financial assistance from the French Ministere de l'Environnement under contract no. 82.136 and the Centre National de la Recherche Scientifique under programme PIREN: Forêts périurbaines. We thank Dr J.C. Leclerc, Laboratory of cell physiology, Université Paris-Sud, Orsay, for his active scientific guidance and Dr M.A. Letrouit-Galinou and Dr S. Déruelle for their interest.

REFERENCES

Adam, N.K., 1930, "The physics and chemistry of surface," Clarendon Press, Oxford.

Baddeley, S., Ferry, B.W., and Finegan, E.J., 1971, A new method of measuring lichen respiration: response of selected species to temperature and sulfur dioxide, The Lichenologist, 5: 18-25.

Baddeley, S., Ferry, B.W., and Finegan, E.J., 1972, The effects of sulphur dioxide on lichen respiration, The Lichenologist, 5: 283-291.

Clark, L.C.J., 1956, Monitor and control of blood and tissue oxygen tension, Transaction of the American Society for Artificial Internal Organs, 71: 201-225.

De Kouchkowsky, Y., 1963, L'induction photosynthétique des chloroplastes isolés, Physiologie végétale, 1: 25-76.

Déruelle, S., 1983, "Ecologie des lichens du Bassin Parisien. Impact de la pollution atmosphérique (engrais, SO_2, Pb) et relation avec les facteurs climatiques," Thèse de Doctorat d'Etat. Université Pierre et Marie Curie, Paris, France.

Déruelle, S., and Petit, P.J.X., 1983, Preliminary studies of the net photosynthesis and respiration responses of some lichens to automobile pollution, Cryptogamie, Bryologie et Lichénologie, 4: 269-278.

Gordon, D.C., Prouse, J., and Prouse, N.J., 1973, The effects of three oils on marine phytoplankton photosynthesis, Marine Biology, 22: 329-333.

Goyal, R., and Seaward, M.R.D., 1981a, Lichen ecology of the Scunthorpe Heathland II. Industrial metal fallout pattern from lichen and soil assay, The Lichenologist, 13: 289-300.

Goyal, R., and Seaward, M.R.D., 1981b, Metal uptake in terricolous lichens. I. Metal localisation within the thallus, New Phytologist, 89: 631-643.

James, P.W., 1973, The effect of air pollutants other than hydrogen fluoride and sulphur dioxide on lichens, in: "Air Pollution and Lichens," B.W. Ferry, M.S. Baddeley, and D.L. Hawksworth, eds, pp. 143-175, Athlone Press, London.

Karydis, M., 1979, Short term effects of hydrocarbon on the photosynthesis and respiration of some phytoplankton species, Botanica Marina, 22: 281-285.

Leclerc, J.C., 1983, Intérêt de la mesure de la photosynthèse avec une électrode à oxygène en milieu aérien. Application à quelques algues subaériennes et lichens, Oceanis, 9: 195-203.

Lemaistre, V., 1981, "Influence de la pollution sur les lichens de quelques forêts périurbaines," DEA d'Ecologie. Université Pierre et Marie Curie, Paris, France.

Lemaistre, V., 1983, "Mise au point d'une électrode à oxygène en milieu aérien. Application à l'étude de l'influence de la pollution automobile globale sur les échanges gazeux de deux lichens," Thèse de Doctorat de troisième cycle. Université Pierre et Marie Curie, Paris, France.

Pakarinen, P., 1981, Nutrient and trace metal content and retention in reindeer lichen carpets of Finnish ombrophilic bogs, Annales Botanici Fennici, 18: 265-274.

Parson, T.R., and Waters, R., 1976, Some preliminary observations on the enhancement of phytoplankton growth by low levels of mineral hydrocarbons, Hydrobiologia, 51: 85-89.

Pearson, L., and Skye, E., 1965, Air pollution affects pattern of photosynthesis in Parmelia sulcata, a corticolous lichen, Science, 148: 1600-1602.

Puckett, K.J., 1976, The effect of heavy metal on some aspects of lichen physiology, Canadian Journal of Botany, 54: 2695-2703.

Punz, W., 1979a, Der einfluss isolierter und kombinierte schadstoffe auf die flechten photosynthese, Photosynthetica, 13: 428-433.

Punz, W., 1979b, The effect of single and combined pollutants on lichen water content, Biologia Plantarum, 21, 472-474.

MINERAL ELEMENT ACCUMULATION IN BOG LICHENS

P. Pakarinen

Department of Botany, University of Helsinki
Helsinki 10, SF-00100 Finland

INTRODUCTION

Lichens are slow growing and the lichen component of ecosystems is usually small. Lichens have, however, ecological and biogeochemical significance in some types of boreal heaths, forests and bogs (Wein and Speer, 1975; Rencz and Auclair, 1978), and as epiphytes in humid, subtropical-temperate swamps (Bosserman and Hagner, 1981). Furthermore, species of <u>Cladonia</u> (generally = <u>Cladina</u>*) have been used as monitoring organisms in order to survey the deposition of atmospheric metals in northern woodlands, bogs and tundra (Tomassini et al., 1976; Pakarinen et al., 1978; Puckett and Finegan, 1980; Glooschenko et al., 1981; Pakarinen, 1981a).

Physiological aspects of mineral uptake and metal accumulation in lichens have been discussed and reviewed by several authors during the past decade (Tuominen and Jaakkola, 1974; Brown, 1976; Nieboer et al., 1978; Nieboer and Richardson, 1981). In this paper, the lichen studies carried out in recent years in Finnish, and to some extent also Canadian, bogs are summarised. Attention has centred on ombrotrophic (ombrogenic) sites, mainly raised bogs, where atmospheric fallout is the ultimate source of mineral nutrients. The main topics which are briefly discussed here include a comparison of the mineral element composition and mineral nutrition of epiphytic versus terricolous lichens, vertical distribution of elements in lichen carpets, with a comparison to their peat substratum, and a review of some estimates concerning the importance of lichens as accumulators of mineral elements in bogs.

* See introductory editorial note on nomenclature.

INTERSPECIFIC AND INTERGENERIC COMPARISONS

According to Canadian studies, interspecies differences in
mineral element concentrations among the reindeer lichens (Cladina)
are relatively small (Tomassini et al., 1976; Puckett & Finegan,
1980). Statistically significant differences, however, have been
observed for some elements, e.g. between C. arbuscula and C. stellaris
(Pakarinen 1981a). There are indications that C. rangiferina might
also take up more major cations, such as K and Ca, than the other
Cladina species (Pakarinen et al., unpublished data). The nutritional
differences between terricolous and corticolous lichens are much more
distinct. Hypogymnia is usually the predominant epiphytic lichen on
bog conifers and, for example, Finnish studies (Pakarinen & Mäkinen,
1976; Pakarinen & Häsänen, 1983; Pakarinen et al., 1983) suggest that
the concentrations of Pb, Zn, Hg, V and Al are several-fold higher in
Hypogymnia physodes compared to C. arbuscula. Data in Table 1 show
that a similar difference exists for a number of other elements, the
concentration ratio (Hypogymnia/Cladina) being highest for Ca, Fe and
Pb (cf. Folkeson, 1979), but relatively low for N, P, K and Mn.

VERTICAL DISTRIBUTION PATTERNS

Reindeer lichens may form patches over 10 cm thick or continuous
carpets in bogs or muskegs. The living part of thalli is commonly 3
to 6 cm long, representing on average about 6 to 10 years growth
(Pakarinen, 1981a), and the dead base may retain its structure for
several years. Besides laboratory studies (e.g. Tuominen, 1968; Brown
and Slingsby, 1972), accumulation of elements by Cladina has been
studied by analysing the vertical distributions of minerals from field
collections. The elements that are actively taken up by the lichen
are usually enriched in the living parts of podetia. Figure 1
illustrates vertical patterns found in C. arbuscula in southern
Finnish raised bogs (Pakarinen, 1981a). The concentration decrease
from live top to dead base was most distinct for the macronutrients
(K, N, P), while the heavy metals (Fe, Pb) accumulated towards the
base.

A common species in oceanic bogs of eastern Canada, C. terrae-
novae, has also been studied recently (Pakarinen and Wetmore,
unpublished) and, based on the vertical concentration patterns, three
behavioural categories of elements were distinguished:
  (1) K, P, N, Cd, Zn, Na, Cu and Mg were enriched in the top
       segment,
  (2) Mn and Ca showed no significant difference in concentration
       between live and dead segments,
  (3) Pb, Al and Fe showed a distinct increase towards the dead
       parts.
In general, vertical patterns appeared to be fairly similar in
Canadian and Finnish bogs.

Table 1. Elemental Composition of Two Lichens from South Finnish
          Raised Bogs: Hypogymnia physodes (Corticolous Species on
          Pine) and Cladonia stellaris (Terricolous Species on Peat).

	Hypogymnia	Cladina	H/C
N %	1.52 + 0.32 (1.04 − 1.96)	0.81 + 0.12 (0.69 − 1.09)	1.9
P mg g^{-1}	0.63 + 0.16 (0.43 − 0.87)	0.38 + 0.05 (0.26 − 0.47)	1.7
K mg g^{-1}	1.85 + 0.22 (1.64 − 2.28)	1.32 + 0.23 (1.02 − 1.70)	1.4
Ca mg g^{-1}	3.00 + 1.21 (2.04 − 5.32)	0.19 + 0.06 (0.10 − 0.30)	15.8
Mg mg g^{-1}	0.47 + 0.11 (0.29 − 0.62)	0.17 + 0.04 (0.11 − 0.23)	2.8
Fe mg g^{-1}	1.81 + 0.73 (0.94 − 3.36)	0.30 + 0.14 (0.15 − 0.60)	6.1
Mn μg g^{-1}	30.4 + 9.6 (18.0 − 40.0)	25.7 + 10.3 (17.0 − 48.0)	1.2
Zn μg g^{-1}	74.2 + 9.0 (62.4 − 91.2)	30.3 + 5.7 (22.9 − 39.9)	2.5
Pb μg g^{-1}	65.4 + 21.8 (37.7 − 94.5)	13.4 + 2.7 (9.8 − 18.4)	4.9
Cu μg g^{-1}	6.8 + 1.9 (4.4 − 10.2)	3.3 + 1.5 (2.4 − 7.2)	2.1

Note: Concentrations are expressed on a dry weight (60°C) basis.
       Material collected from virgin bogs between 1976–80, 9 sites
       for Hypogymnia and 11 sites for C. stellaris within 60–62°N
       (see Pakarinen, 1981a)
       H/C = mean concentration ratio.
       Values in parenthesis indicate the extreme values observed.

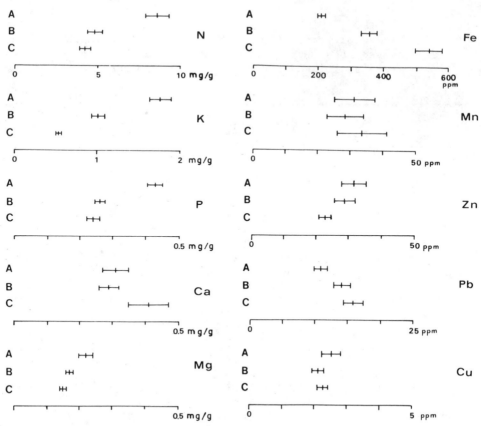

Fig. 1.   Vertical distribution of macronutrients and trace metals in
Cladonia arbuscula in South Finnish raised bogs.   Segments:
A. live top, B. live base, C. dead basal part of the lichen
carpet.   (Reproduced with permission from Pakarinen, 1981a).

BIOGEOCHEMICAL STUDIES

The significance of lichens as nutrient accumulators in northern
bog and forest ecosystems has been assessed recently in studies
performed in interior Alaska (Van Cleve et al., 1981), in northern
Quebec (Auclair and Rencz, 1982) and in southern Finland (Pakarinen,
1981a, 1982).   The results from Finland are from treeless or semi-open
raised bogs, where the annual rate of nutrient accumulation was
compared between the lichen carpets (C. stellaris) and the dominant
Ericaceae shrubs (Ledum palustre, Calluna vulgaris).   The data in
Table 2 show marked differences in the accumulation ratios of chemical
elements: while Fe is accumulated strongly by Cladina, the uptake of
Mn, Ca and Mg by the lichen layer appears to be relatively weak.   Also
when the annual uptake rates of Cladina (C. stellaris and C. arbuscula)

Table 2.   Comparison of the Net Accumulation Rates of Elements by the
           Annual Production of Terricolous Lichen Layer (Cladonia
           stellaris) and Dwarf-shrub Layer (Ericaceae) in South
           Finnish raised bogs (data from Pakarinen 1981a, 1982).

	Cladonia stellaris	Dwarf shrubs	Lichen/shrub accumulation ratio
	$mg\ m^{-2}\ yr^{-1}$		
N	408	394	1.04
P	21.0	30	0.70
K	67	143	0.43
Ca	8.6	159	0.05
Mg	7.9	50	0.16
Fe	14.0	3.0	4.66
Mn	1.4	31.1	0.05
Zn	1.5	1.1	1.36
Cu	0.21	0.20	1.05

Note: For C. stellaris the estimates are for closed lichen
        carpets (100% coverage); only the above-ground shoot production
        is included for dwarf shrubs (mean cover: 35%).

in Finnish bogs were related to the mean annual input of elements by
atmospheric deposition, Ca and Mg showed very low retention rates
(Pakarinen, 1981a).

    The proportion of the terricolous lichens to the total nutrient
capital in phytomass of a spruce bog ecosystem ('muskeg') in Alaska
was estimated to be less than 2% for the macronutrients studied (Van
Cleve et al., 1981).   In a representative Finnish pine bog, the mean
annual production of Cladina was only 3 g m^{-2}, which is less than 1%
of the total primary production (Vasander, 1981).   Thus the importance
of lichens in the nutrient ecology of bog sites is generally small,
although there can be considerable variation between microsites.

CONCLUDING REMARKS

    Of the elements studied, N, P, and K show, in their distribution
patterns and retention rates, strong indication of active uptake
while, particularly in reindeer lichens, the retention of Ca and Mg
was low.   This is consistent with Buck and Brown (1979) who report
that in Cladonia rangiformis the intracellular fraction is dominant
(98%) for K, intermediate (49%) for Mg and low (9%) for Ca.   Thus in
Cladonia, and probably also Cladina, a very high proportion of Ca

appears to be extra-cellularly bound. It is therefore possible that
continued leaching by acid rain, for example, results first in a loss
of divalent cations (Ca, Mg) from the extracellular binding sites,
while loss of intra-cellular K is probably a result of more severe
exposure to acid compounds (cf. the experiments of Tomassini et al.
(1977) and Moser et al. (1980)).

The fact that in Hypogymnia the concentration of Ca was
relatively high is also in agreement with the studies of Buck and
Brown (1979) as well as Lang et al. (1980). Tuominen (1967)
demonstrated that all epiphytic lichen species studied possessed a
significantly greater uptake capacity for strontium than species
growing on the ground and suggests that this correlated with the
uronic acid content of thalli (but see Richardson et al., this
volume). Other possible factors favouring the accumulation of
elements in corticolous species include the inputs from stemflow and
canopy leachates (Pike, 1978).

Bog lichens and bog mosses have been used to detect regional
pollution patterns, and the concentrations for most elements seem to
be consistently higher in Sphagnum mosses and in the surface peat than
in the fruticose lichen layer of the same habitat (Glooschenko et al.,
1981; Pakarinen, 1981a,b). Thus at least the terricolous lichens
(Cladina spp.) are not particularly large 'sinks' of minerals compared
to the other components of a bog ecosystem. As in bryophytes, some
heavy metals (Fe, Pb) are efficiently retained by lichens (Puckett et
al., 1973; Goyal and Seaward, 1982), while some others (e.g. Zn and
Mn) appear to be more mobile and their concentrations may not
necessarily be in direct relationship with the atmospheric deposition
rates. The value of lichens as monitoring organisms is evident, but
many problems associated with the uptake and tolerance of various
mineral elements and with the inter-species variability require
further field and laboratory studies (cf. Nieboer and Richardson,
1981).

REFERENCES

Auclair, A.N.D., and Rencz, A.N., 1982, Concentration, mass and
        distribution of nutrients in a subarctic Picea mariana - Cladonia
        alpestris ecosystem, Canadian Journal of Forest Research, 12:
        947-968.
Bosserman, R.W., and Hagner, J.E., 1981, Elemental composition of
        epiphytic lichens from Okefenokee Swamp, The Bryologist, 84:
        48-58.
Brown, D.H., 1976, Mineral uptake by lichens, in: "Lichenology:
        Progress and Problems," D.H. Brown, D.L. Hawksworth, and R.H.
        Bailey, eds, pp. 419-439, Academic Press, London, New York.
Brown, D.H., and Slingsby, D.R., 1972, The cellular location of lead
        and potassium in the lichen Cladonia rangiformis (L.) Hoffm., New
        Phytologist, 71: 297-305.

Buck, G.W. amd Brown, D.H., 1979, The effect of desiccation on cation
    location in lichens, Annals of Botany, 44: 265-277.
Folkeson, L., 1979, Interspecies calibration of heavy-metal concen-
    trations in nine mosses and lichens: applicability to deposition
    measurements, Water, Air, and Soil Pollution, 11: 253-260.
Glooschenko, W.A., Sims, R., Gregory, M., and Mayer, T., 1981, Use of
    bog vegetation as a monitor of atmospheric input of metals, in:
    "Atmospheric Pollutants in Natural Waters," S.J. Eisenreich, ed.,
    pp. 389-399, Ann Arbor Science Publ., Ann Arbor.
Goyal, R., and Seaward, M.R.D., 1982, Metal uptake in terricolous
    lichens. III. Translocation in the thallus of Peltigera canina,
    New Phytololgist, 90: 85-98.
Lang, G.E., Reiners, W.A., and Pike, L.H., 1980, Structure and
    biomass dynamics of epiphytic lichen communities of balsam fir
    forests in New Hampshire, Ecology, 61: 541-550.
Moser, T.J., Nash III, T.H., and Clark, W.D., 1980, Effects of a
    long-term field sulfur dioxide fumigation on Arctic caribou
    forage lichens, Canadian Journal of Botany, 58: 2235-2240.
Nieboer, E., and Richardson, D.H.S., 1981, Lichens as monitors of
    atmospheric deposition, in: "Atmospheric Pollutants in Natural
    Waters," S.J. Eisenreich, ed., pp. 339-388, Ann Arbor Science
    Publ., Ann Arbor.
Nieboer, E., Richardson, D.H.S., and Tomassini, F.D., 1978, Mineral
    uptake and release by lichens: An overview, The Bryologist, 81:
    226-246.
Pakarinen, P., 1981a, Nutrient and trace metal content and retention
    in reindeer lichen carpets of Finnish ombrotrophic bogs, Annales
    Botanici Fennici, 18: 265-274.
Pakarinen, P., 1981b, Metal content of ombrotrophic Sphagnum mosses
    in NW Europe, Annales Botanici Fennici, 18: 281-292.
Pakarinen, P., 1982, Ombrotrofisten soiden pohjakerroskasvien
    hivenaine- ja ravinne-ekologiasta (Summary: On the trace element
    and nutrient ecology of the ground layer species of ombrotrophic
    bogs), Publications from the Department of Botany, University of
    Helsinki, 10: 1-32.
Pakarinen, P., and Häsänen, E., 1983, Suosammalten ja -jäkälien
    elohopeapitoisuuksista (Summary: Mercury concentrations of bog
    mosses and lichens), Suo, 34: 17-20.
Pakarinen, P., Kaistila, M., and Häsänen, E., 1983, Regional
    concentrations levels of vanadium, aluminium and bromine in
    mosses and lichens, Chemosphere, 12: 1477-1485.
Pakarinen, P., and Mäkinen, A., 1976, Suosammalet, -jäkälät ja männyn
    neulaset raskasmetallien kerääjinä (Summary: Comparison of Pb, Zn
    and Mn contents of mosses, lichens and pine needles in raised
    bogs), Suo, 27: 77-83.
Pakarinen, P., Mäkinen, A., and Rinne, R.J.K., 1978, Heavy metals in
    Cladonia arbuscula and Cladonia mitis in eastern Fennoscandia,
    Annales Botanici Fennici, 15: 281-286.
Pike, L.H., 1978, The importance of epiphytic lichens in mineral
    cycling, The Bryologist, 81: 247-257.

Puckett, K.J., and Finegan, E.J., 1980, An analysis of the element content of lichens from the Northwest Territories, Canada, Canadian Journal of Botany, 58: 2073-2089.

Puckett, K.J., Nieboer, E., Gorzynski, M.J., and Richardson, D.H.S., 1973, The uptake of metal-ions by lichens: a modified ion-exchange process, New Phytologist, 72: 329-342.

Rencz, A.N., and Auclair, A.N.D., 1978, Biomass distribution in a subarctic Picea mariana – Cladonia alpestris woodland, Canadian Journal of Forest Research, 8: 168-176.

Tomassini, F.D., Lavoie, P., Puckett, K.J., Nieboer, E., and Richardson, D.H.S., 1977, The effect of time of exposure to sulphur dioxide on potassium loss from and photosynthesis in the lichen Cladina rangiferina (L.) Harm., New Phytologist, 79: 147-155.

Tomassini, F.D., Puckett, K.J., Nieboer, E., Richardson, D.H., and Grace, B., 1976, Determination of copper, iron, nickel, and sulphur by X-ray fluorescence in lichens from the Mackenzie Valley, Northwest Territories, and the Sudbury District, Ontario, Canadian Journal of Botany, 54: 1591-1603.

Tuominen, Y., 1967, Studies on strontium uptake of the Cladonia alpestris thallus, Annales Botanici Fennici, 4: 1-28.

Tuominen, Y., 1968, Studies on the translocation of caesium and strontium ions in the thallus of Cladonia alpestris, Annales Botanici Fennici, 5: 102-111.

Tuominen, Y., and Jaakkola, T., 1974, Absorption and accumulation of mineral elements and radioactive nuclides, in: "The Lichens," V. Ahmadjian and M.E. Hale, eds, pp. 185-223, Academic Press, New York, London.

Van Cleve, K., Barney, R., and Schlentner, R., 1981, Evidence of temperature control of production and nutrient cycling in two interior Alaska black spruce ecosystems, Canadian Journal of Forest Research, 11: 258-273.

Vasander, H., 1981, The length growth rate, biomass and production of Cladonia arbuscula and C. rangiferina in a raised bog in southern Finland, Annales Botanici Fennici, 18: 237-243.

Wein, R.W., and Speer, J.E., 1975, Lichen biomass in Acadian and boreal forest of Cape Breton Island, Nova Scotia, The Bryologist, 78: 328-333.

THE DISTRIBUTION OF URANIUM AND COMPANION ELEMENTS IN LICHEN HEATH

ASSOCIATED WITH UNDISTURBED URANIUM DEPOSITS IN THE CANADIAN ARCTIC

J.H.H. Looney[a], K.A. Kershaw[a], E. Nieboer[b],
C. Webber[c] and P.I. Stetsko[b]

[a] Department of Biology, McMaster University
Hamilton, Ontario L8S 4K1 Canada
[b] Department of Biochemistry, McMaster University
Hamilton, Ontario L8N 3Z5 Canada
[c] Department of Nuclear Medicine, McMaster University
Hamilton, Ontario L8N 3Z5 Canada

INTRODUCTION

Nieboer et al. (1978) and Nieboer and Richardson (1981) have
given a comprehensive summary of the range of element levels found in
lichens. The data they present show tremendous concentration ranges
correlating with different or contrasting environments and in
particular 10-100 fold increases in metal concentrations near ore
smelters. The origin of the elements found in lichen thalli is
twofold: from particulate atmospheric "fallout" and from ionic
solutions. The latter may be delivered to the thallus either directly
as rainfall or indirectly as surface runoff.

Nieboer et al. (1978, 1982) provide convincing evidence of the
widespread occurrence of trapped particulates in lichens, not only
adjacent to industrial development but also in remote areas. The data
were derived from a number of studies from two geologically diverse
areas of Canada, remote from industrial activity (Mackenzie Valley,
Northwest Territories and southern New Brunswick, see also Pakarinen,
this volume). The geology of these areas includes sandstones,
limestones, shales and granites. The iron/titanium (Fe/Ti) quotient
for a wide range of rock types is remarkably constant, presumably
because their respective oxides originally crystallised together. As
a result, although the absolute concentrations of the two elements
varies for the Canadian shield rock types, the Fe/Ti quotient has an
average value of 7.2. The Fe and Ti contents for a range of species
of the genus <u>Cladonia</u> collected from these non-industrial areas were

193

strongly correlated and had a slope of 7.0 + 0.2. This clearly points
to the presence of these two elements largely in the form of
particulates in many lichens.

    Detailed confirmation of particulate metallic fall-out is given
by Garty et al. (1979). Their study demonstrated, by means of
scanning and transmission electron microscopy, the extensive extracellular deposition of particulates, mainly in the medulla of the
lichen Caloplaca aurantia. The elemental composition of particles
integrated into the lichen tissue was also compared with that of dust
particles collected from the surface of the lichen colonies using
energy dispersive x-ray analysis and a close correspondence was
established.

    It is, therefore, now possible to monitor effectively the
potential impact on the environment of any industrial operation, such
as a mining operation. In the present context, a proposed uranium
mining site is a point source of both wind- and water-borne
particulate contaminants as well as ionic and particulate materials
derived from in situ bedrock weathering. Subsequent alterations,
coupled to mineral cycling within an ecosystem, can substantially
mobilise these secondary products over considerable distances.
Accordingly, permanent transects radiating from the central location
of the ore body and oriented with the direction of the prevailing wind
will yield samples from which levels of particulate and ionic emission
can be determined on a continuous basis.

    The intention of this research was to establish the background
levels of selected elements within the thalli of a number of lichen
species, located in permanent plots established on transects radiating
from the uranium deposit at Lone Gull Lake near Baker Lake, N.W.T.
(64°18'N, 96°3'W) prior to any mining activity. At the same time, the
analytical technology was selected and developed in such a way as to
enable subsequent continuous monitoring of lichen samples, if and when
mining operations commence. Equally, the development of this
programme has been structured so that it is applicable to the
monitoring of other potential arctic mining or industrial operations
across Canada.

METHODS

Field Work at Lone Gull, July-August, 1982

    A field camp was located slightly remote from and north of the
main Urangesellschaft mining camp. Using the main camp as the focal
point, four transects running approximately N, S, E and W were
established; with sample plots (c. 100 x 100 m) located at 0.5, 1, 2,
4, 8 and 12 km along the east-west transects, and at 0.5, 1, 2, 4, 8
and 16 km along the north-south transects, with an additional plot at
32 km on the southern transect. Since the prevailing wind is from the

north and north-west it is anticipated that major particulate fall-out
will be in a southerly direction and the design of the sampling layout
reflects this (Fig. 1).

    Following a general survey of the commoner lichen species in the
area, Dactylina arctica, Cetraria cucullata, and C. nivalis were
chosen as representative of the local vegetation as a whole but also
representative of arctic Canada in general. Approximately 30 g of
each species was collected from each permanent plot, labelled and
returned to the laboratories at McMaster. Following careful hand
sorting and washing, the dried lichen material was powdered with the
aid of liquid nitrogen and 2 g quantities of sieved material (70 mesh)
were mixed with 0.5 g of Hoechst "C" wax and compressed into pellets
of 32 mm diameter and 2-3 mm thickness for elemental analysis.
Pellets were made using a semi-automatic hydraulic press and applying
a pressure of 12 tonnes for 1 minute.

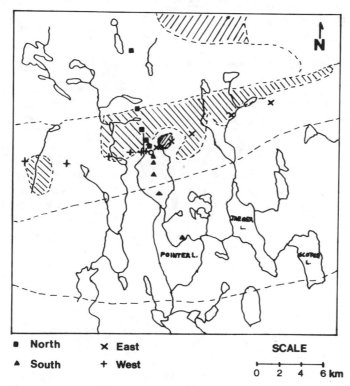

■ North	✕ East
▲ South	+ West

SCALE

0 2 4 6 km

Fig. 1. Area map indicating collection sites for the Lone Gull
         transects. The hatched areas denote orthoquartzite (\\\\\)
         and intrusive rock (/////) formations.

Elemental Analyses

The analyses of the plant tissues for Ti, Fe, Cu, Ni, Pb, S, Ca
and K were achieved by X-ray fluorescence spectrometry (XRF)
(Tomassini et al., 1976; Beckett et al., 1982; Boileau et al., 1982).
The preparative work for the U analysis involved ashing 5 g of
powdered lichen material in a programmable furnace (1 h at 150°C; 2 h,
250°C; 1 h, 350°C; 20 h, 500°C). The U analyses were carried out by
delayed neutron counting (Ernst and Hoffman, 1982), where delayed
neutrons are emitted from induced fission of, principally, ^{235}U.
Samples (30-100 mg of ash) were irradiated in plastic vials for 60 s
at a thermal neutron flux of 5 x $10^2 n$ $cm^{-2}s^{-1}$, and were counted for 60
s after a 10 s delay. A detection limit of 0.03 g of U is possible
with this technique. Appropriate inter-method checks on
representative standard and actual samples of the selected elements
employed electrothermal atomic absorption spectrometry.

Gamma-Ray Spectrometry. ^{137}Cs was determined using from 6 - 12 g
of dried, ground lichen material in a 25 ml vial, using a GeLi
detector supported by a PDP11/20 multichannel analyzer. Each sample
was counted for 15 h. The resulting spectrum was recorded on magnetic
tape or floppy disc and subsequently compared to that for the empty
vial.

Statistical Analysis of Data

Pearson's linear correlation coefficient statistic (r) and
scatter diagrams were assessed using modified programs of the Social
Sciences Statistical Package of the University of Pittsburg (Nie et
al., 1975). In addition, the combined elemental concentration data
for all three lichen species were explored for meaningful patterns of
relationships using detrended correspondence analysis (DCA, Cornell
University Multivariate Analysis Package; Hill, 1979). Currently, DCA
is the most successful and efficient multivariate analysis technique
available and is useful at revealing gradients or relationships within
a data set (Hill and Gauch, 1980; Gauch, 1982). DCA was designed
primarily as a technique for investigating vegetation distribution
data but is also eminently suitable for the identification of
relationships among elemental distributions such as in soils and
vegetation.

Scanning Electron Microscopy

Sputter-coated (Au) surface and cross sections of all three
species were examined for morphological differences.

RESULTS

Ash Content of Samples

The percentage ash content of the lichen samples indicate that <u>D. arctica</u> had considerably less ash than <u>C. cucullata</u> and <u>C. nivalis</u> (Table 1). The percentage ash content did not show a relationship with distance along any transect, nor with the local geology and can best be viewed as resulting from morphological differences.

Calcium Concentration

The data for Ca does not show any consistent distance relationships, with the concentration differences between species (Table 1) being the most interesting of the phenomena found. The significant correlation between Ca and percentage ash for all three species (results for <u>C. nivalis</u> are given in Table 2), supports the importance of Ca as a major component of the ash.

Caesium Concentration and Gamma-Ray Spectrometry

No gamma-emitting, daughter isotopes derived from the decay of either ^{238}U, (^{214}Pb, ^{214}Bi) or ^{232}Th (^{208}Tl) were detected in the samples. ^{137}Cs was detected at very low levels (Table 1), but this radioisotope is normally assigned to nuclear fallout (Svoboda and Taylor 1979; Taylor et al., 1979). During the period when <u>C. nivalis</u> samples were analyzed (August, 1982 to February, 1983), the background was measured four times. The standard deviation for the four measurements (+ 80 counts) was not different from the standard deviation due to counting statistics (+ 85 counts), indicating that the background count rate was the same for all samples. The number of counts accumulated in 24 h for each of the <u>C. nivalis</u> samples ranged from 1865 to 2248. No result was more than two standard deviations above background. Fourteen of the twenty-one samples measured were within one standard deviation of the background counts.

Copper Concentration

The Cu concentration of the three species (Table 1) is comparable to other arctic studies, but Cu is definitely accumulated preferentially by the lichens, as expected for an essential element. This is shown by the Cu/Ti quotient of the lichens being in excess of those of the field rocks (Table 3), i.e. the enrichment factor (Puckett, this volume).

Potassium and Sulphur Concentrations

The thallus concentrations of K and S (Table 1 and Fig. 2) are best interpreted as reflecting physiological requirements. The very

TABLE 1. The Means (X) and Standard Deviations (s) for the Results of the Lichen Analyses by Species, with F-values Indicating Significant Differences between Means.

Species		Ash %	Ca µg g^{-1}	Cs nCi Kg^{-1}	Cu µg g^{-1}	Fe µg g^{-1}	K µg g^{-1}	Ni µg g^{-1}	Pb µg g^{-1}	S µg g^{-1}	Ti µg g^{-1}	U µg g^{-1}
Cetraria cucullata	X	1.66	4380	4.33	3.05	88.9	1460	1.99	3.76	302	13.49	0.020
	s	(0.51)	(2300)	(1.64)	(1.56)	(35.5)	(475)	(0.81)	(1.39)	(69)	(4.53)	(0.018)
Cetraria nivalis	X	2.06	7970	5.00	2.92	74.0	1250	1.83	4.49	263	12.73	0.018
	s	(0.61)	(3860)	(2.86)	(4.10)	(50.2)	(366)	(1.11)	(2.90)	(73)	(8.62)	(0.010)
Dactylina arctica	X	0.89	2100	3.73	3.88	41.5	1380	1.67	5.52	254	4.60	0.013
	s	(0.18)	(860)	(1.80)	(1.50)	(14.5)	(450)	(0.63)	(3.09)	(48)	(1.43)	(0.006)
	F*	40.88 ***	31.46 ***	1.64	0.94	11.05 ***	1.51	0.86	2.99	3.99 *	18.80 ***	2.19

Values are expressed per air-dry weight, and are based on $n = 25$ for each species except for Cs: C. cucullata, $n = 17$; C. nivalis, $n = 21$; D. arctica, $n = 19$.

* F test for difference between means: *, $p < 0.05$; ***, $p < 0.001$.

TABLE 2. Pearsons's (r) Correlation Coefficients for Cetraria nivalis from all Locations, for Elemental Concentrations (per air-dry weight) and % Ash Content.

	Ca	Cs	Cu	Fe	K	Ni	Pb	S	Ti	U
% Ash	0.6946 (.000) ***	-0.1533 (.254)	0.1548 (.230)	-0.2844 (.084)	-0.0044 (.492)	0.2208 (.144)	-0.0167 (.468)	0.1931 (.178)	-0.3359 (.050) *	0.4535 (.011) *
Ca		-0.1169 (.307)	0.2714 (.095)	-0.4871 (.007) **	-0.0103 (.481)	0.0926 (.330)	-0.0638 (.381)	0.2682 (.097)	-0.4805 (.008) **	0.1336 (.262)
Cs			0.3994 (.036) *	0.6964 (.000) ***	-0.2035 (.118)	0.4353 (.024) *	0.5678 (.004) **	-0.1492 (.259)	0.6758 (.000) ***	0.6653 (.000) ***
Cu				-0.0735 (.364)	-0.3577 (.040) *	0.0802 (.352)	0.2658 (.100)	-0.3507 (.043) *	0.0130 (.475)	-0.0169 (.468)
Fe					0.0185 (.465)	0.3135 (.064)	0.5285 (.003) **	-0.0458 (.414)	0.9848 (.000) ***	0.5049 (.005) **
K						-0.2863 (.083)	-0.3085 (.067)	0.8865 (.000) ***	0.0212 (.460)	-0.0995 (.318)
Ni							0.3836 (.029) *	-0.2856 (.083)	0.2439 (.120)	0.3710 (.034) *
Pb								-0.2456 (.118)	0.5290 (.003) **	0.3789 (.031) *
S									-0.0430 (.419)	-0.0072 (.486)
Ti										0.4466 (.013) *

n = 25 except for Cs where n = 21.
One-tailed probability values are given in parentheses beneath the correlation coefficient:
*, $p < 0.05$; **, $p < 0.01$; ***, $p < 0.001$.

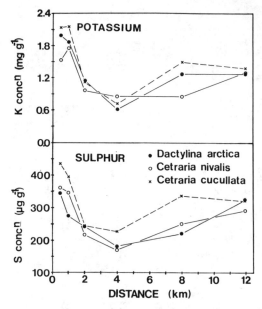

Fig. 2.   Potassium and sulphur concentrations in lichen samples
collected along the West Transect.

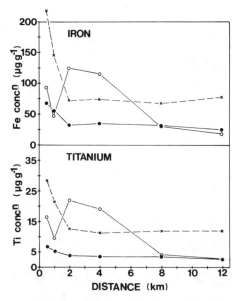

Fig. 3.   Iron and titanium concentrations in lichen samples collected
along the West Transect.
Species key as in Fig. 2.

strong correlations between these two metals (Table 2 and Fig. 2) further supports this.

Iron and Titanium Concentrations

It is obvious from the distribution and correlations (Tables 1 and 2, and Fig. 3) of Fe and Ti that there is a similarity between the accumulation of these two metals. The virtual intraspecies invariance of the Fe/Ti quotient (Table 3) indicates a single source of the trapped particulates that may be assumed to account for much of the observed Fe and all of the Ti concentrations (Nieboer et al., 1978, 1982). It is interesting that this concentration ratio is consistently higher for D. arctica compared to the two Cetraria species (Table 3). This difference may result from the contrasting morphologies of these two lichen genera (Nieboer et al., 1982).

Nickel Concentrations

Concentrations of Ni were relatively low along three transects (2 μg g^{-1}), with slightly higher concentrations (3 μg g^{-1}) for the North Transect; these values are typical of arctic lichens (Tomassini et al., 1976). A minor physiological requirement for nickel is suggested by the enrichment factor (thallus quotient/rock quotient); compare the Ni/Ti quotients for the lichens and the field rocks (Table 3).

Lead Concentrations

The occurrence of higher thallus Pb concentrations than expected, as shown by the Pb/Ti quotients (Table 3), probably reflects an anthropogenic source.

Uranium Concentration

The U concentrations in the lichen samples are extremely low (Table 1) and do not show a concentration by the thalli (Table 3). When the correlations of U with Ti and Fe are considered for C. nivalis (Table 2), it seems probable that the U in the thalli is predominantly in particulate form.

Inter-Element Correlations and Ordination Results

All correlations (Table 2; only those for C. nivalis are shown) have been interpreted with reference to scatter diagrams. Several significant correlations were indeed invalidated by outlier points. Only significant correlations (p > 0.01) that are not affected in this way are considered.

The results of the DCA ordination (Figs 4 and 5) show the clear differences of nutrient uptake or particle-trapping mechanisms between genera, which is discussed below.

TABLE 3. Concentration Quotients of Selected Metals to Ti Based upon Lichen/Lichen or Rock/Rock Values.

Species or Rock Type		Ca/Ti	Cu/Ti	Fe/Ti	K/Ti	Ni/Ti	Pb/Ti	U/Ti
Cetraria cucullata	X	368	0.234	6.53	115	0.161	0.302	0.0014
	s	(259)	(0.135)	(0.54)	(46)	(0.082)	(0.137)	(0.0008)
Cetraria nivalis	X	1090	0.313	5.80	153	0.197	0.402	0.0018
	s	(1140)	(0.517)	(0.91)	(117)	(0.142)	(0.202)	(0.0015)
Dactylina arctica	X	524	0.921	9.02	318	0.392	1.319	0.0029
	s	(322)	(0.474)	(1.00)	(118)	(0.201)	(0.871)	(0.0015)
Dirty-Quartzite		0.80	0.007	6.38	13.63	0.0003	0.011	0.0022
Ortho-Quartzite		1.19	0.042	34.95	41.50	0.2500	0.042	0.0083
Granite		6.70	0.004	9.92	42.21	0.0009	0.089	0.0625
Thelon Sandstone		0.12	0.004	30.63	3.12	0.0142	0.004	0.0026

Lichen values are from this study, x = mean, s = standard deviation, n = 25 for each species. Geological values were calculated from data supplied by Urangesellschaft, Canada Limited.

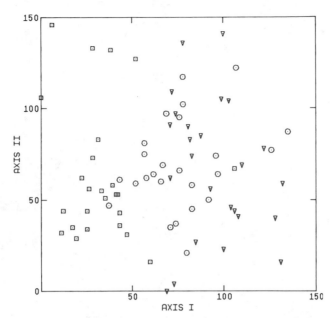

Fig. 4.   Ordination diagram (with species overlays) for Axes I and II
          for the DCA analysis of the lichen elemental content data
          (including ash content) for all species and all 25 sites.
          Axes scaling are relative values: O , Cetraria cucullata;
          ▽ , Cetraria nivalis; □ , Dactylina arctica.
          The interpretation of the position of the points is best
          represented by generic differences shown here by the clear
          separation of the Dactylina data from those of the two
          Cetraria species. Eigenvalues show that Axis I explains more
          variability than Axis II.

DISCUSSION

     From the data (Tables 1 and 3) and the correlations (Table 2),
several important points emerge. Firstly, D. arctica has much lower
levels of percent ash, Ca, Fe, and Ti than the two Cetraria species.
Secondly, there are very pronounced positive correlations between
percentage ash and Ca concentrations, between K and S as well as for
Fe, Ti and U. Thirdly, ^{137}Cs is also strongly correlated with Fe and
Ti. The correlation between Fe and Ti has been shown to result from
their occurrence together in particulates trapped in lichen thalli
(Nieboer et al., 1978, 1982). Further, it is concluded that since U
is also strongly correlated with Fe and Ti, it too is present
primarily in particulate form. The apparent absorption of ^{137}Cs on
trapped particulates and thus its grouping with Fe, Ti and U is
explained below.

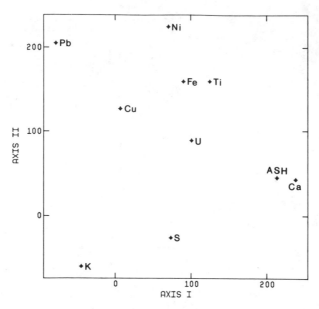

Fig. 5. Ordination diagram (with element overlays) for Axes I and II
for the lichen elemental content data (including ash) used in
the DCA analysis. Axes scaling are in relative units.
Proximity of different elements or ash shows a similarity in
their distribution in the lichens analysed.

The observed levels of Fe, Ti, S, Pb, Ca, K and Ni in C. nivalis
and C. cucullata are comparable to those previously reported for the
same lichen species from the Mackenzie Valley, N.W.T., although the
highest values do suggest some contribution from a mineral enriched
substratum (Tomassini et al., 1976; Nieboer et al., 1978). The Cu
concentrations appear typical of arctic sites without mineralization
of the substratum but with some selective uptake. Comparable
elemental data for D. arctica are less available (Tomassini et al.,
1976), although it is evident from the data reported here that this
species is a less efficient accumulator than the genus Cetraria. The
low levels of U found in the three lichen species are in good
agreement with other reported natural background concentrations of
this metal (reviewed by Beckett et al., 1982). The data indicate that
[137]Cs accumulated by lichens in Canada is similar to that found in
Finland (Tillander et al., 1979; Nieboer and Richardson, 1981).

The magnitude of the Fe/Ti concentration ratios is clearly
diagnostic of a lichen species. In addition, lichens also differ in
their percent inorganic ash content. On average, the Cetraria species
contained about twice as much incombustible ash as D. arctica.
Although genus- and species- dependent, ash levels of 1-2% are typical

of fruticose lichens growing in clear air environments (Lounamaa,
1965). In this situation, the macronutrients Ca, Mg, K, Na and S, and
perhaps Fe, are expected to contribute the largest share since the
observed concentrations of trace elements such as of Ti, Cu, Pb and
Ni, even when trapped as particulates, would only slightly add to the
total (Nieboer et al., 1982). This interpretation draws support from
the lack of association between the observed elemental concentrations
(except Ca) and the percent ash (Table 2).

    Nuclear fallout of ^{137}Cs, and more recently that from the
crashing of nuclear-powered satellites, is presumably particulate.
Although ^{137}Cs has a short isotope half-life (30.1 y), its presence in
lichen samples collected in the N.W.T. and other parts of Canada as
recently as 1978 remains well above the detection limit of gamma-ray
spectrometry. Particulates of minerals such as metal oxides, hydrous
oxides, metal sulphides and clays are often charged, especially when
moistened (Stumm and Morgan, 1981). In fact, the solid/solution
interfaces and the associated colloidal properties provide a vehicle
for physiochemical interactions that have important ecological
repercussions. Forces acting at interfaces are extensions of forces
acting within the two separate phases. Surface adsorption of ions,
organic molecules and, no doubt, small particulates by inorganic
particulate matter is a known phenomena and is the subject of intense
study (Stumm and Morgan, 1981). We propose therefore, on the basis of
the strong association of ^{137}Cs with Fe, Ti and U, that this
radioisotope is adsorbed either as the ion or very small particulates
onto inorganic dust, which is available for trapping by lichens. To
our knowledge, this mode of distribution of radioisotopes resulting
from nuclear fallout has not been recognised previously.

    It is evident that the elemental concentration differences
between D. arctica and the two Cetraria species reflect markedly
different abilities to trap particulates in the lichen thallus. This
difference is very clearly shown in the ordination of the lichen
samples using DCA (Fig. 4). The samples of D. arctica, with one
exception, have very low loadings on the first ordination axis of Fig.
4. The average Fe/Ti concentration quotients were considerably higher
for this lichen (Table 3). Since composition is known to vary with
the size of particulates (Nieboer et al., 1982), this observation
suggests differential trapping according to size. Scanning electron
microscopic studies of individual thalli of the three species exhibit
contrasting surface irregularities and quite different internal
organization, including differences in the prominence and size of
interstitial spaces. These distinct features (Fig. 6) agree well with
the apparently contrasting, particulate-trapping abilities of
Dactylina to the Cetraria species.

    The DCA ordination confirms the correlations discussed above:
between Ca and ash; Fe, Ti, and U; and K and S (Fig. 5). On the basis
of this joint evidence it is indeed reasonable to conclude that the

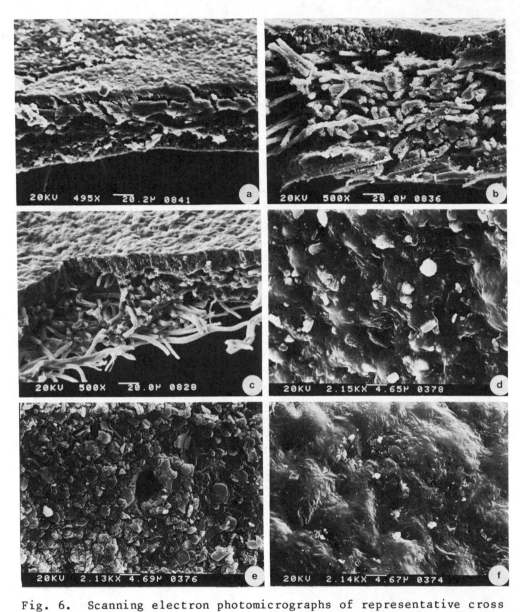

Fig. 6.   Scanning electron photomicrographs of representative cross
           sections (a–c) and areas of the upper surface (d–f) of the
           three lichen species showing contrasted morphologies: (a,d)
           Cetraria cucullata; (b,e) Cetraria nivalis; (c,f) Dactylina
           arctica.
           Note the lack of an interior (lower) cortex in D. arctica (c)
           and the apparent absence of available interstitial space in
           C. cucullata (a).
           The presence of the pore in C. nivalis (e) was a constant
           feature of this species. Bars indicate scale in µm

background concentrations of U are primarily particulate in nature and
that S and K are predominantly located intracellularly. The weaker
correlations (p < 0.05) of S and K with Fe and Ti, as well as the
loading along Axis I in Fig. 5, suggest that there is a minor
contribution to the S and K contents by particulate trapping.
Further, the grouping of Cu, Pb and Ni with Fe, Ti and U in Fig. 5
receives some support from a number of significant inter- elemental
correlations in Table 2. It is possible that the exact position of Ni
and Pb in Fig. 5 has been determined by the relatively high
concentrations observed in a small number of samples. The latter
values may reflect local geology.

While we have noted that the distribution of the different
elements in the lichen thalli are not related to distance from the ore
body, distance diagrams do reveal similarities between elemental
distributions. The strong correlation between K and S is shown in
this manner in Fig. 2. Here a distance-dependent relationship does
not exist but a similar distribution pattern of elemental concentrations between sample sites does. This type of relationship is clearly
shown for Fe and Ti (Fig. 3), which no doubt reflects their known
association in particulates.

The data show a negative correlation between Ca (or % ash) and Fe
or Ti (Table 2). Interestingly, Erdman et al. (1977) reported a
strong positive correlation between Ca and % ash but a strong negative
association for silicon and ash as well as Al and ash. In this study
Ca was also found to be the main contributor to the non-combustible,
inorganic ash. X-ray diffraction studies by Erdman et al. (1977)
revealed that Ca was present as calcium oxalate in the intact lichen
thallus and as calcium carbonate (calcite) in the ash. Mass balance
calculations from our own data do indeed indicate that Ca is the
primary lichen constituent contributing to the ash. Collectively,
these observations suggest that calcium-rich lichen thalli appear to
exclude exogenous dust particulates, perhaps because calcium oxalate
of endogenous origin fills the extracellular interstitial spaces.

In Table 3, a comparison is made of the element-to-titanium
quotients observed in the three lichen species with those reported for
the four main rock types found in the study area. It is evident, from
the good agreement for both the Fe/Ti and U/Ti concentration ratios,
that dirty quartzite most closely resembles the composition of the
particulates trapped in the lichens. Perhaps this rock type weathers
readily and the resulting particulates are distributed across the
study area by wind and surface-water run off. As expected, the
contencentration ratios Ca/Ti, Cu/Ti and K/Ti are elevated in the
lichens and reflect the physiological need for Ca, Cu and K by these
plants. While the data for Ni is equivocal, the enhancement of the
Pb/Ti ratio in the three lichen species would appear to be consistent
with the known global Pb fallout from automobile and related emissions
(Nieboer et al., 1978).

ACKNOWLEDGEMENTS

    Financial assistance to several of us from the Department of
Indian and Northern Affairs (DIAND), and particularly the program
Arctic Land Use Research (ALUR), is gratefully acknowledged as is
continued support from The Natural Sciences and Engineering Research
Council of Canada (NSERC). We gratefully acknowledge the technical
assistance of Mr. Brian Smith in the field. The help in field
logistics and the provision of geochemical data by Urangesellschaft
Canada Ltd. are also appreciated.

REFERENCES

Beckett, P.J., Boileau, L.J.R., Padovan, D., Richardson, D.H.S., and
        Nieboer, E., 1982, Lichens and mosses as monitors of industrial
        activity associated with uranium mining in northern Ontario,
        Canada - Part 2: Distance dependent uranium and lead
        accumulation patterns, Environmental Pollution (Series B), 4:
        91-107.
Boileau, L.J.R., Beckett, P.J., Lavoie, P., Richardson, D.H.S., and
        Nieboer, E., 1982, Lichens and mosses as monitors of industrial
        activity associated with uranium mining in northern Ontario,
        Canada - Part 1: Field procedures, chemical analysis and
        interspecies comparisons, Environmental Pollution (Series B), 4:
        69-84.
Erdman, J.A., Gough, L.P., and White, R.W., 1977, Calcium oxalate as
        source of high ash yields in the terricolous lichen Parmelia
        chlorochroa, The Bryologist, 80: 334-339.
Ernst, P.C., and Hoffman, E.L., 1982, Development of automated
        analytical systems for large throughput, Journal of
        Radioanalytical Chemistry, 70: 527-537.
Garty, J., Galun, M., and Kessel, M., 1979, Localization of heavy
        metals and other elements accumulated in the lichen thallus, New
        Phytologist, 82: 159-168.
Gauch, H.G. Jr., 1982, "Multivariate Analysis in Community Ecology,"
        Cambridge University Press, Cambridge.
Hill, M.O., 1979, "DECORANA - A Fortran Program for Detrended
        Correspondence Analysis and Reciprocal Averaging," Cornell
        University, Ithaca, New York.
Hill, M.O. and Gauch, H.G. Jr., 1980, Detrended correspondence
        analysis, an improved ordination technique, Vegetatio, 42:
        47-58.
Lounamaa, K.J., 1965, Studies on the content of iron, manganese and
        zinc in macrolichens, Annales Botanici Fennici, 2: 127-137.
Nie, N.H., Hull, C.H., Jenkins, J.G., Steinbrenner, K., and Bent,
        D.H., 1975, "SPSS, statistical package for the social
        sciences," 2nd Edition, McGraw-Hill, New York.
Nieboer, E., and Richardson, D.H.S., 1981, Lichens as monitors of
        atmospheric deposition, in: "Atmospheric Pollutants in Natural
        Waters," S.J. Eisenreich, ed., pp. 339-388, Ann Arbor Science

Publications Inc.

Nieboer, E., Richardson, D.H.S., Boileau, L.J.R., Beckett, P.J., Lavoie, P., and Padovan, D., 1982, Lichens and mosses as monitors of industrial activity associated with uranium mining in northern Ontario, Canada - Part 3: Accumulations of iron and titanium and their mutual dependence, Environmental Pollution (Series B), 4: 181-192.

Nieboer, E., Richardson, D.H.S., and Tomassini, F.D. 1978, Mineral uptake and release in lichens: an overview, The Bryologist, 81: 226-246.

Stumm, W., and Morgan, J.J., 1981, "Aquatic Chemistry: An Introduction Emphasizing Chemical Equilibria in Natural Waters," 2nd Edition, Wiley, New York.

Svoboda, J., and Taylor, H.W., 1979, Persistence of Cesium-137 in arctic lichens, Dryas integrifolia, and lake sediments, Arctic and Alpine Research, 11: 95-108.

Taylor, H.W., Hutchinson, E.A., McInnes, K.L., and Svoboda, J., 1979, Cosmos 954: Search for airborne radioactivity on lichens in the crash area, Northwest Territories, Canada, Science, 205: 1383-1385.

Tillander, M., Jaakkola, T., and Miettinen, J.K., 1979, Cesium-137 in the foodchain lichen-reindeer-man during 1976-1978, in: "Radioactive Food-chains in the Subarctic Environment," Paper No. 97. Technical progress report, U.S. Energy Research and Development Administration.

Tomassini, F.D., Puckett, K.J., Nieboer, E., Richardson, D.H.S., and Grace, B., 1976, Determination of copper, iron, nickel and sulphur by x-ray fluorescence in lichens from the Mackenzie Valley, Northwest Territories, and the Sudbury District, Ontario, Canadian Journal of Botany, 54: 1591-1603.

TEMPORAL VARIATION IN LICHEN ELEMENT LEVELS

K.J. Puckett

Air Quality Research Branch, Atmospheric Environment
Service, Environment Canada, 4905 Dufferin Street
Downsview, Ontario, Canada M3H 5T4

INTRODUCTION

Lichens have certain features which have resulted in their
intensive use as indicators of air quality and as monitors of the
atmospheric deposition of various elements. These salient features
include the lack of roots or structures which have the absorptive
function of roots and thus some lichens are dependent for their
mineral nutrients to a large extent on material landing on the lichen
thallus as the result of wet and dry deposition from the atmosphere.
Also, lichens unlike higher plants, do not have a well-developed
cuticle and hence there is no comparable physical barrier to impede
exchange with the environment. Consequently, lichens can accumulate
mineral elements to levels far greater than their expected physiological needs. Lichens are perennial and this feature together with
the other characteristics has led to the use of these plants as long-
term integrators of deposition from the atmosphere of elements
originating from both natural and man-made sources.

In interpreting these deposition patterns around various line,
point and area element sources, considerable attention has been given
to understanding the reasons for the spatial variation in lichen
element levels while comparatively little attention has been given to
the temporal aspects of element accumulation. Element accumulation by
lichens can be considered in terms of several time scales. Laboratory
experiments have shown that when lichens come into contact with
solutions containing metals the response is immediate (Puckett et al.,
1973; Wainwright and Beckett, 1975). Thus, the lichen responds in
seconds to changes in its immediate environment. The other extreme is
where increasing metal levels in lichens have been demonstrated by
collections over a period of 80 years, where the lichen is reflecting

211

changes in the overall environment which occur over decades (Rao et
al., 1977; Schutte, 1977; Lawrey and Hale, 1981). Crittenden (1983)
monitored the gains and losses in N and K levels in Stereocaulon
paschale and Cladonia stellaris over a period of days. Changes in
metal concentrations in terms of weeks and months (Steinnes and
Krog, 1977; Garty and Fuchs, 1982; Johnsen et al., 1983) have been
described in transplant experiments where lichens collected from non-
contaminated areas are resited in the vicinity of industrial
operations which emit metals to the atmosphere. On a longer time
scale, lichens in tundra communities have been utilised to monitor
over years the deposition and loss of radionuclides originating from
atmospheric nuclear weapon testing (Hanson, 1982).

Another time scale is that imposed on lichens by the changing
physical environment throughout the year. Information on seasonal
variation in lichen element levels is limited but shows several
response patterns. Kovács-Lang and Verseghy (1974) followed seasonal
changes in Ca and K levels in European terricolous lichens over a
period of two years. They noted that element levels were highest in
the autumn and winter while concentrations were uniformly lower in the
spring and summer. Similarly, Lewis Smith (1978) observed seasonal
fluctuations in Ca, K and Na concentrations in maritime Antarctic
lichens. Differences between summer and winter concentrations of the
three cations suggested the possibility of an annual cycle of mineral
element concentrations in lichens, with relatively high spring to
summer levels and winter minima.

In contrast, Scotter and Miltimore (1973) sampled Cladonia and
Cetraria species from a site in the Northwest Territories, Canada, on
3 occasions over an 11 month period. They found no seasonal
differences in levels of the 7 elements studied which included Mg, Mn,
and K. Values differed slightly between collection periods in the
above study but since no variance estimate was quoted it is difficult
to determine whether the differences noted were real or merely
represented variability within the sampling site. This caveat also
applies to other studies where seasonal variation has been described.
Similarly, Carstairs and Oechel (1978) found no changes in nutrient
concentrations of C. stellaris collected on five occasions during the
growing season in the Canadian subarctic. Further, Chapin et al.,
(1980) found only small differences in the N and P content of six
lichen species collected in June and September from an Alaskan tundra
ecosystem.

The objective of the work described in this paper was to determine whether lichens show changes in their element levels in response
to changing seasonal conditions and whether such changes can be
described in terms of varying element deposition rates throughout the
year.

MATERIALS AND METHODS

Samples of <u>Cladonia rangiferina</u> were collected on a monthly basis (November, 1979 – November, 1983) from a site 160 km north of Toronto, Ontario, but only data from November, 1979 to June, 1981 will be discussed in this paper. The samples were collected from thin mineral soil overlying bedrock and all samples (10 replicates/monthly collection) were taken from the same area (approx 2 sq. m.). The samples (podetial tips, 1.0 cm in length) were cleaned of debris, crushed in liquid nitrogen and then oven-dried at 75°C for 24 hours. The samples were analysed for Al, Ca, Cl, Mg, Mn, K, Na, Ti and V by neutron activation (Puckett and Finegan, 1980). Rock fragments which had accumulated near the lichen collection site were sieved and fragments which passed through a 202 μm grid (sieve size 70) were analysed by neutron activation.

RESULTS AND DISCUSSION

In considering temporal variation, it is common practice to examine each element separately even though variation in some optimal combination of elements might be of greater importance to the lichen. Insight into the factors controlling lichen element concentration can be gained from examining the relationships between the element concentrations within the lichen.

Correlations between Elements
<u>   </u>

Correlation coefficients between pairs of elements are shown in

Table 1. Correlation between Aluminium and Other Elements in <u>Cladonia rangiferina</u>

Element	Correlation coefficient	N
Vanadium	0.87	245
Sodium	0.86	249
Titanium	0.86	191
Calcium	0.58	249
		*
Manganese	0.21	249
Magnesium	0.20	151
Chlorine	0.15	177
Potassium	0.02	249

* Correlation coefficients above line significant at the 0.01 level

Table 2. Correlation between Potassium and Other Elements in <u>Cladonia</u>
          <u>rangiferina</u>

Element	Correlation coefficient	N
Magnesium	0.54	151
Manganese	0.27	249
		*
Calcium	0.24	249
Vanadium	0.23	245
Chlorine	0.21	199
Titanium	0.19	196
Sodium	0.08	249
Aluminium	0.02	249

* Correlation coefficients above line significant at the 0.01 level

Tables 1 and 2. Aluminium is very highly correlated with V, Na, Ti
(Fig. 1) and, to a lesser extent, Ca but there is no relationship,
positive or negative, with K (Fig. 2). Potassium is significantly
correlated with Mg (Fig. 3) and, to a much lesser extent, Mn. Similar
correlations were noted between the same elements in <u>C. rangiferina</u>
collected from a different area (Northwest Territories, Canada) where

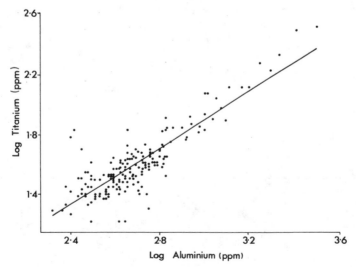

Fig. 1. Relationship between Al and Ti levels in <u>Cladonia</u>
         <u>rangiferina</u>.

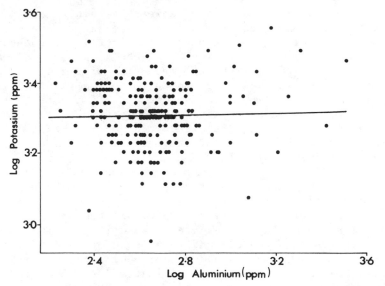

Fig. 2.   Relationship between Al and K levels in <u>Cladonia</u>
          <u>rangiferina</u>.

Al was again highly correlated with Ti, V and Na and K was correlated
to Cl, Ca, Zn, Mg and Na, in order of decreasing r values (Puckett and
Finegan, 1980).   Thus, these groups of elements are correlated both in
space and time.

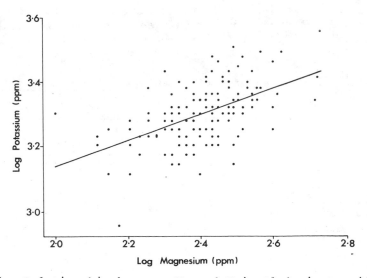

Fig. 3.   Relationship between Mg and K in <u>Cladonia rangiferina</u>.

The significant correlations between certain elements may be indicative of a common source of these elements. Lichens found in southern Ontario will be subject to both natural and anthropogenic sources of elements. For terricolous lichens, a major element source will be material derived from the weathering of the local bedrock and soils. One of the simplest approaches to source identification is by comparing the lichen element content to that of a potential source. This comparison can be approached by calculating an enrichment factor, where the concentration of the element in question in the lichen is compared to the concentration of the same element in the potential source and the resulting ratio is normalised by the use of an element which is considered to be the most unambiguous indicator of that particular source. Enrichment factors of unity imply that the element in question is not enriched relative to the potential source and that the potential source in question is probably the main source of that element. Enrichment factors greater than unity, in practice greater than 5, indicate that the element is enriched relative to the potential source and implies additional sources of the element in question although enrichment also indicates differing accumulation mechanisms and metabolic requirements (Puckett and Finegan, 1980). Enrichment factors for the elements in C. rangiferina using local bed-rock as the potential source and Al as the reference element are given in Table 3. Sodium, Ti and V have enrichment factors less than 5 whereas K, Mn and Cl have enrichment factors greater than 5 with Cl having a very large enrichment factor. The Ca enrichment factor of 4.9 is borderline. The enrichment factors quoted are for July 1980 but are representative of the entire period.

Material derived from local bedrock would, therefore, appear to be the main source of Al, Na, Ti and V. Particulate entrapment would seem to be most likely accumulation mechanism for these elements (Nieboer and Richardson, 1981). Such entrapped particulates, unless water-soluble, would be of limited metabolic significance and obser-

Table 3.   Enrichment Factors of Elements in Cladonia rangiferina
            (July 1980)

Element	Enrichment Factor
Sodium	0.7 ± 0.1
Titanium	1.4 ± 0.3
Vanadium	4.1 ± 0.5
Calcium	4.9 ± 1.4
Potassium	7.1 ± 2.1
Manganese	18.8 ± 8.8
Chlorine	85.1 ± 20.3

Table 4.   Elution of Metals from <u>Cladonia rangiferina</u>

Metal	Metal in water-eluted lichen ($\mu g^{-1}g$)	Metal in acid-eluted lichen ($\mu g^{-1}g$)	% Loss
Aluminium	404 $\pm$ 23	376 $\pm$ 48	7
Sodium	102 $\pm$ 19	89 $\pm$ 53	13
Titanium	31 $\pm$ 5	28 $\pm$ 9	11
Vanadium	0.63 $\pm$ 0.10	0.53 $\pm$ 0.06	16

Lichen material was washed in deionised water to remove any loosely
adhering particles and then samples (500 mg air dry weight) were
incubated in either deionised water or in HNO_3 (1N, 50 ml) for 3 h.
Water-eluted and acid-eluted material was analysed by neutron
activation (Puckett and Finegan, 1980). The values quoted are the
mean $\pm$ 1 SD of 5 replicates.

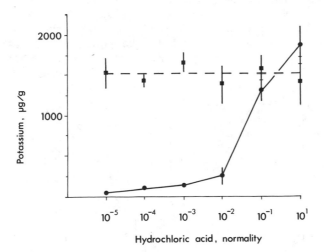

Fig. 4.   Potassium loss from <u>Cladonia rangiferina</u> as a function of
          increasing acidity.
          Samples (150 mg) of <u>Cladonia rangiferina</u> were incubated in
          solutions of increasing hydrogen-ion concentration (25 ml)
          for 3 h after which the solution was filtered and the K in
          solution determined. Heat pre-treated samples (100°C for 24
          hours) were treated as above. ●——● = Live thalli;
          ■---■ = Heat-treated thalli.

vations on their chemistry imply limited solubility (Gough and Erdman, 1977; Garty et al., 1979; Johnsen, 1981). Elution with nitric acid only removed a small proportion of the total Al, Na, Ti and V (Table 4), an observation supporting the limited availability of these elements. In contrast, 90% of the total K can be removed by acid elution (Fig. 4), and a similar proportion can be removed from heat-treated lichens eluted with deionised water, indicating the mainly intracellular location of K (see also Puckett, 1976; Buck and Brown, 1979).

Temporal Variation

Aluminium is an element of limited metabolic significance whose concentration varies primarily as a function of source strength. In contrast, K is an essential element whose mainly intracellular location will be influenced by both source strength fluctuations and differing metabolic requirements throughout the year.

The month to month variation in the Al and K concentrations in C. rangiferina is shown in Figs 5 and 6. Initially, the Al levels show only minor fluctuations from month to month but there are some major deviations from the norm later in the period. In contrast, the K levels are uniform throughout the period and only show a small month to month variation. To examine the variation in more detail, each monthly collection was compared to the overall mean value for Al and K (Figs 7 and 8). The largest positive values relative to the mean value for Al were found in mid-winter; low negative deviations were found in early spring with a tendency to positive values, relative to the mean, again in early summer. For K the largest positive devia-

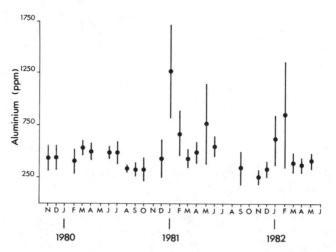

Fig. 5.   Monthly variation in the Al concentration in Cladonia rangiferina.   Values given as mean ± 1 S.D.

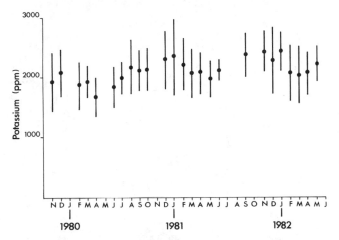

Fig. 6.   Monthly variation in the K concentration in <u>Cladonia</u>
<u>rangiferina</u>. Values given as mean ± S.D.

tions from the mean were found in mid-winter, small negative values in
early spring, a tendency to positive values in early summer but,
unlike Al, relatively high positive values were seen in late summer-
early autumn.   This pattern in element concentration was seen in 1981
and 1982 but missing data precludes its identification in the winter
of 1979-1980.   In addition, although Al and K levels showed a similar
pattern in their fluctuation through the year, it should be emphasised
that the concentrations of these elements were not correlated in the
lichen thallus.

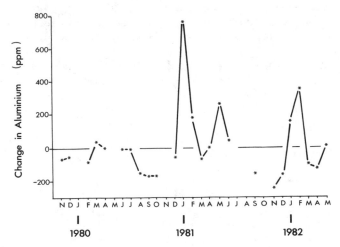

Fig. 7.   Monthly variation in Al concentrations in relation to the
mean aluminium concentration over the period.

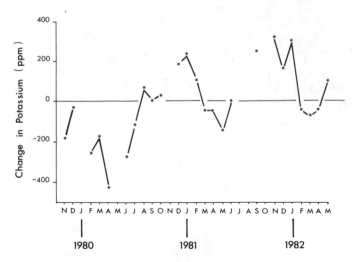

Fig. 8. Monthly variation in the K concentrations in relation to the mean K concentration over the period.

One factor that will influence both the Al and K levels will be the variations in source strengths throughout the year. These elements will be delivered to the lichen surface either dissolved in precipitation or as particles. In terms of reflecting atmospheric concentrations, the relationship between measured atmospheric metal concentrations and in situ lichen (or moss) metal concentrations has not been rigorously examined. Parallelism has been observed between the geometric means of element concentrations of atmospheric particulates and the arithmetic means of lichen metal concentrations in the urban area of Sendai, Japan (Saeki et al., 1977). Several studies have attempted to describe the link between cryptogram element levels and measured deposition. A general indication that lichens reflect atmospheric deposition was the observed tendency for element concentrations in a particular lichen species to remain in a narrow range even though substrates varied in their element composition, indicating some other factor influencing lichen element composition. Jenkins and Davies (1966) suggest that this factor would have to be regional in nature and suggested atmospheric particulates as a major source based on a close correlation between element values in lichen ash and those of deposited atmospheric particulates. Comparisons of metal levels in cryptograms and precipitation have also shown empirical relationships. Linear correlations between Cu, Pb and Zn concentrations in the moss Brachythecium rutabulum and bulk precipitation have been described by Andersen et al., (1978) while linear relationships were inferred between annual deposition estimates (in precipitation) and metal concentrations in the moss Hylocomium splendens (Hanssen et al., 1980 and see also Thomas et al., 1983). In a more mechanistic study, metal levels in transplanted Hypogymia physodes and Dicranoweisia cirrata and bulk precipitation were

compared on a monthly basis (Pilegaard, 1979). Metal levels in bulk
precipitation followed a decreasing power curve when the distance from
the source was increased and metal levels in transplanted lichens and
mosses and <u>in situ</u> lichens also showed the same relationship.

The deposition of water-soluble elements in precipitation in
Ontario has been described by Chan et al., (1983). Precipitation was
collected in a Sangamo collector which allows for the collection of
wet but not dry deposition. There was no significant correlation
between the deposition of Al and V in precipitation (Fig. 9), although
the lack of correlation may be explained by these V concentrations
being very close to the detection limits of the methods used. In
contrast, these elements were highly correlated in the lichen,
providing further evidence that Al and V are accumulated by
particulate entrapment. Also, there was no significant correlation
between the deposition of K and Mg in precipitation (Fig. 10) whereas
these elements were significantly linked in the lichen. As indicated,
a large proportion of the total K is intracellular and the differing
relationship between these two elements in the lichen and precipitation may reflect a preferential accumulation of K or Mg thereby
maintaining a pre-determined intracellular ratio between these two
elements.

Seasonal variation in Al and K deposition are shown in Figs 11
and 12. For the period studied, deposition for both elements in
winter was lower than the previous autumn or the subsequent spring or
summer while, in contrast, both the Al and K levels in the lichen were
at their highest during the winter period. However, precipitation
during the period November-April is mainly in the form of snow and

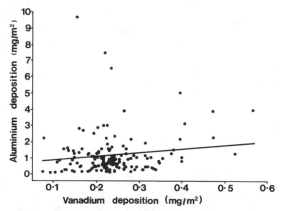

Fig. 9.   Relationship between Al and V deposition in precipitation.
Each point represents the deposition over a 3 month period
for each station in the network. Data from Chan et al.
(1983).

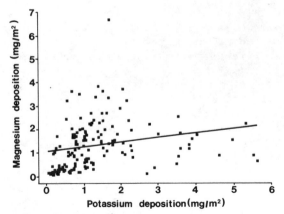

Fig. 10. Relationship between Mg and K deposition in precipitation.
Each point represents the deposition over a 3 month period
for each station in the network. Data from Chan et al.
(1983).

material deposited will be accumulated in the snowpack which covers
the lichens and will be unavailable unless there are thaw periods.
Measured temperatures and predicted snowpack melt periods indicated
that minor thaws occurred in the area in mid-December 1980 and late
January 1981 (Louie et al., 1985), at which time the lichen Al and K
concentrations were at their highest. Several studies have shown that
the initial meltwater leaches a large proportion of the accumulated
chemical species (Johannessen and Henriksen, 1978; Johannes et al.,
1981; Cadle et al., 1984). Thus the lichen element levels may have
been enhanced by this relatively low volume of meltwater rich in
deposited material. In contrast, in April when the entire snowpack

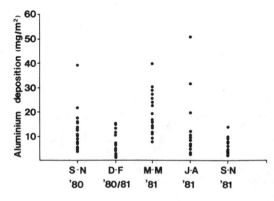

Fig. 11. Aluminium deposition for the period November 1980-1981.
Data from Chan et al. (1983).

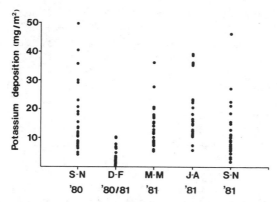

Fig. 12. Potassium deposition for the period November 1980-1981.
Data from Chan et al. (1983).

melts, the remaining material accumulated throughout the winter
becomes available to the lichen, yet the lichen showed the lowest
element levels at this time. In this instance, the large volume of
meltwater flowing over the lichens in a short time period may serve to
leach existing Al and K from the apical portion of lichen rather than
act as a source of additional material.

REFERENCES

Andersen, A., Hovmand, M.F., and Johnsen, I., 1978, Atmospheric heavy
    metal deposition in the Copenhagen area, Environmental Pollution,
    17: 133-151.
Buck, G.W., and Brown, D.H., 1979, The effect of desiccation on cation
    location in lichens, Annals of Botany, 44: 265-277.
Cadle, S.H., Muhlbaier Dasch, J., and Grossnickle, N.E., 1984,
    Northern Michigan Snowpack - a study of acid stability and
    release, Atmospheric Environment, 18: 807-816.
Carstairs, A.G., and Oechel, W.C., 1978, Effects of several
    microclimatic factors and nutrients on net carbon dioxide
    exchange in Cladonia alpestris (L.) Rabh. in the subarctic,
    Alpine and Arctic Research, 10: 81-94.
Chan, W.H., Tang, A.J.S., and Lusis, M.A., 1983, Precipitation
    concentration and wet deposition fields of pollutants in Ontario,
    September 1980 to December 1981. Report ARB-61-83 ARSP. Ontario
    Ministry of the Environment, Toronto, Ontario, Canada. pp.79.
Chapin, F.S., Johnson, D.A., and McKendrick, J.D., 1980, Seasonal
    movement of nutrients in plants of differing growth form in an
    Alaskan tundra ecosystem. Implications for herbivory, Journal of
    Ecology, 68: 189-209.
Crittenden, P.D., 1983, The role of lichens in the nitrogen economy of
    subarctic woodlands: Nitrogen loss from the nitrogen-fixing
    lichen Stereocaulon paschale during rainfall, in: "Nitrogen as an

ecological factor," J.A. Lee, S. McNeill and I.H. Rorison, eds, pp. 43-68. Blackwell Scientific Publication, Oxford.

Garty, J., Galun, M., and Kessel, M. 1979, Localization of heavy metals and other elements accumulated in the lichen thallus, New Phytologist, 82: 159-168.

Garty, J., and Fuchs, C., 1982, Heavy metals in the lichen Ramalina duriaei transplanted in biomonitoring stations, Water, Air and Soil Pollution, 17: 175-183.

Gough, L.P., and Erdman, J.A., 1977, Influence of a coal-fired powerplant on the element content of Parmelia chlorochroa, The Bryologist, 80: 492-501.

Hanssen, J.E., Rambaek, J.P., Semb, A., and Steinnes, E., 1980, Atmospheric deposition of trace elements in Norway. Proceedings International Conference "Ecological impact of acid precipitation," Norway, 1980, pp. 116-117. SNSF project.

Hanson, W.C., 1982, ^{137}Cs concentrations in northern Alaskan eskimos, 1962-1979: Effects of ecological, cultural and political factors, Health Physics, 42: 433-447.

Jenkins, D.A., and Davies, R.I., 1966, Trace element content of organic accumulations, Nature, 210: 1296-1297.

Johannes, A.H., Galloway, J.N., and Troutman, D.E., 1981, Snowpack storage and ion release, in: "Integrated Lake Watershed Acidification Study (ILWAS)," EPRI report EA-1825, pp. 6-1.

Johannessen, M., and Henriksen, A., 1978, Chemistry of snow meltwater: changes in concentration during melting, Water Resources Research, 14: 615-619.

Johnsen, I., 1981, Heavy metal deposition on plants in relation to emission and bulk precipitation, Silva Fennica, 15: 444-445.

Johnsen, I., Pilegaard, K., and Nymand, E., 1983, Heavy metal uptake in transplanted and in situ yarrow (Achillea millefolium) and epiphytic cryptogams at rural, urban and industrial localities in Denmark, Environmental Monitoring and Assessment, 3: 13-22.

Kovacs-Lang, E., and Verseghy, K., 1974, Seasonal changes in the K and Ca contents of terricolous xerophyton lichen species and their soils, Acta Agronomica Academiae Scientiarum Hungaricae, 23: 325-333.

Lawrey, J.D., and Hale, M.E., 1981, Retrospective study of lichen lead accumulation in the Northeastern United States, The Bryologist, 84: 449-456.

Lewis Smith, R.I., 1978, Summer and winter concentrations of sodium, potassium and calcium in some maritime Antarctic cryptogams, Journal of Ecology, 66: 891-909.

Louie, P.Y.T., Johnstone, K., and Barrie, L.A., 1985, Acidic snowmelt shock potential model for basin studies, in: "Proceedings of Canadian Hydrology Symposium," Quebec City, June 10-12, 1984 (In press).

Nieboer, E., and Richardson, D.H.S., 1981, Lichens as monitors of atmospheric deposition, in: "Atmospheric pollutants in natural waters," S.J. Eisenreich, ed., pp. 339-388. Ann Arbor Science, Ann Arbor.

Pilegaard, K., 1979, Heavy metals in bulk precipitation and transplanted Hypogymnia physodes and Dicranoweisia cirrata in the vicinity of a Danish steelworks, Water, Air & Soil Pollution, 11: 77–91.

Puckett, K.J., 1976, The effect of heavy metals on some aspects of lichen physiology, Canadian Journal of Botany, 54: 2695–2703.

Puckett, K.J., and Finegan, E.J., 1980, An analysis of the element content of lichens from the Northwest Territories, Canada, Canadian Journal of Botany, 58: 2073–2089.

Puckett, K.J., Nieboer, E., Gorzynski, M.J., and Richardson, D.H.S., 1973, The uptake of metal ions by lichens: a modified ion-exchange process, New Phytologist, 72: 329–342.

Rao, D.N., Robitaille, G., and LeBlanc, F., 1977, Influence of heavy metal pollution on lichens and bryophytes, Journal of the Hattori Botanical Laboratory, 42: 213–239.

Saeki, M., Kunii, K., Seki, T., Sugiyama, K, Suzuki, T., and Shishido, S., 1977, Metal burden of urban lichens, Environmental Research, 13: 256–266.

Schutte, J.A., 1977, Chromium in two corticolous lichens from Ohio and West Virginia, The Bryologist, 80: 279–283.

Scotter, G.W., and Miltimore, J.E., 1973, Mineral content of forage plants from the reindeer preserve, Northwest Territories, Canadian Journal of Plant Science, 53: 263–268.

Steinnes, E., and Krog, H., 1977, Mercury, arsenic and selenium fall-out from an industrial complex studied by means of lichen transplants, Oikos, 28: 160–164.

Thomas, W., Riess, W., and Herrmann, R., 1983, Processes and rates of deposition of air pollutants in different ecosystems, in: "Effects of Accumulation of Air Pollutants in Forest Ecosystems," B. Ulrich and J. Pankrath, eds, pp. 65–82. D. Reidel Publisher.

Wainwright, S.J., and Beckett, P.J., 1975, Kinetic studies on the binding of zinc ions by the lichen Usnea florida (L.) Web., New Phytologist, 75: 91–98.

LEAD AND URANIUM UPTAKE BY LICHENS

D.H.S. Richardson[a], S. Kiang,[a],
V. Ahmadjian[b] and E. Nieboer[c]

[a] Botany School, Trinity College
Dublin, Ireland
[b] Department of Biology, Clarke University
Worcester, MA 01610, U.S.A.
[c] Department of Biochemistry, McMaster University
Hamilton, Ontario, Canada

INTRODUCTION

Interest in the uptake of Pb and U stems from both practical and academic considerations. Lead is currently one of the most widespread environmental contaminants owing to its use as an anti-knock petrol additive. Lead emitted by vehicle exhausts is taken up by both plants and animals and poses a particular health hazard for young children (Richardson, 1982). Uranium is more limited as a pollutant but can be important where this element is mined or refined. Uranium from such operations is accumulated by lichens, mosses and other organisms (Beckett et al., 1982). Monitoring such contamination by using lichens is of value because the environmental impact of U is uncertain but of concern. This element and particularly its decay product, radon, is thought to induce lung cancer in miners (Van Hook, 1979; Band et al., 1982; Gottlieb and Husen 1982).

The academic interest in Pb and U uptake by lichens is related to the interesting chemistry of these two metals. Lead is a borderline element with considerable class B character. Thus Pb^{2+} exhibits a preference for binding to sulphur and nitrogen centres, although it will also bind to oxygen-containing ligands. In contrast U is termed a class A element, as its cationic species UO_2^{2+} binds most readily to the latter ligand type (Nieboer and Richardson, 1980). Indeed the contrasting behaviour of these two makes them leading stains in electron microscopy (see Boileau et al., 1985b). Uranium has the ability to form both neutral (UO_2L) and anionic species ($UO_2L_2^{2-}$) in solution. By dissolving uranyl nitrate in either distilled water,

227

Table 1.   Exchange of Nickel or Copper for Strontium in the Lichen
            Cladonia rangiferina

Metal available μmol 50 ml^{-1}	Sr^{2+} released during Ni^{2+} or Cu^{2+} Incubation, μmol g^{-1}	Ni^{2+} or Cu^{2+} taken up in treatment μmol g^{-1}	Sr^{2+}: metal molar ratio
Ni^{2+}			
25	14.9	13.9	1:0.9
250	29.7	28.8	1:1.0
25#	14.6	15.3	1:1.0
Cu^{2+}			
2	1.8	1.8	1:1.0
12.5	10.7	11.5	1:1.1
25	18.3	19.9	1:1.1
50	28.5	31.9	1:1.1
50#	24.7	25.5	1:1.0

* Values are the mean of three replicates.   Standard deviation of
  mean $\pm$ 10% for all experiments.
\# Experiment done with acid-treated lichen.
  (From Burton et al., 1981)

phthalate buffer or oxalate buffer it is possible to provide
a lichen predominantly or exclusively with the cationic species, the
neutral species or the anionic species (Boileau et al., 1985a).   Thus
U uptake from solution provides a model system by which to study
cation and anion uptake in lichens.

     The uptake of cations by lichens is initially via an ion exchange
process which occurs rapidly and is not inhibited by metabolic
poisons.   A slower intracellular uptake of much smaller capacity may
also take place in the case of physiologically-important ions such as
Mg^{2+} or Zn^{2+} or even toxic ions, for example Cd^{2+} (Puckett et al.,
1973; Nieboer and Richardson, 1981; Brown and Beckett, 1984).   Studies
on the ion exchange uptake system in lichens have revealed that molar
exchange ratios are conserved when one metal ion is used to replace
another on the metal binding sites (Table 1, Burton et al., 1981).

     In contrast to the uptake of cations, that of anions requires the
lichen to be alive and may be an active energy-dependent process.
Based on studies of the uptake of arsenate and earlier investigations
on phosphate it seems that two uptake systems are probably involved
for some anions (Fig. 1) but much still remains to be learned
(Richardson et al., 1984).   It is interesting that phosphate and
sulphite compete with arsenate for uptake whereas the uptake of

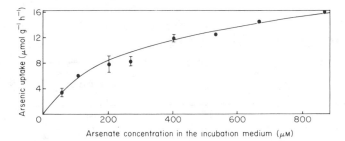

Fig. 1.   The dependence of arsenate uptake by <u>Umbilicaria muhlenbergii</u>
          on the concentration supplied.
          Note that a line joining all points would resolve uptake into
          two parts and further analysis with other mathematical
          techniques suggests two uptake systems may be involved.
          (From Richardson et al., 1984).

sulphate has an enhancing effect (Nieboer et al., 1984a).

      The present review is concerned with two aspects of ion uptake by
lichens which are not well understood. Firstly, we present results of
recent experiments on lead uptake and secondly, using the uranyl ion
as an example, show how the nature of the ionic species (whether
cation, anion or neutral) affects the amount and pattern of uptake.
We then discuss what is currently known about where and to what U and
Pb ions bind in lichens.

LEAD ACCUMULATION

Location of Metals Within the Thallus

      Lichens growing near major roads exhibit elevated levels of lead
(Laaksovirta et al., 1976; Lawrey and Hale, 1981). However it is
still not clear where metal particulates are accumulated within thalli
or to what molecules metal ions bind. Asta and Garrec (1980) found
that Ca accumulated in the thick-walled tissue of the upper and lower
cortices, while K and Mg were abundant in the algal layer. The
medulla was found to be poor in all mineral elements. Garty et al.
(1979) and Lawrey (1980) concluded that metals of anthropogenic origin
accumulated in the medulla. This also seems to be true for naturally-
occurring elements that are potentially toxic. Thus Purvis (1984)
found that lichens growing on copper mineralized rocks had their
medullary hyphae encrusted with crystals containing this metal.
Further analysis determined that these crystals were copper oxalate
and confirmed the suggestions by Richardson and Nieboer (1980) that
oxalates, although not involved in the ion exchange uptake process,
could provide metabolically-controlled sinks for metal ions,
especially where particular metal ions are present in great excess

Table 2.   Nickel Uptake into Different Parts of the Thallus of
           Umbilicaria muhlenbergii

	Algal zone	Medulla	Plates
Percentage composition by weight of each fraction	15	46	39
Ni content as $\mu g\ g^{-1}$ dry weight	1314	1041	646
The Ni content of each zone as a percentage of the total	21	52	27

(From Richardson and Nieboer, 1983)

(see also Wilson and Jones, 1984). This may also apply to free-
living fungi, as Stokes and Lindsay (1979) found blue-green material
in Penicillium sp. which they considered to be copper oxalate.

       In an experimental study of the uptake of Ni^{2+} ions, it was found
that they bound to both the algal and fungal components of Umbilicaria
muhlenbergii (Richardson and Nieboer, 1983b). The algal zone had the
highest uptake per gram dry weight but the medulla, forming a larger
proportion of the thallus, accumulated more metal overall. The lower
surface plates, that replace rhizinae in this lichen were intermediate
in uptake (Table 2). Goyal and Seaward (1981, 1982) found that, in
Peltigera, the rhizinae were able to accumulate high metal levels and
hypothesized that these structures could play a role in preventing the
accumulation of toxic amounts of metals within the thallus, as this
genus has a very high metal uptake capacity (Richardson and Nieboer,
1983b).

Experimental Studies on Lead Uptake

       Washed samples of lichens equivalent to 0.1 g dry weight (16,
5 mm discs or appropriate samples) were incubated in 10 ml of solution
containing 10 mol of Pb^{2+} as the nitrate salt with the pH adjusted to
4.5 using NaOH. Samples were then washed for 0.5 h in deionised
distilled water followed by treatment with 0.1 N HCl for 1 h, to
displace Pb from the lichen (Kiang, 1984). After filtering off the
lichen samples, using sintered glass filters, the solutions were
analysed for their lead content using differential pulse polarography
or anodic stripping voltammetry (Richardson, 1982). Preliminary
experiments using Peltigera established that approximately 6 µmoles
remained in solution following incubation, 0.5 µmol was unbound within
the lichen and was removed by the washing step and 3 µmoles could be

Table 3.   Lead Uptake by Live and Heat-killed Lichen Thalli

| Lichen species | Pb Uptake (μmol g^{-1} dry weight) | | |
	Live	Dead	Live/Dead
*Peltigera horizontalis	45.1 + 2.8	58.4 + 1.2	0.77
*Hypogymnia physodes	40.5 + 2.7	60.2 + 4.7	0.67
+Lobaria pulmonaria	33.7 + 4.4	41.8 + 5.5	0.81
+Xanthoria parietina	13.9 + 1.4	26.4 + 1.4	0.53

* 16, 5 mm discs
+ Lobes equivalent to c. 0.1 g dry wt.

eluted by 0.1 M HCl which displaced the Pb from the lichen binding
sites. The remainder was not recovered and presumably lost by
adsorption onto glass surfaces or taken up intracellularly and hence
not displaceable.

 Further experiments confirmed observations by previous workers
that dead lichens take up more metal ions than live lichens and that
the uptake capacity varies between different genera (Table 3).

 In an attempt to locate the major metal binding sites, lichens
were pretreated to remove different components. Some samples were
immersed in water at 60°C for 30 minutes to remove isolichenan, and
other water-soluble polysaccharides. Other samples were submerged in
acetone for the same period to remove lichen substances (Fig. 2a & b).
A further set were shaken for 24, 36 or 48 hours (depending on the
lichen) in a 7.5% w/v solution of "Ariel". This biological detergent
has been found to remove much of the extracellular matrix of the
cortical regions of lichen thalli (Anglesea et al., 1982). A

Table 4.   The Effect of Chemical Treatments on Pb Uptake

| Treatment | Pb Uptake (μmol g^{-1} dry weight) | | |
	Lobaria	Hypogymnia	Peltigera
control	33.7 + 4.4	40.5 + 2.6	32.0 + 2.2
heat kill	41.8 + 5.5	60.2 + 4.7	41.2 + 3.9
hot water	44.0 + 4.1	49.4 + 4.3	43.0 + 1.2
acetone	59.1 + 4.5	51.4 + 5.7	43.9 + 1.9
detergent	68.8 + 8.5	125.5 + 7.8	72.4 + 2.6

final set was heat killed by oven drying at 70°C for 12 hours. The
effect of these pretreatments on Pb uptake is shown in Table 4. Hot
water and heat killing both increased uptake, with heat killing being
the more effective probably because of the more complete destruction
of cell membrane systems. Acetone treatment enhanced uptake even
further. It effectively destroys membranes by dissolving the lipid

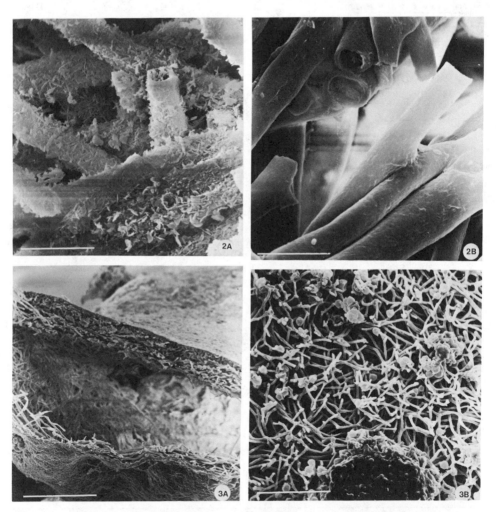

Fig. 2.   Hyphae of <u>Peltigera horizontalis</u> showing A) untreated thallus
          with extracellular crystals; B) acetone-treated thallus.

Fig. 3.   <u>Hypogymnia physodes</u> thallus A) section showing intact matrix
          on upper cortex and hollow thallus; B) upper cortex, after
          detergent treatment showing patches of residual cortical
          material and large areas free of extracellular material.

Table 5.   Lead Uptake by Cultured Symbionts and Re-synthesised Lichens
           of Cladonia cristatella

| Sample | Pb Uptake (μmol g^{-1} dry weight) | | |
	Live	Dead	Live/Dead
Phycobiont	4.6 $\pm$ 0.5	87.1 $\pm$ 6.4	0.05
Mycobiont	35.9 $\pm$ 1.7	93.1 $\pm$ 8.6	0.39
Re-synthesised Lichen	34.4 $\pm$ 0.9	45.3 $\pm$ 3.2	0.76

component and also removes the lichen substances which are relatively
water-insoluble. Surprisingly detergent treatment which removes the
extracellular matrix (Fig. 3a & b) resulted in the maximum increase in
Pb uptake. These data indicate that the main metal binding sites of
lichens are located on the cell wall rather than the extracellular
matrix. They also confirm previous observations that lichen
substances are not involved as major metal-binding components (Brown,
1976).

To examine the cell wall components in further detail, cultures
of isolated symbionts of Cladonia cristatella were used. Colonies of
this lichen have been resynthesised on small mica sheets soaked in
Bolds Basal Medium, positioned so that one end was embedded in an agar
medium (Ahmadjian and Jacobs, 1981, 1983). The isolated symbionts
were grown in culture, centrifuged and washed and then incubated in

Table 6.   Lead Uptake by Cell Wall Fractions of the Isolated Mycobiont
           from Cladonia cristatella

Fraction*	Pb Uptake (μmol g^{-1} DW fraction)	% of Total Cell Wall†	Pb Uptake (μmol g^{-1} DW cell wall)
Fraction I	188.2	7	13.2
Fraction II	1530.6	5	76.5
Fraction III	96.9	5	4.8
Fraction IV	6.3	83	5.2
Total		100	99.7

† 1 g DW cell wall $\simeq$ 1 g DW mycelium
* Fractions include I = isolichenan and protein; II = protein and
  glucosamine; III = lichenan; IV = chitin.

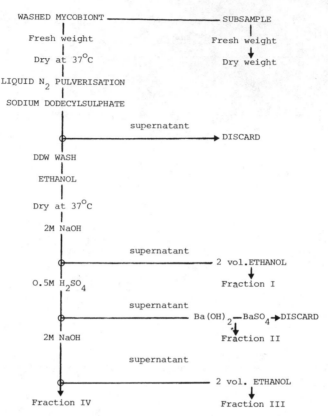

Fig. 4. Cell wall fractionation sequence for the mycobiont of
         <u>Cladonia cristatella</u>.
         Adapted from Mahadevan and Tatum (1965).

Pb solution in the same way as lichen samples. The resynthesised
cultures were treated similarly to whole thalli. Results (Table 5)
showed that the characteristics of Pb uptake by the isolated mycobiont
and by the resynthesized lichen were comparable to those of naturally-
collected lichens. The low figure for uptake by live algae contrasted
markedly with the high dead uptake and this is discussed later.

A study was carried out on cultures of the mycobiont of <u>C.
cristatella</u> to investigate the nature of the binding molecule. Hyphae
were washed and then fractionated using a technique adapted from
Mahadevan and Tatum (1965) (Fig. 4). The Pb uptake capacity of each
of the fractions was then examined. The fraction which was found to
bind the greatest amount of Pb was that which included protein (Table
6).

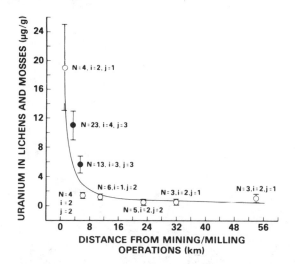

Fig. 5.   Uranium content of lichens (<u>Cladonia rangiferina</u> and <u>C.</u>
<u>mitis</u>) and mosses (<u>Pleurozium schreberi</u> and <u>Sphagnum</u> sp.)
from Elliot Lake transects.
N = total number of samples at site; i = number of lichen
species and j = number of moss species. Vertical bars =
standard error. (From Beckett et al., 1982).

URANIUM ACCUMULATION

     Field studies have shown (Fig. 5) that enhanced levels of U occur
in lichens up to 9 km from the major U mining centre of Elliot Lake in
Ontario, Canada (Beckett et al, 1982).   Studies on the Fe/Ti content
of these lichens, as well as levels of other elements, led to the
conclusion that particulate trapping was the most important
accumulation mechanism for U fallout associated with the mining and
milling of this metal (Nieboer et al., 1982).

     Laboratory investigations on the uptake of uranyl ions showed
that lichens are also capable of accumulating U from rainwater or
surface run-off.   The total uptake capacity varied depending on
whether U was present as the cationic species (Fig. 6 curve A), the
neutral species (Fig. 6 curve B) or the anionic species (Fig. 6
curve C) (Boileau et al., 1985a).   The uptake capacity for the
cationic species, about 50 μmol g^{-1} was similar to that for cations
such as Ni^{2+} or Cu^{2+} while that for the neutral and anionic species
was respectively close to 20 and 2 μmol g^{-1} (Boileau et al., 1985a).
The uptake of the neutral species was complicated by the fact that at
higher available U levels there was a significant amount of
uncomplexed cationic species present in the incubation medium.

     Toxicity studies revealed that the anionic species in oxalate

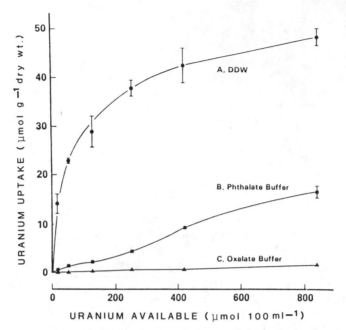

Fig. 6.   Uptake of uranium by <u>Cladonia rangiferina</u> from uranyl
          solutions in A) deionised water, B) phthalate buffer and C)
          oxalate buffer.
          Standard deviation from 3 - 5 replicates.
          (From Boileau et al., 1985a).

buffer inhibited photosynthetic ^{14}C-fixation to a greater extent than
the uncomplexed cation UO^{2+}, even though the latter was taken up in
much greater amounts.   The uptake of U also affected the pattern of
fixation in that less $NaH^{14}CO_3$ was incorporated into the ethanol-
soluble fraction, which includes the sugar alcohol ribitol that is
transported to the fungus (Richardson, 1985).   More radio-activity was
found in insoluble components which include proteins and cell wall
materials (Boileau et al., 1985b).   It is interesting that none of the
ionic species of U induced the release of large amounts of K^+ ions.
Such a release is a feature of exposure to moderate or high
concentrations of toxic cations or anions, e.g. Pb^{2+} or HSO^-_3.   If
loss of K^+ occurs, it is interpreted as due to harmful effects on the
lichen cell membranes (Richardson and Nieboer, 1983a).

     It is clear from the above that the effects of U uptake on
photosynthesis and the pattern of fixation do not correlate with the
absolute amounts of U taken up by the lichen.   The cationic species is
thought to bind to cell membranes as well as the cell walls.   The
observed toxic effects are probably due to its ability to reduce

membrane permeability to the bicarbonate ion and also interfere with
sugar transport between the symbionts (Boileau et al., 1985b). The
small amount of the dianionic U species, absorbed from the oxalate
buffer, had the greatest effects on total photosynthetic fixation.
The reason for this is that anions are mostly taken up into the cell,
rather than binding extracellularly to cell walls and membranes. Once
in an intracellular position, the anionic species of the uranyl ion
would be strategically placed to induce injury.

DISCUSSION

Under natural conditions lichens accumulate metals both by
particulate trapping and via ion exchange. The latter can be shown by
displacing the bound cations with hydrogen ions or other metal ions
(see Nieboer and Richardson, 1981; Brown and Beckett, 1984). Scanning
electron microscopy coupled with an electron probe can help to
determine where metals accumulate within a lichen. Jones et al.
(1982) found that Pb in the thalli of Stereocaulon vesuvianum was
located both on hyphae (where it was presumably bound by ion exchange)
and as discrete particles in the intracellular spaces. Studies on Cu
uptake by the free-living fungus Penicillium indicated that, after 72
hours in a Cu-enriched medium, the metal was located mainly on the
cell wall and cell membrane but there were a few Cu- rich granules in
the cytoplasm (Fukami et al., 1983).

Metal Uptake Capacity

Dead lichens take up more Pb^{2+} than living ones (Table 3), and
this is also the case for U and other elements. Thus Boileau et al.
(1985a) showed that dead Cladonia rangiferina took up 30% more UO_2^{2+}.
This phenomenon is not restricted to lichens as boiled samples of the
free-living fungus Penicillium digitatum also exhibited greatly
enhanced U uptake (Galun et al., 1982). It seems that killing the
cells of these organisms destroys the cell membranes so exposing
intracellular metal-binding sites which are not normally available.
While this no doubt accounts for the major part of such increases in
binding capacity, scanning electron microscopy revealed that dry heat
killing also causes shrinkage of thalli and cracks in the
extracellular matrix (Kiang, 1984). This may make some cell wall
binding sites more accessible.

Ultrastructural studies on resynthesized lichens have shown that
they differ from thalli growing naturally in having a reduced amount
of extracellular matrix and by the presence of crystalline deposits on
the hyphae. Evidently these differences play a minor role as regards
metal uptake capacity since living, naturally-collected or
resynthesised lichens and cultured mycobionts all have an uptake
capacity for Pb of about 40 μmoles g^{-1} (see Tables 3 & 5). The
studies of Boileau et al. (1985a) showed an uptake capacity for U in
C. rangiferina of close to 50 μmoles g^{-1}. These values are much less

than recorded for free-living fungi. Thus <u>Aureobasium pullulans</u> took
up about 100 moles g^{-1} of Cu^{2+} in the case of resistant strains while
sensitive strains took up twice as much (Gadd and Griffiths, 1980).
Even higher values have been recorded for U uptake in <u>Rhizopus
arrhizus</u> and <u>Penicillium chrysogenum</u>, which accumulated in excess of
600 μmoles g^{-1} (Tsezos and Volesky, 1981). The reason why lichen
fungi, both in symbiosis and in culture, have a lower uptake capacity
for metals is obscure. It may be related to the fact that isolated
lichen fungi grow very slowly, even in liquid media, or to a
difference in the composition of the cell wall since Ahmadjian (1982)
noted that lichen mycobionts develop gelatinous coverings on their
hyphae. Thus the cell walls of lichen fungi may either have fewer
binding sites or these may be somehow masked.

With regard to lichen algae, it is known that these develop
sheaths or a gelatinous outer layer when growing in culture (Drew and
Smith, 1967; Richardson and Smith, 1968). Their presence may reduce
the binding capacity of living cells by protecting the sites since, as
shown in Table 5, the uptake capacity more than doubles on heat
killing. It is interesting to note that Horikoshi et al. (1979)
observed that the uptake capacity of free-living <u>Chlorella regularis</u>
for UO$_2^{2+}$, was increased from 66 to 280 μmoles g^{-1} by scalding.
Calculating the live to dead uptake ratio, their value of 0.07 is
similar to that for <u>Trebouxia</u> (Table 5). Other workers have also
found that dead unicellular algae adsorb more metal ions (Khummongkol
et al., 1982). In a later study, Horikoshi et al., (1981) found that
the total uptake capacity was reduced if the cells were grown
heterotrophically. Lichen algae, particularly <u>Trebouxia</u>, are usually
grown on glucose-supplemented media (as in the present study) so that
the <u>Trebouxia</u> cells within lichen thalli may take up more metal ions
than indicated by the data in Table 5.

Nature of the Metal Binding Group

The nature of the cation binding groups in lichens has been
speculated upon by many authors. The possibility that uptake of metal
ions could be by sorption onto the cellulose network of the cell walls
via Van Der Waals forces has been considered. However, both studies
on lichens and recent experiments using unicellular algae discount
such a mechanism (Puckett et al., 1973; Geisweid and Urbach, 1983).

It is most likely that cation exchange involving carboxylic acid-
or hydrocarboxylic acid-binding sites is involved (Nieboer et al.,
1978). There are three types of evidence for this. Firstly, the
selectivity sequence or order of binding observed in lichens
(Cu,Pb>Zn>Ni>>Mg>Sr) is similar to the sequence for a 1:1 complex with
acetic acid (Richardson et al., 1980). Secondly, titration curve
studies show that uptake of metal ions by lichens is a function of pH
and that the determined pK$_a$ values, e.g. 4.8 and 6.2 in the case of <u>C.
rangiferina</u>, are indicative of carboxylic acid sites (Tuominen, 1967;

Burton et al., 1981; Nieboer et al., 1984b). Thirdly, pH measurements
on thalli covered by a thin water film are also consistent with this
interpretation. Türk et al. (1974) observed thallus pH values of
4.2+ 0.6 for members of the Parmeliaceae and higher values of 5.9+ 0.5
for Xanthoria parietina, a species of eutrophicated habitats.

   Richardson and Nieboer (1980) suggested that the metal-binding
sites on lichens might be located within the extracellular matrix composed of polymers such as lichenan. They calculated, based on the
analysis of native lichenan by Schmidt et al. (1934) which indicated
that there was one carboxylic acid unit per 47 glucose units, that the
extracellular matrix could provide sufficient binding capacity as it
made up about 30% of the thallus volume in X. parietina (Collins and
Farrar, 1978). Such binding seemed feasible since, in brown algae,
the primary binding site for metal ions such as Zn^{2+} is the extracellular carbohydrate although, when the tide recedes, the metals can
be transferred to internal pools within the algal cells (Skipnes et
al., 1976). The fact that the removal of much or all of the
extracellular matrix of lichens using the technique of Anglesea et al.
(1982) did not reduce the metal-binding capacity indicates that the
sites are not primarily located on this component. Incidentally,
prolonged detergent treatment also removes the crystals of lichen
compounds from the medullary hyphae though this is more easily
achieved by immersion in acetone.

   Other candidates for the metal-binding molecules in lichens are
the uronic acids, components of pectin and pectin-like compounds.
Since Tuominen's early assertion in 1967 that uronic acids could be
important, little progress has been made to confirm or refute this
idea (Tuominen, 1967; Tuominen and Jaakkola, 1973). Certainly in
higher plants the negative charges of the carboxylate groups of
pectins are stated to account for the tremendous cation-binding
capacity of higher plant cell walls (Nobel, 1974). Furthermore uronic
acids have been identified in mosses, algae and bacteria (Craigie and
Maass. 1966; Percival, 1966; Beveridge et al., 1982). Unfortunately,
a very recent study on the lichen algae Trebouxia and Pseudotrebouxia
reported negative results from a test for uronic acids and amino
sugars (König and Peveling, 1984). They found that the cell wall was
composed of cellulose and probably a β-1.4-mannan, together with a
small amount of protein and sporopollenin. The latter may be a common
feature of all green lichen symbionts and is an oxidative polymer of
carotenoids and carotenoid esters to which hydroxyl-, C-methyl-ester
and ether groups are bound (but see Honegger, 1984). As regards the
fungal component of lichens, there is little evidence that their cell
walls contain uronic acids (Brown, 1976). Indeed the only fungus in
which uronic acids have been demonstrated is Mucor rouxii which is a
member of the Zygomycotina rather than the Ascomycotina to which
lichen fungi belong (Rogers et al., 1980). In a recent study of the
cell walls of free-living fungi, chitin, chitosan, protein,
non-nitrogenous carbohydrate, lipid and phosphorus were identified

(Cuthbert, 1984).

     In the preliminary study reported here on cultured isolated
lichen fungi, the fraction exhibiting greatest binding capacity was
that which included protein as a component. Furthermore the fractions
containing lichenan, isolichenan and chitin seemed to be relatively
unimportant as regards metal binding. Perhaps this is not surprising
as the latest analyses of lichenan and other polysaccharides in
lichens have not confirmed the presence of carboxylic acid residues in
these polymers (Yokota et al., 1979; Gorin and Iacomini, 1984). The
integral part played by proteins in fungal cell walls is well known
(Bartnicki-Garcia, 1968; Hunsley and Burnett, 1970). Proteins can be
sources of anionic functional groups capable of binding metal ions
and, while this has been discussed, only limited experimental work has
been performed. In a recent study of U uptake by the yeast,
Saccharomyces cerevisiae, Strandberg et al. (1981) concluded that the
carboxylic acid residues of amino acids which make up wall proteins
are implicated with U binding. However, they proposed that
phosphomannans, which are another component of the walls of these
fungi, may be even more important. The phosphate groups of these
mannans were considered to be the sites of U complexation since the
amount of binding was related to the phosphate content of the polymer.

     Galun et al. (1982) proposed that chitin, which is a polymer of
N-acetyl-D-glucosamine, was an important metal-binding molecule in
fungi. This was based on experiments in which micro-columns of P.
digitatum hyphae or chitin were perfused with UO_2^{2+}. The hyphae took
up some 40 µmoles g^{-1} of U while the chitin columns absorbed some 99%
of the available metal. However, on a dry weight basis the chitin
only took up about 1.5 µmoles g^{-1} which is in reasonable agreement
with the binding capacity of the chitin-containing fraction reported
in Table 5. Thus the role of chitin as a major metal-binding
component in fungal and particularly lichen cell walls remains to be
confirmed.

CONCLUDING REMARKS

     It is evident from the foregoing paragraphs that while cation
binding by lichens most likely involves carboxylic acid-containing
ligands, the exact nature of the binding molecule or molecules has yet
to be determined. The preliminary study of mycobiont cell walls,
using fractionation coupled with metal uptake studies, suggested that
cell-wall proteins might be important in metal binding. It would now
be interesting to extend such studies to whole or dissected lichens
and to apply other procedures, e.g. of König and Peveling (1984), to
help find out more about the nature of the main metal binding
component.

     The methods now available for obtaining algae direct from lichen
thalli (Ascaso, 1980) should help to determine whether such algae

indeed have a higher uptake capacity when in symbiosis. Detailed
studies of the cell walls of these directly-isolated cells coupled
with experiments on metal uptake would be of great interest. Studies
on fractionated whole thalli, together with investigations on lichen
algae such as Trebouxia, Nostoc and Coccomyxa, might help to explain
why the total metal uptake capacity of different genera of lichens
varies so much. In this context, further studies on the extracellular
matrix of lichens may contribute to an understanding of the variation
in metal levels found in different species collected from the same
general area. The extracellular matrix and lichen acids appear not to
take part in metal-ion binding, but may be important as regards the
total metal content of naturally-collected thalli. Where thalli are
filled, e.g. with metal oxalates (formed by some lichens in response
to the presence of excess metal ions), there seems to be little
capacity for trapping particulates. Thus, lichens growing on
base-rich substrates have high calcium oxalate contents and exhibit
low particulate-trapping ability (Looney et al., this volume).

In conclusion, there is still much to be learned about the
extracellular uptake of metal ions by lichens. The variety of
biochemical techniques that can be applied as well as the availability
of electron microscopes with electron probes should ensure a
fascinating study for anyone who decides to take up the challenge.

ACKNOWLEDGEMENTS

The authors acknowledge the help of the Trinity College Electron
Microscope Unit for providing facilities, training and photographic
assistance connected with the scanning electron microscope studies.

REFERENCES

Ahmadjian, V., 1982, Algal and fungal symbioses, in: "Progress in
    Phycological Research," Volume 1, F.E. Round and V.J. Chapman
    eds, pp. 179-233, Elsevier, Amsterdam.
Ahmadjian, V., and Jacobs, J.B., 1981, Relationship between fungus
    and alga in the lichen Cladonia cristatella Tuck., Nature
    (London), 289: 169-172.
Ahmadjian, V., and Jacobs, J.B., 1983, Algal-fungal relationships in
    lichens: recognition, synthesis and development, in: "Algal
    Symbiosis," L.J. Goff ed., pp. 147-172, Cambridge University
    Press, Cambridge.
Anglesea, D., Veltkamp, C., and Greenhalgh, G.N., 1982, The upper
    cortex of Parmelia saxatilis and other lichen thalli, The
    Lichenologist 14: 29-38.
Ascaso, C., 1980, A rapid method for the quantitative isolation of
    green algae from lichens, Annals of Botany, 45: 483.
Asta, J., and Garrec, J.P., 1980, Etude de la repartition du
    calcium, potasium, magnesium et phosphore dans les differentes
    couches anatomiques de dix lichens par analyse directe a la

microsonde electronique, Cryptogamie, Bryologie et Lichenologie,
    1: 3-20.
Band, P., Fledstein, M., Watson, I., King, G., and Saccomanno, G.,
    1982, Lung cancer screening programs in Canadian uranium miners,
    Recent Results in Cancer, 82: 153-158.
Bartnicki-Garcia, S., 1962, Cell wall chemistry, morphogenesis
    and taxonomy of fungi, Annual Review of Microbiology, 22:
    87-108.
Beckett, P.J., Boileau, L.J.R., Padovan, D., and Richardson, D.H.S.,
    1982, Lichens and mosses as monitors of industrial activity
    associated with uranium mining in Northern Ontario, Canada - Part
    2 : Distance dependent uranium and lead accumulation patterns,
    Environmental Pollution (Series B), 4: 91-107.
Beveridge, T.J., Forsberg, C.W., and Doyle, R.J., 1982, Major sites
    of metal binding in Bacillus licheniformis walls, Journal of
    Bacteriology, 150: 1438-1448.
Boileau, L.J.R., Nieboer, E., and Richardson, D.H.S., 1985a, Uranium
    accumulation in the lichen Cladonia rangiferina (L.) Wigg. Part
    I: Uptake of cationic, neutral and anionic forms of the uranyl
    ion, Canadian Journal of Botany, 63: 384-389.
Boileau, L.J.R., Nieboer, E., and Richardson, D.H.S., 1985b, Uranium
    accumulation in the lichen Cladonia rangiferina (L.) Wigg. Part
    II. Toxic effects of cationic, neutral and anionic forms of the
    uranyl ion, Canadian Journal of Botany, 63: 390-397.
Brown, D.H., 1976, Mineral uptake by lichens, in: "Lichenology:
    Progress and Problems," D.H. Brown, D.L. Hawksworth, and R.H.
    Bailey, eds, pp. 419-440, Academic Press, London.
Brown, D.H., and Beckett, R.P., 1984, Uptake and effect of cations on
    lichen metabolism, The Lichenologist, 16: 173-188.
Burton, M.A.S., LeSueur, P., and Puckett, K.J., 1981, Copper, nickel
    and thallium uptake by the lichen Cladonia rangiferina, Canadian
    Journal of Botany, 59: 91-100.
Collins, C.R., and Farrar, J.F., 1978, Structural resistances to mass
    transfer in the lichen Xanthoria parietina, New Phytologist, 81:
    71-83.
Craigie, J.S., and Maass, W.S.G., 1966, The cation exchanger in
    Sphagnum species, Annals of Botany, 30: 153-154.
Cuthbert, A., 1984, Cell wall composition and resistance in haustorial
    mycoparasites, Bulletin of the British Mycological Society, 17
    (Suppl. 3): 1.
Drew, E.A., and Smith, D.C., 1967, Studies in the physiology of
    lichens. VII. The physiology of the Nostoc component of
    Peltigera polydactyla compared with cultured and free living
    forms, New Phytologist, 66: 379-388.
Fukami, M., Yamazaki, S., and Toda, S., 1983, Distribution of copper
    in the cells of heavy metal tolerant fungus Penicillium
    ochro-chloron, cultured in concentrated copper medium,
    Agricultural Biological Chemistry (Japan), 47: 1367-1369.
Gadd, G.M., and Griffiths, A.J., 1980, Influence of pH on toxicity
    and uptake of copper in Aureobasidium pullulans, Transactions of

the British Mycological Society, 75: 91-96.

Galun, M., Keller, P., Malki, D., Feldstein, H., Galun, E., Siegel, S.M., and Siegel, B.Z., 1982, Removal of uranium (VI) from solution by fungal biomass and fungal wall-related biopolymers, Science, 219: 285-286.

Garty, I., Galun, M., and Kessel, M., 1979, Localisation of heavy metals and other elements accumulated in the lichen thallus, New Phytologist, 82: 159-168.

Geisweid, H.J., and Urbach, W., 1983, Sorption of cadmium by the green microalgae Chlorella vulgaris, Ankistrodesmus braunii and Eremosphaera viridis, Zeitschrift fur Pflanzenphysiologie, 109: 127-141.

Gorin, P.A.J., and Iacomini, M., 1984, Polysaccharides of the lichens Cetraria islandica and Ramalina usnea, Carbohydrate Research, 128: 119-132.

Gottlieb, L.S., and Husen, L.A., 1982, Lung cancer among Navajo uranium miners, Chest, 81: 449-552.

Goyal, R., and Seaward, M.R.D., 1981, Metal uptake in terricolous lichens. I. Metal localisation within the thallus, New Phytologist, 89: 631-645.

Goyal, R., and Seaward, M.R.D., 1982, Metal uptake in terricolous lichens. III. Translocation in the thallus of Peltigera canina, New Phytologist, 90: 85-98.

Honegger, R., 1984, Cytological aspects of the mycobiont - phycobiont relationship in lichens, The Lichenologist, 16: 111-127.

Horikoshi, T., Nakajima, A., and Sakaguchi, T., 1979, Uptake of uranium by Chlorella regularis, Agricultrual and Biological Chemistry (Japan), 43: 617-623.

Horikoshi, T., Nakajima, A., and Sakaguchi, T., 1981, Accumulation of uranium by Chlorella cells grown under auxotrophic heterotrophic and mixotrophic culture conditions, Agricultural and Biological Chemistry (Japan), 45: 781-783.

Hunsley, D., and Burnett, J.H., 1970, The ultrastructural architecture of the walls of some hyphal fungi, Journal of General Microbiology, 62: 203-218.

Jones, D., Wilson, M.J., and Laundon, J.R., 1982, Observations on the location and form of lead in Stereocaulon vesuvianum, The Lichenologist, 14: 281-286.

Khummongkol, D., Canterford, G.S., and Fryer, C., 1982, Accumulation of heavy metals in unicellular algae, Biotechnology and Bioengineering, 24: 2643-2660.

Kiang, S., 1984, "A Study of the Uptake of Lead by Lichens," Honours B.A. Moderatorship Thesis, Trinity College, Dublin.

Konig, J., and Peveling, E., 1984, Cell walls of the phycobionts Trebouxia and Pseudotrebouxia: Constituents and their localization, The Lichenologist, 16: 129-144.

Laaksovirta, K., Olkkonen, H., and Alakuijala, P., 1976, Observations on the lead content of lichen and bark adjacent to a highway in southern Finland, Environmental Pollution, 11: 247-255.

Lawrey, J.D., 1980, Calcium accumulation by lichens and transfer to lichen herbivores, Mycologia, 72: 586-594.

Lawrey, J.D., and Hale, M.E., 1981, Retrospective study of lichen lead accumulation in the northeastern United States, The Bryologist, 84: 449-456.

Mahadevan, P.R., and Tatum, E.L., 1965, Relationship of the major constituents of the Neurospora crassa cell wall to wild-type and colonial morphology, Journal of Bacteriology, 90: 1073-1081.

Nieboer, E., and Richardson, D.H.S., 1980, The replacement of the nondescript term "heavy metals" by a biologically and chemically significant classification of metal ions, Environmental Pollution (Series B), 1: 3-26.

Nieboer, E., and Richardson, D.H.S., 1981, Lichens as monitors of atmospheric deposition, in: "Atmospheric Pollutants in Natural Waters," S.J. Eisenreich, ed., pp. 339-388, Ann Arbor Science, Ann Arbor.

Nieboer, E., Richardson, D.H.S., and Tomassini, F.D., 1978, Mineral uptake and release by lichens: An overview, The Bryologist, 81: 226-246.

Nieboer, E., Richardson, D.H.S., Boileau, L.J.R., Beckett, P.J., Lavoie, P., and Padovan, D., 1982, Lichens and mosses as monitors of industrial activity associated with uranium mining in Northeastern Ontario, Canada - Part 3: Accumulations of iron and titanium and their mutual dependence, Environmental Pollution (Series B), 4: 181-192.

Nieboer, E., Padovan, D., Lavoie, P., and Richardson, D.H.S., 1984a, Anion accumulation by lichens. II. Competition and toxicity studies involving arsenate, phosphate, sulphate and sulphite, New Phytologist, 96: 83-93.

Nieboer, E., MacFarlane, J.D., and Richardson, D.H.S., 1984b, Modification of plant cell buffering capacities by gaseous air pollutants, in: "Gaseous Air Pollutants and Plant Metabolism," M.J. Koziol, and F.R. Whatley, eds, pp. 313-330, Butterworths, London.

Nobel, P.S., 1974, "Introduction to Biophysical Plant Physiology," W.H. Freeman, San Francisco.

Percival, E., 1966, The natural distribution of plant polysaccharides, in: "Comparative Phytochemistry," T. Swain, ed., pp. 139-158, Academic Press, London.

Puckett, K.J., Nieboer, E., Gorzynski, M.J., and Richardson, D.H.S., 1973, The uptake of metal ions by lichens: A modified ion-exchange process, New Phytologist, 72: 329-342.

Purvis, O.W., 1984, The occurrence of copper oxalate in lichens growing on copper sulphide-bearing rocks in Scandinavia, The Lichenologist, 16: 197-204.

Richardson, D.H.S., 1985, Carbohydrate Transfer, in: "Surface Physiology of Lichens," C. Vicente, D.H. Brown, and M.E. Legaz, eds, pp. 25-55 Universidad Complutense de Madrid, Madrid.

Richardson, D.H.S., and Smith, D.C., 1968, Lichen physiology. X. The isolated algal and fungal symbionts of Xanthoria aureola, New Phytologist, 67: 69-77.

Richardson, D.H.S., and Nieboer, E., 1980, Surface binding and
    accumulation of metals in lichens, in: "Cellular Interactions in
    Symbiosis and Parasites," C.B. Cook, P.W. Pappas, and E.D.
    Rudolph, eds, pp. 75-94, Ohio State University Press, Columbia.

Richardson, D.H.S., and Nieboer, E., 1983a, Ecophysiological responses
    of lichens to sulphur dioxide, Journal of the Hattori Botanical
    Laboratory, 54: 331-351.

Richardson, D.H.S., and Nieboer, E., 1983b, The uptake of nickel ions
    by lichen thalli of the genera Umbilicaria and Peltigera, The
    Lichenologist, 15: 81-88.

Richardson, D.H.S., Beckett, P.J., and Nieboer, E., 1980, Nickel in
    lichens, bryophytes, fungi and algae, in: "Nickel in the
    Environment," J.D. Nriagu, ed., pp. 367-406, John Wiley, New
    York.

Richardson, D.H.S., Nieboer, E., Lavoie, P., and Padovan, D., 1984,
    Anion accumulation by lichens. I. The characteristics and
    kinetics of arsenate uptake by Umbilicaria muhlenbergii, New
    Phytologist, 96: 71-82.

Richardson, R.M., 1982, Blood-lead concentrations in three to eight
    year old school-children from Dublin City and rural County
    Wicklow, Irish Journal of Medical Science, 151: 203-210.

Rogers, H.J., Perkins, H.R., and Ward, J.B., 1980, "Microbial Cell
    Walls and Membranes," Chapman and Hall, London.

Schmidt, E., Schnegg, R., and Wurzner, E., 1934, Die Kettenlange des
    Lichenins Hativer Zusammensettzung, Naturwissenschaften, 22: 172.

Skipnes, O., Roalad, T., and Haug, A., 1976, Uptake of zinc and
    strontium by brown algae, Physiologia Plantarum, 34: 314-320.

Strandberg, G.W., Shumate, S.E., and Parrott, J.R., 1981, Microbial
    cells as biosórbants for heavy metals: Accumulation of uranium by
    Saccharomyces cerevisiae and Pseudomonas aeruginosa, Applied and
    Environmental Microbiology, 41: 237-245.

Stokes, P.M., and Lindsay, J.E., 1979, Copper tolerance and
    accumulation in Penicillium ochro-chloron isolated from
    copper-plating solution. Mycologia, 71: 796-806.

Tsezos, M., and Volesky, B., 1981, Bisorption of uranium and thorium,
    Biotechnology and Bioengineering, 23: 583-604.

Tuominen, Y., 1967, Studies on the strontium uptake of the Cladonia
    alpestris thallus, Annales Botanici Fennici, 4: 1-28.

Tuominen, Y., and Jaakkola, T., 1973, The absorption and accumulation
    of mineral elements and radioactive nuclides, in: "The Lichens,"
    V. Ahmadjian, and M.E. Hale, eds, pp. 185-224, Academic Press,
    New York.

Türk, R., Wirth, V., and Lange, O.L., 1974, Carbon dioxide exchange
    measurements for the determination of sulphur dioxide resistance
    in lichens. Canadian Translation Bureau, Ottawa, Translation no.
    366007; Oecologia (Berlin), 15: 33-64.

Van Hook, R.I., 1979, Potential health and environmental effects of
    trace elements and radionuclides from increased coal utilization,
    Environmental Health Perspectives, 33: 227-247.

Wilson, M.J., and Jones, D., 1984, The occurrence and significance of

manganese oxalate in <u>Pertusaria corallina</u> (Lichenes),
   <u>Pedobiologica</u>, 26: 373–379.
Yokota, I.,. Shibata, S., and Saito, H., 1979, A ^{13}C–N.M.R. analysis
   of linkages in lichen polysaccharides: an approach to chemical
   taxonomy of lichens, <u>Carbohydrate Research</u>, 69: 252–258.

THE ROLE OF THE CELL WALL IN THE INTRACELLULAR UPTAKE OF CATIONS BY

LICHENS

D.H. Brown and R.P. Beckett

Department of Botany
The University
Bristol BS8 1UG, U.K.

INTRODUCTION

When lichens or bryophytes are placed in solutions containing cations, an equilibrium is rapidly achieved between free cations in solution and cations bound to the exchange sites on the cell wells. For a given cation the extent of this process depends on the nature of the exchange sites, the affinity of the ion for these sites and the nature and number of pre-existing cations (Nieboer et al., 1978; Nieboer and Richardson, 1981; Brown and Beckett, 1984, 1985). With low concentrations of "heavy" metals in limited volumes of solution lichens and bryophytes can significantly reduce the cation concentration in solution. As heavy metal cations are effectively bound outside the plasma membrane, which surrounds the metabolically active cytoplasm, such binding might represent an apparent means of detoxifying the solution.

Early workers studying heavy metal resistance in higher plants postulated that modifications to cation binding sites in the roots might account for differences in sensitivity to specific heavy metals and the retention of such elements in the roots of tolerant plants. Some experimental evidence was provided in support of this suggestion (Peterson, 1969; Turner and Marshall, 1971; Hogan and Rauser, 1981) but later workers were unable to repeat these observations (Mathys, 1973). It has also been suggested that the cell walls of bryophytes and lichens could act as buffers against heavy metal damage (Brown and Bates, 1972; Brown and Slingsby, 1972). This was initially put forward as an explanation of how these plants might survive the very high concentrations of elements which have frequently been detected in plants from heavy metal contaminated habitats (Laaksovirta and Olkkonen, 1977; Seaward et al., 1978). However, there is now evidence

to show that metal-rich particles can become trapped by lichens and
bryophytes and may represent a substantial proportion of the total
metal content (Garty et al., 1979; Nieboer and Richardson, 1981; Jones
et al., 1982; Nieboer et al., 1982). As many heavy metals may be only
slowly dissolved from these particles, it is very probable that the
elevated metal contents previously noted are tolerated because the
potentially toxic elements are physically associated with the plant
rather than freely available to influence cell metabolism. Evidence
now suggests that tolerance mechanisms in bryophytes (Brown and House,
1978) and lichens (Beckett and Brown, 1984b, Brown and Beckett, 1984,
1985) may partly involve differences in intracellular cation uptake
capacity.

     Recent work has shown that heavy metals, such as Zn (Brown and
Beckett, 1983) and Cd (Beckett and Brown, 1984a, b) can be taken into
the cell where they may or may not influence cell metabolism. Thus it
has been shown that lichens with cyanobacterial photobionts (e.g.
Peltigera) are more photosynthetically sensitive to added Cd, Cu and
Zn than are lichens with chlorophycean photobionts (e.g. Cladonia)
(Brown and Beckett, 1983). However, when cyanobacterial lichen populations have been exposed to Zn or Cd in nature their sensitivity to
addition of either element in laboratory tests is substantially
reduced (Beckett and Brown, 1983a) and, for at least one population
from a heavy metal-contaminated mine site, resistance is correlated
with a reduced rate of Cd uptake into the cell without any alteration
to extracellular cation binding capacity (Beckett and Brown, 1984b).

     As cation binding to the cell wall is due to a passive physio-
chemical process, it is possible for such cations to be released back
into solution when the external cation supply is removed. Hence it
might be possible for the cell wall exchange sites to act as a
reservoir of cations for intracellular uptake rather than as a buffer
against such uptake. Such a transfer from the cell wall into the cell
has been implied in a number of other plant groups (e.g. seaweeds
(Skipnes et al., 1976), higher plant roots (Page and Dainty, 1963) and
bryophytes (Heywood, 1982; Beckett and Brown, 1983b). This paper
reports the first attempt to specifically demonstrate such a movement.
We used the element Cd because, although potentially toxic, its direct
intracellular uptake has already been characterised (Beckett and
Brown, 1984a,b).

MATERIALS AND METHODS

     Lichen samples were freshly collected from near Bristol.
Peltigera horizontalis (Goblin Combe) and Cladonia portentosa (Priddy)
were obtained from uncontaminated sites while P. membranacea was collected from a disused Zn mine (Shipham). In all experiments five replicate samples, each containing either six 7 mm discs from Peltigera
or c. 100 mg fresh weight of 1 cm apices of Cladonia, were shaken for
10 min in 50 ml CdSO$_4$ solution followed by washing in 10 ml deionised

water for 20 min. Further storage was in 10 ml deionised water under continuous fluorescent illumination (135 μE m^{-2} s^{-1}) at 23°C, unless otherwise stated.

Samples were fractionated to determine the cellular location of Cd and K by the sequential elution method of Beckett and Brown (1984a), involving the displacement of wall-bound exchangeable cations in 0.02M NiCl$_2$ and detection of intracellular cations by a subsequent complete digestion in concentrated HNO$_3$. Cations were quantified by atomic absorption spectrophotometry in an air/acetylene flame. Photosynthetic and respiratory measurements were made by infrared gas analysis using the method of Snelgar et al. (1980). Data is presented ± standard deviation unless otherwise stated.

RESULTS AND DISCUSSION

Transfer of Cadmium from Cell Wall to Cell Interior

Preliminary experiments showed that uptake of Cd into the cell of C. portentosa was detectable following a 10 min incubation period (35 ± 7 μg g^{-1}) but substantially greater uptake occurred to the extracellular exchangeable sites (6680 ± 470 μg g^{-1}). After removal of the 300 μm Cd solution and incubation in deionised water, the intracellular Cd concentration rose progressively to 174 ± 38 μg g^{-1} after 96 h (Fig. 1). With P. horizontalis, over a similar time period, a more rapid transfer of Cd into the cell was demonstrated (Fig. 1), the intracellular Cd concentration rising from 31 ± 2 μg g^{-1} to 434 ± 94 μg g^{-1} after 123.5 h. It should be noted that although the intracellular Cd concentrations were similar after treatment for 10 min in the two species, Cd uptake by Peltigera was from a 50 μM Cd solution.

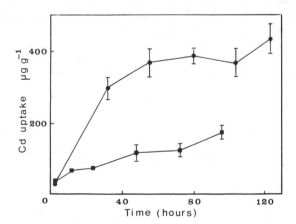

Fig. 1. Transfer of a pulse of Cd from the cell wall to the cell interior in Cladonia portentosa (■) and Peltigera horizontalis (●) from uncontaminated sites. Bar = standard error.

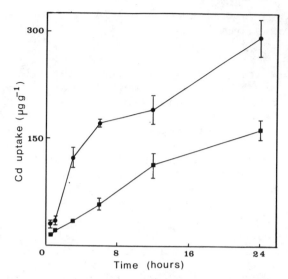

Fig. 2. Transfer of a pulse of Cd from the cell wall to the cell
       interior in <u>Peltigera horizontalis</u> (uncontaminated site = ●)
       and <u>P. membranacea</u> (contaminated site = ■). Bar = standard
       error.

Calculations have shown that the quantity of additional Cd taken
into the cell exceeds that which might be retained in the Cd solution
trapped between cells on the removal of lichens from the Cd treatment
solution. Additional uptake is therefore only possible from material
previously bound to the lichen's extracellular exchange sites.

With both species a loss of Cd from the extracellular exchangeable sites was detectable (<u>Cladonia</u> from 6680 + 470 to 5870 + 580 µg
g^{-1} in 96 h and <u>Peltigera</u> 2120 + 100 to 1020 + 160 µg g^{-1} in 123.5 h).
The decline in Cd recovered from the cell wall was always greater than
the increase in intracellular Cd, i.e. in both cases net loss of Cd
from the lichen occurred. In the <u>Peltigera</u> experiment this was into
moist filter paper, on which the samples were incubated at 16°C under
35 µE m^{-2} s^{-1} illumination, and for the <u>Cladonia</u> experiment into deionised water (see methods section).

The rate of Cd transfer into the cells of <u>Peltigera</u> was initially
much faster than with <u>Cladonia</u>, despite the greater uptake of Cd to
the cell walls in the latter species. Uptake is therefore more
closely related to differences in the initial intracellular Cd uptake
rate than the size of the extracellularly-bound Cd supply.

Later experiments with <u>Peltigera</u> spp. involved shorter incubation
periods following the initial 100 µM Cd pulse. The rate of intracellular Cd increase in <u>P. horizontalis</u> from the uncontaminated site
was approximately double that of <u>P. membranacea</u> from the disused mine

site (Fig. 2). Beckett and Brown (1984b) have previously shown that when a solution of Cd was supplied throughout the incubation period, Peltigera samples from the disused mine site used in the present study had lower intracellular Cd uptake rates than material from uncontaminated sites. This difference can be seen here after the 10 min treatment period (Fig. 2). However, it is not certain whether the subsequent rate of intracellular Cd uptake, derived from the cell wall Cd pool, is related to different intracellular uptake rates or to species differences in cell wall Cd-binding ability, i.e. P. horizontalis (uncontaminated site) = 3410 ± 660 µg g^{-1} and P. membranacea (mine site) = 1940 ± 230 µg g^{-1}, as noted by Beckett and Brown (1984a). Calculations based on the additional intracellular Cd uptake from 0.5 to 3 h (including an allowance for different extracellular Cd concentrations) indicate that the intracellular uptake rate was 2.7 times higher in the uncontaminated Peltigera sample when using Cd from the cell wall and 2.3 times higher when taking Cd directly from solution.

It is not possible to conclude from the above experiments whether the uptake of a pulse of extracellularly-bound Cd into the cell interior is the result of the passive establishment of a new equilibrium across the plasma membrane or the active displacement of Cd from the cell wall exchange sites by released intracellular cations.

Cellular Damage

References to the effects of heavy metals on cell physiology have often emphasised inhibition of photosynthesis and damage to membranes (Puckett, 1976; Rao et al., 1977; Puckett and Burton, 1981). Many of these studies used unnaturally high concentrations of toxic elements in short-term studies, thereby tending to affect both processes similarly. Evidence now indicates that photosynthesis may be a more sensitive monitor of physiological damage because it is either inhibited at lower element concentrations or after shorter incubation periods (Nieboer et al., 1979; Brown and Beckett, 1983, 1984, 1985). Membrane damage will obviously influence intracellular cation uptake but photosynthetic inhibition may also be important if light energy is involved, either directly in ion uptake or indirectly via carbohydrate products, in maintaining a suitable metabolic state in algal or fungal cells.

While the uptake of a pulse of Cd from the cell wall to the cell interior is very approximately linear with C. portentosa, a declining rate was found with Peltigera, especially with P. horizontalis from the uncontaminated site. These examples cover the range of sensitivities to Cd-induced inhibition of photosynthesis. It is, therefore, possible that this decline in Cd transfer rate may reflect damage caused to the cell by a build-up of potentially toxic Cd cations.

There was a rapid decline in gross photosynthesis following a 10

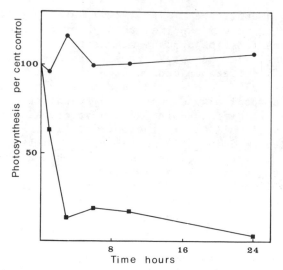

Fig. 3.   Changes in photosynthetic rates during the transfer of a
           pulse of Cd from the cell wall to the cell interior in
           Peltigera horizontalis (uncontaminated site = ■) and
           P. membranacea (contaminated site = ●).

min exposure to 100 μM Cd in the uncontaminated Peltigera sample (Fig.
3). Most damage occurred within the first 3 h after the Cd pulse.
The material from the contaminated site was unaffected throughout the
24 h period (Fig. 3), even though the intracellular Cd concentration
rose to more that present in the P. horizontalis sample from the
uncontaminated site after 3 h (Fig. 2). Material from the uncontaminated Peltigera population also showed an initial 40% increase in
respiration, followed by a decline to 70% of the rate observed in
material treated with deionised water only (Fig. 4). The respiration
rate of P. membranacea from the contaminated site was not significantly altered by the original Cd that entered the cells during the 10
min exposure period or any subsequent transfer.

     A substantial rate of Cd transfer to the cell interior occurred
during the first 3 - 6 h after the initial pulse of Cd to the cell
wall. This uptake occurred in P. horizontalis during the period when
photosynthesis declined to a minimum. Hence Cd uptake can occur in
lichens which have severely damaged photosynthetic capacities.
Although Beckett and Brown (1984a, b) found intracellular Cd uptake to
be light-stimulated in material from uncontaminated sites, they also
showed that a substantial uptake rate occurred in the dark in material
from both uncontaminated and polluted sites. The latter mechanism may
be the dominant process in these transfer experiments, as preliminary
results show that the rate of transfer is unaffected by light.
However, the rate-limiting step might be the release of Cd from the

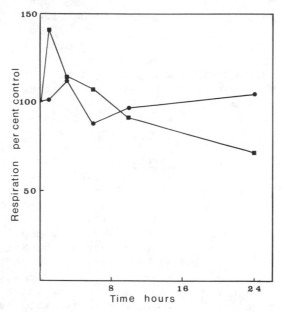

Fig. 4. Changes in respiratory rates during the transfer of a pulse
of Cd from the cell wall to the cell interior in <u>Peltigera
horizontalis</u> (uncontaminated site = ■) and <u>P. membranacea</u>
(contaminated site = ●).

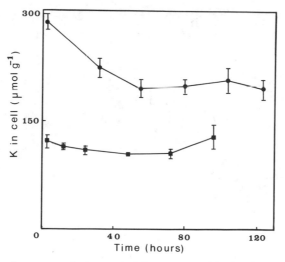

Fig. 5. Potassium remaining within the cell during the transfer of a
pulse of Cd from the cell wall to the cell interior in
<u>Cladonia portentosa</u> (■) and <u>Peltigera horizontalis</u> (●) from
uncontaminated sites. Bar = standard error.

cell wall exchange sites and this may not be affected by light.

While photosynthesis is damaged more rapidly and at lower Cd concentrations than is membrane damage, monitored by loss of soluble intracellular K (Brown and Beckett, 1983, 1984), both processes are correlated with the sensitivity of different lichen species and populations to added heavy metals (Beckett and Brown, 1984b; Brown and Beckett, 1984). No obvious loss of intracellular K occurred from the green alga-containing C. portentosa, when maintained in the absence of additional Cd, beyond that supplied in a 10 min pulse, whereas the cyanobacteria-containing P. horizontalis lost a significant proportion of its intracellular K over a 123.5 h period (Fig. 5). Short-term studies using Peltigera spp. with contrasting Cd tolerance showed a substantial loss of intracellular K from the sensitive P. horizontalis population but, apart from a small initial loss of K, no net loss over a 24 h period from the resistant P. membranacea population (Fig. 6). This is in agreement with the results obtained by Beckett and Brown (1984b) for material continuously supplied with Cd. Loss of intracellular K from the sensitive P. horizontalis population sample is only substantial 6 h after the initial Cd pulse was applied (Fig. 6), whereas the greatest decline in photosynthesis (Fig. 3) occurred within 3 h.

The conspicuous decline in intracellular Cd uptake rate observed after 6 h in P. horizontalis from the uncontaminated site is likely to be due to cellular damage, especially to membrane systems. It must be remembered that the data presented here merely show the amount of Cd present within the intracellular pool and rates of intracellular Cd up-

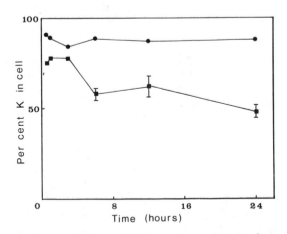

Fig. 6. Potassium remaining within the cell during the transfer of a pulse of Cd from the cell wall to the cell interior in Peltigera horizontalis (uncontaminated site = ■) and P. membranacea (contaminated site = ●). Bar = standard error.

take are the sum of opposing influx and efflux rates. As K loss may
be a measure of membrane damage, enhanced release of Cd may have
occurred through damaged plasma membranes, resulting in a lower
retention of Cd within the cell. Alternatively, it is possible that
during Ni displacement of cell wall-bound Cd some of the Ni may enter
damaged cells and displace intracellular Cd. Such a process may
account for the apparent increase in extracellular Cd uptake by
desiccated Peltigera that was noted by Brown and Beckett (1985).

The rate of intracellular uptake of Cd derived from the cell wall
also shows some decline with increasing time in the Cd-tolerant population of P. membranacea. This decline occurs without substantial
damage to the lichen and may only represent a reduced uptake rate due
to depletion of the source of Cd on the cell wall or saturation of the
intracellular Cd pool. Similar arguments might apply to
C. portentosa, although there the size of the cell wall exchangeable
pool is much larger.

Concluding Remarks

The data presented here clearly show that a pulse of Cd applied
to the cell wall of three different populations of lichens can be subsequently partially transferred to the interior of the lichen cells.
However, the rate of transfer to the cell interior is approximately
one tenth of that observed when Cd is supplied in solution throughout
the period of study (Brown and Beckett, 1984). Although it has not
been directly demonstrated, the high cation exchange capacity of
lichens has been presumed to be mainly due to fungal cells, as a
result of their greater mass within the lichen thallus (Collins and
Farrar, 1978). Thus for both physiologically essential and metabolically damaging cations the fungal cell wall exchange sites might
act as a way of concentrating cations prior to intracellular uptake.
This does not preclude a direct pathway for intracellular uptake as
long as cations are present in the bathing solution.

Placing desiccated lichens in solution induces losses of cations
and other intracellular soluble chemicals if the plasma membrane has
been damaged by dehydration stress (Farrar and Smith, 1978; Buck and
Brown, 1979). Brown and Buck (1979) showed that loss of intracellular
Mg and K occurred during the rehydration stage and that both elements,
particularly the divalent Mg, can become bound to the cell wall during
this stage. Unless severely damaged, membranes recover their normal
permeability properties when rehydrated. Lang et al. (1976) and
Crittenden (1983) presented data suggesting that, during simulated and
actual rainfall incidents, released monovalent cations may subsequently become reincorporated by lichens. Brown and Beckett (1985)
reported greater intracellular Cd uptake rates from solution by
desiccated lichens, presumably due to entry through damaged plasma
membranes. As element concentrations are highest in the earliest
stages of rainfall episodes (Brown, 1982; Crittenden, 1983), greatest

intracellular uptake may occur at this time. The present work
suggests that material then bound to the cell wall will also remain
available for intracellular uptake while the lichen remains wet.

The data presented here show that while the cell wall cation
exchange sites may accummulate heavy metal cations, binding
potentially toxic elements at these sites does not necessarily prevent
their subsequent uptake into the cell. Specualations which suggested
that the cell wall might act as a buffer against intracellular uptake
and damage by heavy metal cations, can apparently no longer be
sustained. Moreover, Cd may have a relatively low affinity for the
cell wall exchange sites compared to other heavy metals with stronger
Class B properties (Nieboer and Richardson, 1980) and it remains to be
shown whether this cell-wall to cell-interior transfer can occur with
other elements.

REFERENCES

Beckett, R.P., and Brown, D.H., 1983a, Naturally and experimentally
    induced zinc and copper resistance in the lichen genus _Peltigera_,
    Annals of Botany, 52: 43-50.
Beckett, R.P., and Brown, D.H., 1983b, Cellular uptake of heavy metals
    by bryophytes and lichens, _Proceedings of the 4th International
    Conference on Heavy Metals in the Environment_, 1: 447-450.
Beckett, R.P., and Brown, D.H., 1984a, The control of cadmium uptake
    in the lichen genus _Peltigera_, _Journal of Experimental Botany_,
    35: 1071-1082.
Beckett, R.P., and Brown, D.H., 1984b, The relationship between
    cadmium uptake and heavy metal tolerance in the lichen genus
    Peltigera, _New Phytologist_, 97: 301-311.
Brown, D.H., and Bates, J.W., 1972, Uptake of lead by two populations
    of _Grimmia doniana_, _Journal of Bryology_, 7: 187-193.
Brown, D.H., and Beckett, R.P., 1983, Differential sensitivity of
    lichens to heavy metals, _Annals of Botany_, 52: 51-57.
Brown, D.H., and Beckett, R.P., 1984, Uptake and effect of cations on
    lichen metabolism, _The Lichenologist_, 16: 173-188.
Brown, D.H., and Beckett, R.P., 1985, Minerals and lichens:
    acquisition, localisation and effect, _in_: "Surface Physiology of
    Lichens," C. Vicente, D.H. Brown, and M.E. Legaz, eds,
    pp. 127-149, Universidad Computense de Madrid.
Brown, D.H., and Buck, G.W., 1979, Desiccation effects and cation
    distribution in bryophytes, _New Phytologist_, 82: 115-125.
Brown, D.H., and House, K.L., 1978, Evidence of a copper-tolerant
    ecotype of the hepatic _Solenostoma crenulatum_, _Annals of Botany_,
    42: 1383-1392.
Brown, D.H., and Slingsby, D.R., 1972, The cellular location of lead
    and potassium in the lichen _Cladonia rangiformis_ (L.) Hoffm., _New
    Phytologist_, 71: 297-305.
Buck, G.W., and Brown, D.H., 1979, The effects of desiccation on
    cation location in lichens, _Annals of Botany_, 44: 265-277.

Collins J.M., and Farrar, J.F., 1978, Structural resistance to mass transfer in the lichen Xanthoria parietina, New Phytologist, 81: 71–83.

Crittenden, P.D., 1983, The role of lichens in the nitrogen economy of subarctic woodlands: nitrogen loss from the nitrogen-fixing lichen Stereocaulon paschale during rainfall, British Ecological Society Symposium, 22: 43–68.

Farrar, J.F., and Smith, D.C., 1976, Ecological physiology of the lichen Hypogymnia physodes. III. The importance of the rewetting phase, New Phytologist, 77: 115–125.

Garty, J., Galun, M., and Kessel, M., 1979, Localization of heavy metals and other elements accumulated in the lichen thallus, New Phytologist, 89: 631–645.

Heywood, D., 1982, "A possible role of magnesium in preventing intracellular uptake of zinc in the moss Rhytidiadelphus squarrosus," B.Sc. thesis, University of Bristol.

Hogan, G.D., and Rauser, W.E., 1981, Role of copper binding, absorption and translocation in copper tolerance of Agrostis gigantea Roth., Journal of Experimental Botany, 32: 27–36.

Jones, D., Wilson, M.J., and Laundon, J.R., 1982, Observations on the location and form of lead in Stereocaulon vesuvianum, The Lichenologist, 14: 281–286.

Laaksovirta, K., and Olkkonen, H., 1977, Epiphytic lichen vegetation and element contents of Hypogymnia physodes and pine needles examined as indicators of air pollution at Kokkola, W. Finland, Annales Botanici Fennici, 14: 112–130.

Lang, G.E., Reiners, W.A., and Heier, R.K., 1976, Potential alteration of precipitation chemistry by epiphytic lichens, Oecologia (Berlin), 25: 229–241.

Mathys, W., 1973, Vergleichende Untersuchungen der Zinkaufnahme von resistenten und sensitiven Populationen von Agrostis tenuis Sibth., Flora, Jena, 162: 492–499.

Nieboer, E., and Richardson, D.H.S., 1980, The replacement of the non descript term "heavy metal" by a biologically and chemically significant classification of metal ions, Environmental Pollution (Series B), 1: 3–26.

Nieboer, E., and Richardson, D.H.S., 1981, Lichens as monitors of atmospheric deposition, in: "Atmospheric Pollution in Natural Waters," S.J. Eisenreich, ed., pp. 339–388, Ann Arbor Science, Ann Arbor, Michigan.

Nieboer, E., Richardson, D.H.S., Boileau, L.J.R., Beckett, P.J. Lavoie, P., and Padovan, D., 1982, Lichens and mosses as monitors of industrial activity associated with uranium mining in Northern Ontario, Canada – Part 3: Accumulations of iron and titanium and their mutual dependence, Environmental Pollution (Series B), 4: 181–192.

Nieboer, E., Richardson, D.H.S., Lavoie, P., and Padovan, D., 1979, The role of metal-ion binding in modifying the toxic effects of sulphur dioxide on the lichen Umbilicaria muhlenbergii. I. Potassium efflux studies, New Phytologist, 82: 612–632.

Nieboer, E., Richardson, D.H.S., and Tomassini, F.D., 1978, Mineral
uptake and release by lichens: An overview, The Bryologist, 81:
226-246.

Page, E.R., and Dainty, J., 1964, Manganese uptake by excised oat
roots, Journal of Experimental Botany, 15: 428-443.

Peterson, P.J., 1969, The distribution of zinc-65 in Agrostis tenuis
Sibth. and A. stolonifera L. tissues, Journal of Experimental
Botany, 20: 863-875.

Puckett, K.J., 1976, The effect of metals on some aspects of lichen
physiology, Canadian Journal of Botany, 54: 2695-2703.

Puckett, K.J., and Burton M.A.S., 1981, The effects of trace elements
on lower plants, in: "Effect of Heavy Metal Pollution on Plants.
Vol. 2. Metals in the Environment," N.P. Lepp, ed., pp. 213-238,
Applied Science Publishers, London.

Rao, D.N., Robitaille, G., and LeBlanc, F., 1977, Influence of heavy
metal pollution on lichens and bryophytes, Journal of the Hattori
Botanical Laboratory, 42: 213-239.

Seaward, M.R.D., Goyal, R., and Bylinska, E.A., 1978, Heavy metal
content of some terricolous lichens from mineral-enriched sites
in northern England, The Naturalist, 103: 135-141.

Skipnes, O., Roalad, T., and Haug, A., 1976, Uptake of zinc and
strontium by brown algae, Physiologia Plantarum, 34: 314-320.

Snelgar, W.P., Brown, D.H., and Green, T.G.A., 1980, A provisional
survey of the interactions between net photosynthetic rate,
respiratory rate, and thallus water content in some New Zealand
cryptogams, New Zealand Journal of Botany, 18: 247-256.

Turner, R.G., and Marshall, C., 1971, The accumulation of [65]Zn by root
homogenates of zinc-tolerant and non-tolerant clones of Agrostis
tenuis Sibth., New Phytologist, 70: 539-545.

ULTRASTRUCTURAL STUDIES OF DESICCATED LICHENS

C. Ascaso[a], D.H. Brown[b] and S. Rapsch[a]

[a] Instituto de Edafologia y Biologia Vegetal
Consejo Superior de Investigaciones Cientificas
Serrano 115 bis, Madrid 28006, Spain
[b] Department of Botany, The University
Bristol BS8 1UG, U.K.

INTRODUCTION

The association of fungi and algae to produce the unique lichen thallus involves morphological and physiological modifications to both partners (Ahmadjian and Jacobs, 1983). Establishment of the symbiotic relationship is promoted by conditions which do not favour the independent growth of either partner. The range of conditions under which artificial lichen synthesis has been successful is relatively limited and involves an appropriate temperature, suitable substratum, low nutrient levels and alternate periods of wetting and drying (Mattes and Feige, 1983). The natural habitats of lichens are frequently desiccated, either daily or for longer periods. It is not known whether lichens require drying or this is a stress they can adjust to or tolerate (Ahmadjian, 1982).

The resistance of lichens to drought has been regularly reviewed (Kappen, 1973; Bewley, 1979; Mattes and Feige, 1983). Rehydration of severely desiccated lichens may result in elevated respiration rates (Smith and Molesworth, 1973) and the loss of intracellular chemicals through damaged plasma membranes (Farrar and Smith, 1976; Buck and Brown, 1979).

Studies of ultrastructural changes associated with desiccation have often involved using material from extreme habitats. Lichens from deserts show algal cells penetrated by haustoria (Ben-Shaul et al., 1979; Galun et al., 1970). Sometimes intracellular haustoria were only found in mature parts of the thallus, suggesting that the internal environment of the thallus may affect the types of cellular interactions that occur (Malachowski et al., 1980). Studies of

259

seasonal variations in cell structure can be related to climatic
conditions. Variations in the amount of starch and other storage
products (Ahmadjian, 1966; Jacobs and Ahmadjian, 1969) and in the
general organization of the cytoplasm (Ascaso and Galvan, 1977;
Holopainen, 1982) have been observed.

     Ultrastructural changes induced by desiccation stress have been
investigated under laboratory conditions (Brown and Wilson, 1968;
Peveling, 1970; Harris and Kershaw, 1971; Jacobs and Ahmadjian, 1971;
Ascaso and Galvan, 1976; Peveling, 1977; Ascaso, 1979; Peveling and
Robenek, 1980). Some of these studies have involved either prolonged
periods of desiccation or relatively uncontrolled conditions. Observations have often been confined to a single, frequently desiccation
tolerant, species.

     In this study we have begun to investigate the changes produced
by relatively short periods of incubation under defined conditions of
light and relative humidity in growth chambers. Comparisons have been
made between Lasallia pustulata from exposed rock surfaces, where it
is expected to be desiccated for much of the year, and Lobaria
amplissima and Lobaria pulmonaria from the bases of tree trunks in a
relatively open wood by a stream, where both (particularly
L. amplissima) are likely to be moist for much longer periods of the
year. Using loss of intracellular K as a measure of desiccation
damage, Buck and Brown (1979) reported that Lasallia is substantially
more desiccation-resistant than the two Lobaria species. It is also
expected that prolonged storage in a hydrated condition (Farrar, 1976)
will cause damage to the more desiccation-resistant Lasallia than to
the two relatively more desiccation-sensitive Lobaria species (Buck
and Brown, 1979).

     It is suggested that the algal cells, whilst having the same
basic structure, may differ in their sensitivity to water loss and
this may be partly due to lower osmotic potentials being generated in
cells normally subjected to the osmotic stress of desiccation. The
use of phosphate buffer in the fixation medium may result in dis-
tortion of the cells in a way related to their osmotic potential and
thus their desiccation sensitivity.

MATERIALS AND METHODS

     Lichen samples were transported to the laboratory in polythene
bags and either immediately prefixed (field samples) or placed in
small plastic boxes in the growth chamber for 2 days at 100% or 0%
relative humidity, 12:12 hours light:dark regime at 1200 lux with 20°C
and 14°C maximum and minimum temperatures.

     Lobaria amplissima was prefixed by two methods. Small sections
of thalli, from 2-3 mm behind the lobe tips, were cut into small
pieces with a razor blade and placed in 3.25% glutaraldehyde in either

0.05M phosphate buffer (GB), pH 7.1, or in distilled water (GW). The
other species were only fixed in GB. Samples were prefixed at 4°C for
3 hours for sectioning and for 90 minutes for freeze etching. Samples
for sectioning were washed overnight in buffer or water and postfixed
in 1% osmium tetroxide in 0.05 M phosphate buffer, pH 7.1, for 3
hours. The material was dehydrated in graded ethanol solutions,
embedded in Spurr (1969) resin, sectioned and stained with Reynolds
(1963) lead citrate.

    Samples for freeze etching were glycerinated for 1 hour after the
glutaraldehyde fixation. Fracturing was carried out at -110°C and the
temperature raised to -100°C and samples etched for 90 seconds.

    Ultrathin sections and freeze-etched replicas were examined with
a Philips EM 300 electron microscope. Image analysis used a
Mop-Videoplan (Kontron) semi-automatic image analyzer.

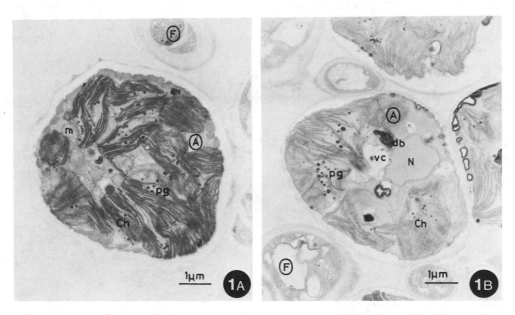

Fig. 1. Phycobiont (<u>Myrmecia</u>) in <u>Lobaria amplissima</u>, field sample.
         A) GW fixed, B) GB fixed.
         Note dark bodies in B.

         Abbreviations: A, alga; ch, chloroplast; db, dark body; F,
         fungus; m, mitochondrion; N, nucleus; P, pyrenoid; pg,
         plastoglobuli (<u>Myrmecia</u>) or pyrenoglobuli (<u>Trebouxia</u>); s,
         starch; sb, storage body; vc, vesicular complex.

RESULTS AND DISCUSSION

     In this study we have only investigated the ultrastructural
changes occurring in the algal cells of these typical foliose lichens.

Lobaria

     Transmission electron micrographs of L. amplissima (Fig. 1) show
that the phycobiont, Myrmecia, contains mitochondria, vesicular
complexes and ribosomes in the cytoplasm, a centrally located nucleus
and several parietal chloroplasts similar to those previously reported
for Myrmecia in Dermatocarpon hepaticum (Galun et al., 1971).
Thylakoids in field samples are associated in groups and there are
plastoglobuli located in lamella-free areas of the chloroplast. Fixation with glutaraldehyde in water (GW) (Fig. 1a) or in phosphate
buffer (GB) (Fig. 1b) produced some changes in the ultrastructure of
field samples; in GB dark bodies appeared in the cytoplasm and a
substantial degree of cell plasmolysis occurred.

     Following storage for 2 days at 100% r.h., an increase in storage
bodies or storage droplets was observed in the phycobiont cytoplasm
(Fig. 2a). Fixation with GB induced an accumulation of dark bodies in
the cytoplasm (Fig. 2b). Normally these dark bodies are situated out-

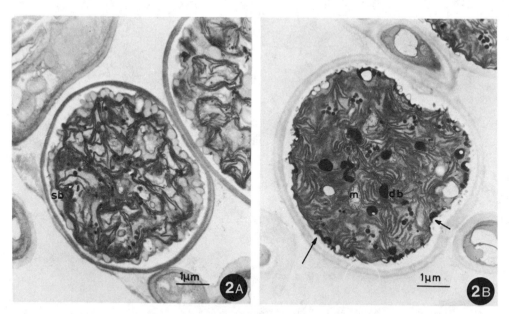

Fig. 2. Phycobiont (Myrmecia) in Lobaria amplissima after 2 days 100%
         r.h. A) GW fixed, B) GB fixed.
         Note abundant storage bodies in A and dark bodies in B.
         Arrows indicate half-moon like structures.

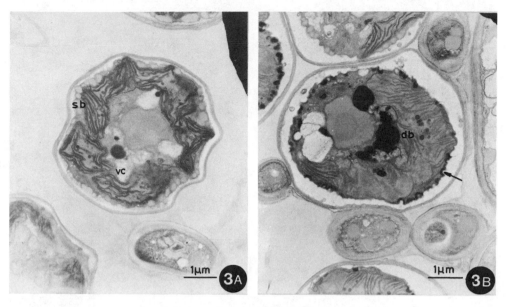

Fig. 3.   Phycobiont (<u>Myrmecia</u>) in <u>Lobaria amplissima</u> after 2 days 0%
          r.h.   A) GW fixed, B) GB fixed.
          Note vesicular complex in both samples. Dark bodies,
          half-moon like structures and plasmolysis in B.

side the chloroplast membrane. Many half-moon like structures
developed in place of the peripheral electron transparent storage
bodies, situated beneath the plasmalemma; some of the storage bodies
now appeared empty.

Cytoplasmic storage bodies were less abundant following storage
at 0% r.h. (Fig. 3a). Vesicular complexes were present between the
chloroplast and central nucleus while the plastoglobuli presented a
perforated appearance. Again, fixation of samples with GB induced the
formation of dark bodies and half-moon like structures (Fig. 3b).

Quantitative measurements of the observed ultrastructural changes
in <u>Myrmecia</u> cells are summarised in Table 1. The area of dark bodies
is signficantly higher in the GB samples. Samples fixed with GW had
similar amounts of dark bodies in the field samples and stored
material. Samples fixed with GB had more dark bodies in stored
material (1.07 and 1.08 μm^2) than in the field samples (0.28 μm^2).
The buffer apparently induced more artifacts, e.g. dark bodies and
plasmolysis, in stored samples. Treatment for 2 days at 100% and 0%
r.h. had several effects on algal cells; changes in protoplast size,
amount and form of storage bodies and appearance of plastoglobuli
(Table 1).

Table 1. Area, Number and Appearance of Storage Bodies in Field or Stored Myrmecia in Lobaria amplissima

| | Field sample | | 100% r.h. | | 0% 5.h. | |
| | | | Stored 2 days* | | | |
	GW	GB	GW	GB	GW	GB
Average area of protoplast#	26.0 ± 7.9	23.0 ± 7.8	14.9 ± 5.4^A	24.7 ± 7.0^A	19.4 ± 56.8	19.2 ± 5.6
Average area of dark bodies per protoplast#	0.02^A	0.28^A	0.02^B	1.07^B	0.11^C	1.08^C
Dark body area as per cent of protoplast area	0.07	1.23	0.12	4.32	0.57	5.64
Half-moon-shaped bodies beneath plasmalemma				++		+
Appearance of plastogobuli	Normal	Normal	Normal	Light	Holed	Destroyed

*Incubated under 12h, 20°C, 1200 lux light: 12h, 14°C dark regime.
GW = fixed in glutaraldehyde in water; GB = fixed in glutaraldehyde in 0.05M phosphate buffer
# Area in μm^2
A B C indicate, within storage conditions, statistically significant differences (T-test) between GW and GB
    (P < 0.01)
+ = arbitrary scale of occurrence.

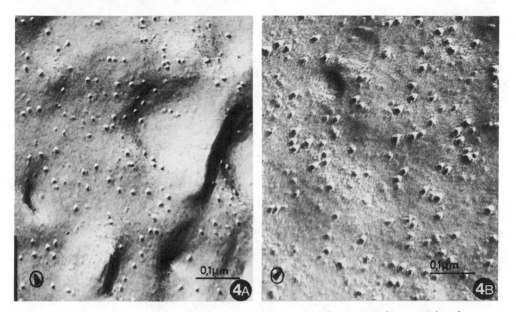

Fig. 4.   Plasmalemma EF-face from <u>Myrmecia</u> in <u>Lobaria amplissima</u>,
field sample.   A) GW fixed, B) GB fixed.

Observations on the freeze-etched appearance of the plasmalemma
of <u>Myrmecia</u> from <u>L. amplissima</u>, were carried out with material fixed
in <u>GW</u> and GB.   Figure 4 shows the EF-face of the plasmalemma from a
field sample (Fig. 4a GW, Fig. 4b GB).   The intramembranous protein
particles (IMPs) detected by freeze-fracture electron microscopy
probably represent integral membrane transport proteins (Verkleij and
Ververgaert, 1978).

A large number of replicas have been measured to determine the
density and diameters of the IMPs.   The frequency distribution of
particle sizes on the EF-face of the plasmalemma is shown in Fig. 5.
Particles with the largest diameters were found in buffer-fixed
samples.   This effect is clear in field samples, the range of
diameters for GW being c.5-15 nm and GB c.8-24 nm.   The effect is
similar, but not so obvious, in the 100% r.h. incubated samples.   We
have been unable to obtain sufficient replicas of the 0% r.h. GB
samples for comparison.   Storage produced an increase in numbers of
small diameter particles (Fig. 5).   This is also true of the EF-face
of the plasmalemma in GB-fixed <u>Myrmecia</u> from <u>L. pulmonaria</u> (Fig. 6).
The wide range of particle diameters (between 9 and 22 nm for field
samples) can be interpreted, as for <u>L. amplissima</u>, as being due to the
presence of the phosphate buffer during fixation.

Variations occur in particle density of from 200 to 550 particles
μm^{-2} with <u>Myrmecia</u> from <u>L. amplissima</u> and from 400 to 600 in

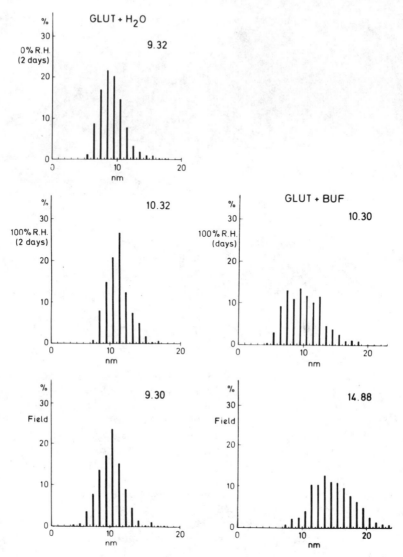

Fig. 5. Particle size frequency distribution of plasmalemma EF-face from _Myrmecia_ in _Lobaria amplissima_.

L. pulmonaria. The buffer increased particle diameter in both species and there is an inverse relationship between diameter and density of particles which indicates that areas with small particles are denser than those with big particles. Average particle diameters on the PF-face of _Myrmecia_ from _L. amplissima_ were smaller than those on the EF-face. Particle density was higher on the PF-face than the EF-face, ranging from 1000 to 3000 particles μm^{-2} .

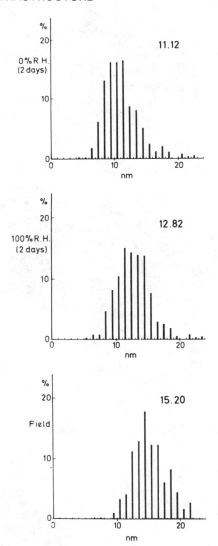

Fig. 6.   Particle size frequency distribution of plasmalemma EF-face
from _Myrmecia_ in _Lobaria pulmonaria_.   Fixed in GB.

Lasallia

    All studies on the phycobiont (_Trebouxia_) of _L. pustulata_ were
carried out after GB prefixation.   Figure 7 shows _Trebouxia_ cells from
field samples.   One large lobate chloroplast occupies the centre of
the cell while the nucleus and mitochondria have a peripheral
position.   The pyrenoid matrix contains osmiophilic globules
(pyrenoglobuli) aligned along the thylakoids.   Small starch granules

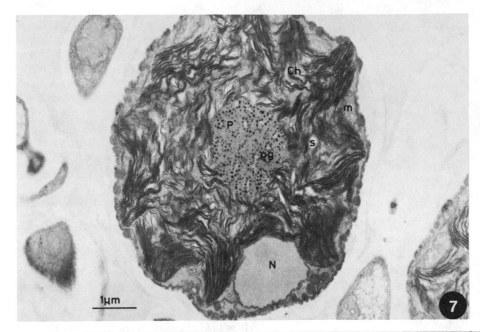

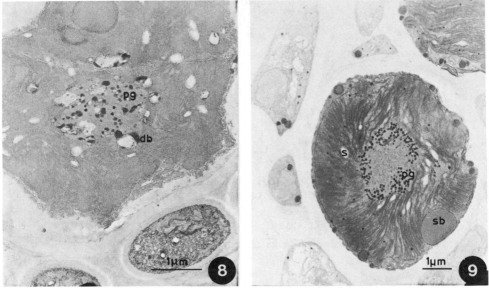

Fig. 7.   Phycobiont (Trebouxia) in Lasallia pustulata, field sample.
Fig. 8.   Phycobiont (Trebouxia) in Lasallia pustulata after 2 days at
          100% r.h.
          Note degenerated protoplast.
Fig. 9.   Phycobiont (Trebouxia) in Lasallia pustulata after 2 days at
          0% r.h.
          Note peripheral position of pyrenoglobuli and presence of
          starch.

were present close to the thylakoid membranes. No dark cytoplasmic
bodies or half-moon like structures were seen. Their absence suggests
that either the buffer had minimal effect on Trebouxia cells or, more
likely, the cells have a genuinely different chemistry and morphology.
Only a little plasmolysis occurred, compared to the Lobaria phycobionts. This probably reflects the lower intracellular osmotic
potential of algal cells from the more desiccation-tolerant Lasallia.

Storage at 100% r.h. for 2 days (Fig. 8) produced obvious
degenerative effects on the protoplast. The collapse of cell contents
is a general feature, producing a granulous mass within the algal cell
wall. Cell organelles disintegrated and only residues of pyrenoglobuli and dark bodies remained discernable.

Several differences were seen after storage at 0% r.h. for 2 days
(Fig. 9). Starch grains appeared around the pyrenoid and pyrenoglobuli were peripherally distributed. Large storage bodies were
present in the cytoplasm between the plasmalemma and the exterior of
the chloroplast but sometimes these had the appearance of myelin
bodies.

Table 2 presents a summary of observations made on field samples
and dry-stored material: no data was obtained from the damaged
material stored at 100% r.h. Phycobiont protoplasts from dried
samples were smaller than control ones. This was not due to an
osmotic effect because plasmolysis was low and similar in both field
and dried material.

Table 2.   Area, Number and Appearance of Storage Bodies in Field or
           Stored Trebouxia in Lasallia pustulata

Average area μm^2	Field sample	Stored 2 days* 0% r.h.
Protoplast	50.3 ± 17.3^A	31.3 ± 14.4^A
Starch	0.19^B	0.51^B
Storage bodies under plasmalemma	0.04	0.11
Dark bodies outside chloroplast	–	0.12
Distribution of pyrenoglobuli in pyrendoid	Random	Peripheral

*Incubation conditions 12h, 20°C light: 12 h, 14°C dark.
Values of a specific feature are statistically different (T-test) at
^{A} P < 0.01; ^{B} P < 0.025.

The presence of significant amounts of starch between the
thylakoids of desiccated samples (also noted by Harris and Kershaw,
1971) may be due to the retention of a greater proportion of
photosynthetic products within the algal cells at low water content
(Tysiaczny and Kershaw, 1979). MacFarlane and Kershaw (1982) reported
that carbohydrate transfer from algal to fungal cells occurred at
lower water contents in desiccation-tolerant species (but see
MacFarlane and Kershaw, this volume) and Cowan et al. (1979) reported
that photosynthesis declined at higher water contents than did the
synthesis of macromolecules. So it is obvious that more work is
required in order to clarify the effect of hydration on the retention
and polymerisation of carbohydrates in autotrophic symbionts.

    In _Trebouxia_, the peripheral distribution of pyrenoglobuli under
dry conditions is a well-documented event (Jacobs and Ahmadjian, 1971;
Ascaso and Galvan, 1976). The present observations show that this
relocation can be a rapid process.

    Freeze-etching studies show that in _Trebouxia_ in _L. pustulata_
(Fig. 10) there are characteristic ridges on the EF-face with
relatively few particles (Fig. 10a), and grooves on the PF-face
(Fig. 10b) with numerous intramembranous particles (Peveling and
Robenek, 1980). All the material was fixed with GB, but the particle

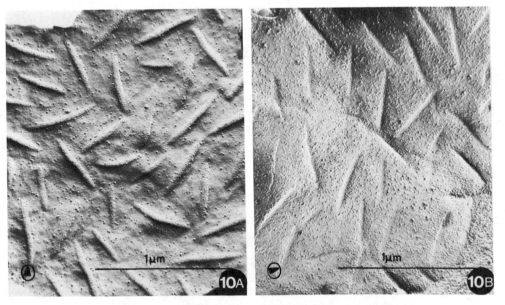

Fig. 10. _Trebouxia_ plasmalemma fracture faces in _Lasallia pustulata_,
          field sample. A) EF- and B) PF-faces.

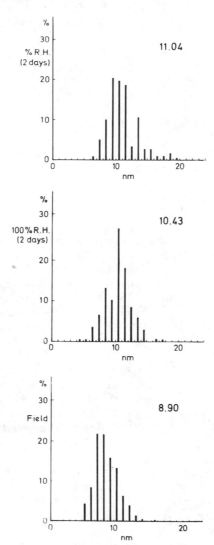

Fig. 11. Particle size frequency distribution of plasmalemma EF-face
from Trebouxia in Lasallia pustulata.

diameters (5-15 nm) (Fig. 11) were not so large as in Myrmecia field
samples in GB. With storage the diameters of the IMPs on the EF-face
of the Trebouxia plasmalemma increased. Diameters were greater in
material stored in the light at 100% r.h. (10.43 nm) and 0% r.h.
(11.04 nm) than in field samples (8.80 nm) (Fig. 11). This increase
in diameter is associated with a decrease in density, the respective
values being 212, 216 and 365 particles μm^{-2}.

The results obtained here with Trebouxia do not agree with those reported by Peveling and Robenek (1980) who observed a decrease in particle number and size on desiccation. In should be noted, however, that particle density in <u>Lobaria</u> species either remained unchanged or increased on storage in a wet or dry condition. These differences are probably due to different experimental conditions. It must be appreciated that although samples were placed at different humidities, the time taken to reach equilibrium may vary and represent a substantial proportion of the total storage period. Bewley (1979) has emphasised that the rate of desiccation may critically influence cell ultrastructure and physiology, particularly in sensitive species. Peveling and Robenek (1980) compared material stored dry for 9 months with samples rehydrated for days.

CONCLUDING REMARKS

This work has shown that the use of a phosphate buffer during prefixation causes a number of alterations and these are different in <u>Myrmecia</u> and <u>Trebouxia</u>. More artifacts are produced after storage. Phosphate buffer produces changes in the size and density of plasmalemma particles in <u>Myrmecia</u>, with a clear inverse relationship between particle density and diameter in GB-treated samples. Holopainen (1982) found that the effect the concentration of phosphate buffer had on ultrastructure varied with the time of year.

Brief storage times have been shown to be sufficient to cause changes in protoplast ultrastructure, changing the quality of storage bodies in the cytoplasm and pyrenoglobuli in the pyrenoid, increasing the amount of starch when dried, destroying the chloroplast structure (<u>L. pustulata</u> at 100% r.h.) and altering the size and density of intramembranous particles.

Many of the changes we have observed have been reported by other workers. This work has shown that changes can occur rapidly and therefore more care must be taken over the conditions the material is exposed to between harvesting and fixation. It is not known whether all of the changes observed here have a metabolic basis. The amount and physical state of water in membranes may influence the size and density of particles and the fact that all samples are being rehydrated during fixation may make comparisons between pretreatments possible but not between workers due to the differences in technique employed.

ACKNOWLEDGEMENTS

We thank Mr F. Pinto and Mrs T. Carnota for technical assistance. Financial support was provided by the Comision Asesora de Ciencia y Tecnologia, Proyecto n°. 502/81 and for D.H.B. a Royal Society/C.S.I.C. European Scientifc Exchange Programme award.

REFERENCES

Ahmadjian, V., and Jacobs, J.B., 1983, Algal-fungal relationships in
    lichens: recognition, synthesis and development, in: "Algal
    Symbiosis", L.J. Goff, ed., pp. 147-172, Cambridge University
    Press, Cambridge
Ahmadjian, V., 1966, Lichens, in: "Symbiosis," Vol. 1, S.M. Henry,
    ed., pp. 35-97, Academic Press, New York.
Ahmadjian, V., 1982, Algal/fungal symbioses, in: "Progress in
    Phycological Research," Vol. 1, F.E. Round, and D.J. Chapman,
    eds, pp. 179-233, Elsevier, Amsterdam.
Ascaso, C., 1979, Variaciones ultrastructurales en Parmelia conspersa
    S.L. provocadas "in vitro" por diferentes grados de humedad
    ambiental, Anales de Edafologia y Agrobiologia, 38: 1409-1419.
Ascaso, C., and Galvan, J., 1976, The ultrastructure of the symbionts
    of Rhizocarpon geographicum, Parmelia conspersa and Umbilicaria
    pustulata growing under dryness conditions, Protoplasma, 87:
    409-418.
Ascaso, C., and Galvan, J., 1977, Variations saisonnieres dans
    l'ultrastructure d'un lichen foliace: Parmelia conspersa S.L.,
    Anales de Edafologia y Agrobiologia, 36: 1157-1166.
Ben-Shaul, Y., Paran, N., and Galun, M., 1969, The ultrastructure of
    the association between phycobiont and mycobiont in three
    ecotypes of the lichen Caloplaca aurantia var. aurantia, Journal
    de Microscopie (Paris), 8: 415-422.
Bewley, J.D., 1979, Physiological aspects of desiccation tolerance,
    Annual Review of Plant Physiology, 30:195-238.
Brown, R.M., and Wilson, R., 1968, Electron microscopy of the lichen
    Physcia aipolia (Ehrh.) Nyl., Journal of Phycology, 4: 230-240.
Buck, G.W., and Brown, D.H., 1979, The effect of desiccation on cation
    location in lichens, Annals of Botany, 44: 265-277.
Cowan, D.A., Green, T.G.A., and Wilson, A.T., 1979, Lichen metabolism,
    1. The use of tritium labelled water in studies of anhydrobiotic
    metabolism in Ramalina celastri and Peltigera polydactyla, New
    Phytologist, 82: 489-503.
Farrar, J.F., 1976, Ecological physiology of the lichen Hypogymnia
    physodes. I. Some effects of constant water saturation, New
    Phytologist, 77: 93-103.
Farrar, J.F., and Smith, D.C., 1976, Ecological physiology of the
    lichen Hypogymnia physodes. III. The importance of the
    rewetting phase, New Phytologist, 77: 115-125.
Galun, M., Paran, N., and Ben-Shaul, Y., 1970, An ultrastructural
    study of the fungus-alga association in Lecanora radiosa growing
    under different environmental conditions, Journal de Microscopie
    (Paris), 9: 801-806.
Galun, M., Paran, N., and Ben-Shaul, Y., 1971, Electron microscopic
    study of the lichen Dermatocarpon hepaticum (Ach.) Th.Fr.,
    Protoplasma, 73: 457-468.
Harris, G.P., and Kershaw, K.A., 1971, Thallus growth and the
    distribution of stored metabolites in the phycobionts of the

lichens <u>Parmelia sulcata</u> and <u>P. physodes</u>, <u>Canadian Journal of</u>
<u>Botany</u>, 49: 1367–1372.

Holopainen, T.H., 1982, Summer versus winter condition of the
ultrastructure of the epiphytic lichens <u>Bryoria capillaris</u> and
<u>Hypogymnia physodes</u> in central Finland, <u>Annales Botanici Fennici</u>,
19: 39–52.

Jacobs, J.B., and Ahmadjian, V., 1969, The ultrastructure of lichens.
I. A general survey, <u>Journal of Phycology</u>, 5: 227–240.

Jacobs, J.B., and Ahmadjian, V., 1971, The ultrastructure of lichens.
II. <u>Cladonia cristatella</u>: The lichens and its isolated
symbionts, <u>Journal of Phycology</u>, 7: 71–82.

Kappen, L., 1973, Response to extreme environments, <u>in</u>: "The Lichens,"
V. Ahmadjian, and M.E. Hale, eds, pp. 310–380, Academic Press,
New York.

MacFarlane, J.D., and Kershaw, K.A., 1982, Physiological–environmental
interactions in lichens. XIV. The environmental control of
glucose movement from alga to fungus in <u>Peltigera polydactyla</u>,
<u>P. rufescens</u> and <u>Collema furfuraceum</u>, <u>New Phytologist</u>, 91:
93–101.

Malachowski, J.A., Baker, K.K., and Hooper, G.R., 1980, Anatomy and
algal–fungal interactions in the lichen <u>Usnea cavernosa</u>, <u>Journal</u>
<u>of Phycology</u>, 16: 346–354.

Matthes, U., and Feige, G.B., 1983, Ecophysiology of lichen symbioses,
<u>Encyclopedia of Plant Physiology, New Series</u>, 12C: 423–467.

Peveling, E., 1970, Das Vorkommen von Starke in Chlorophyceen-
Phycobionten, <u>Planta</u>, 93: 82–85.

Peveling, E., 1977, Die ultrastruktur einiger Flechten nack langen
Trockenzeiten, <u>Protoplasma</u>, 92: 129–136.

Peveling, E., and Robenek, H., 1980, The plasmalemma structure in the
phycobiont <u>Trebouxia</u> at different stages of humidity of a lichen
thallus, <u>New Phytologist</u>, 84: 371–374.

Reynolds, E.S., 1963, The use of lead citrate at high pH as an
electron-opaque stain in electron microscopy, <u>Journal of Cell</u>
<u>Biology</u>, 17: 208–212.

Smith, D.C., 1962, The biology of lichen thalli, <u>Biological Reviews</u>,
37: 537–570.

Smith, D.C., and Molesworth, S., 1973, Lichen Physiology. XIII.
Effects of rewetting dry lichens, <u>New Phytologist</u>, 72: 525–533.

Spurr, A.R., 1969, A low epoxy resin embedding medium for electron
microscopy, <u>Journal of Ultrastructural Research</u>, 26: 31–43.

Tysiaczny, M.J., and Kershaw, K.A., 1979, Physiological–environmental
interactions in lichens. VII. The environmental control of
glucose movement from alga to fungus in <u>Peltigera canina</u>
var. <u>praetextata</u> Hue, <u>New Phytologist</u>, 83: 137–146.

Verkleij, A.J., and Ververgaert, P.H.J.Th., 1978, Freeze-fracture
morphology of biological membranes, <u>Biochimica et Biophysica</u>
<u>Acta</u>, 515: 303–327.

THE ARCHITECTURE OF THE CONCENTRIC BODIES IN THE MYCOBIONT OF

PELTIGERA PRAETEXTATA

E. Peveling[a], H. Robenek[b] and B. Berns[a]

[a] Botanisches Institut, Westfälische Wilhelms-
Universität, Schlossgarten 3, D-4400 Münster
[b] Medizinische Cytobiologie, Westfälische Wilhelms-
Universität, Domagkstrasse 3, D-4400 Münster, Federal
Republic of Germany

INTRODUCTION

In 1968 Brown and Wilson reported for the first time the existence of very characteristic organelles in lichenized fungi. They described these organelles as ellipsoidal in shape, consisting of two shells of different electron density surrounded by a central electron-transparent core. In connection with the outer shell the authors observed double-unit membrane structures forming fingerlike projections. A very similar structure, but without connecting membranes, was described shortly afterwards (Peveling, 1969). In the latter investigation the organelles always appeared as round structures with a central core of two concentrically arranged shells and were therefore termed concentric bodies. During the following years, with the increase in ultrastructural investigations of lichens, such organelles were found in all mycobionts. Griffiths and Greenwood (1972) summarised the different observations and proposed as the final name for these organelles the term "concentric bodies", which is now generally accepted.

There is some evidence to show that, chemically, concentric bodies contain protein (Galun et al., 1974; Boissière, 1979, 1982) and lack polysaccharides (Boissière, 1982). There are no hypotheses as to the origin of concentric bodies (Galun and Bubrick, 1984).

Besides experiments to find out the function of these organelles, there are continuing investigations on the architecture of the concentric bodies. According to the latest detailed

275

observations, concentric bodies are organelles which "consist of
radiating plasma membranes, which are unified by more or less dense
deposits arranged in concentric shells" (Boissière, 1982). This
author distinguished four layers around the electron transparent
core depending on the electron density. In tangential sections
she described the cut membrane as a kind of network with a somewhat
reticulated appearance depending on the cutting level. Close to
the concentric bodies a tubular system has sometimes been observed
(Brown and Wilson, 1968; Jacobs and Ahmadjian, 1971; Griffiths and
Greenwood, 1972; Boissière, 1982).

In this paper we present an improved model of the architecture
of the concentric bodies.

MATERIALS AND METHODS

Peltigera praetextata was collected during May and September
in the Black Forest (900 m NN), stored for no more than 5 days and,
if air dried, moistened before fixation. Pieces, 1 mm^2, were taken
only from the growing regions close to the thallus margin.

Moist samples were infiltrated with 6% glutaraldehyde buffered
with cacodylate, pH 7.2, in which they were kept for 3 h at room
temperature. The fixation procedure was followed by washing in
buffer and transfer to 2% aqueous osmium tetroxide for postfixation
and staining for 2 h. Ethanol and propylene oxide were used for
dehydration before embedding in an epon mixture according to Luft
(1961).

For freeze-fracture studies moist samples, without any
pretreatment, were frozen in liquid Freon 22 at -150°C and stored
in liquid nitrogen. Replicas were made according to standard
techniques (Moor and Mühlethaler, 1963) on a Balzers BA 300
freeze-etching apparatus (Balzers AG, Liechtenstein) equipped with
an electron gun evaporator and an oscillating quartz crystal. The
samples were fractured at -100°C and 2 x 10^{-6} Torr without etching.
Replicas were obtained by shadowing the fracture surface with
platinum and carbon at an angle of 38° followed with carbon at 90°.
The replicas were cleaned overnight in household bleach and washed
in distilled water.

The ultrathin sections, after staining with uranylacetate
(Pease, 1964) and lead citrate (Reynolds, 1963), and the replicas
were examined in a Siemens Elmiscope 102 at 80 kV.

OBSERVATIONS AND DISCUSSION

The concentric bodies in the mycobiont of P. praetextata are
arranged in groups in the cytoplasm (Fig. 1). The number of
concentric bodies within one group may be up to a hundred according

to serial sections. The cytoplasm between the concentric
bodies, and in a small area around them, is less dense than in the
other cell areas. Sometimes a tubular system occurs between the
concentric bodies (Fig. 2). Since serial sections can be cut
through a complete group of concentric bodies without the
appearance of such a tubular system these structures are not
regularly associated with concentric bodies.

The diameter of the concentric bodies ranges from 230 – 300 nm
and sometimes even 400 nm. That this is a real diversity in width
becomes obvious in serial sections, since the smaller concentric
bodies appear only in 3 following sections (see number 6 and 7 in
Figs 5b, c and d) and the larger ones in four sections (see number
8 in Figs 5a – d). If the thickness of the sections is 60 – 80 nm
and the same concentric body appears three or four times we have
support for the spherical shape of this organelle.

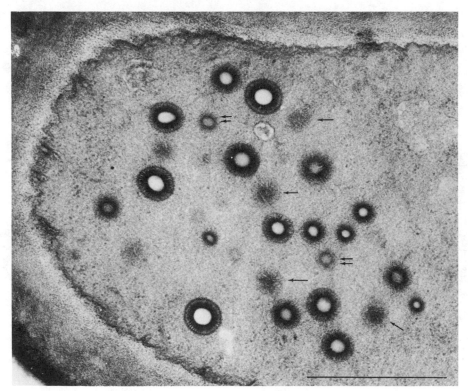

Fig. 1   Part of a mycobiont hypha with a group of concentric
          bodies. Arrows indicate concentric bodies which have been
          cut through their periphery. Double arrows point to
          organelles which are incomplete. In all figures the
          scale, if not otherwise indicated, is 1 μm.

According to transverse sections through these organelles the concentric bodies consist of an electron transparent core and two concentrically arranged shells, which differ by their electron density (Figs 1 and 2). While these main structures have been mentioned in all investigated concentric bodies (Peveling, 1969; Griffith and Greenwood, 1972; Galun et al., 1974), Boissière described in addition a "very thin edge" surrounding the core and another "dark line" about in the middle of the outer shell (Boissière, 1977, 1982; see Fig. 7).

Peripheral sections through the concentric bodies show numerous ray-like structures. They are arranged in more or less curved lines, which are usually interrupted (single arrows in Fig. 1). Some of such rays seem to traverse the periphery of a concentric body as parallel lines. Occasionally a core is still visible, but distinct shell structures are missing (double arrows in Fig. 1).

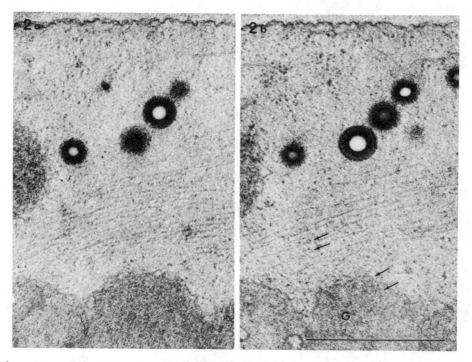

Fig. 2a and b Two sequential sections through a part of a hypha with some concentric bodies and the tubular system. The tubules seem to be connected with unknown structures (G) shown at the bottom of the picture. Arrows point to small circles within the tubules and the unknown structures.

High resolution electron micrographs of thin sections (Fig. 3) and of freeze-fractured preparations (Fig. 4) reveal that the appearance of the two shells of the concentric bodies is caused by very closely arranged rays. The length of the rays varies from 60 to 100 nm with a width of 8 - 9 nm but some are only half of this length. From some images it may be concluded that there are alternating long and short rays (arrows in Figs 3 and 4b). With such an arrangement of rays, the difference in contrast between the two shells seems plausible. The very high number of closely arranged rays forms the inner "dark" shell, while the fewer longer rays creates a "lighter" area in thin sections. All the rays start at the border of the central core. When viewed from the convex side of the central core (Fig. 4d), the closely packed rays present a membrane-like aspect. Concave views are without such an appearance (Figs 4a and b). The boundary between the central core and the shells appears as a dotted line. The continuous structures reported earlier (Peveling, 1969; Griffiths and Greenwood, 1972; see Fig. 7) were probably due to poor preparation and resolution. The dotted line proposed by Boissière (1977; see Fig. 7) is more appropriate.

The rays consist of almost round particles of 8 - 9 nm diameter arranged in lines (Fig. 4). Images of thin sections show small circles at the outer end of the rays (Fig. 3 and Fig. 5, arrow to number 3). Occasionally such circles can also be seen as rays of a concentric body, which had been cut through its periphery

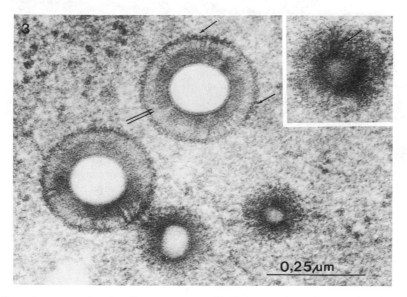

Fig. 3   Concentric bodies cut at different levels. Single arrows
          point to circles at the end of the rays.

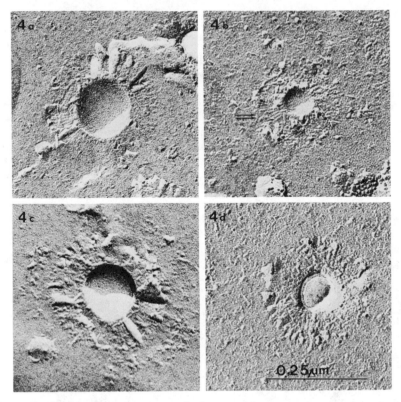

Fig. 4a - d Different images of freeze-fractured concentric
            bodies. The double arrows in Fig. 3 and Fig. 4d point to
            ending rays at the border of the "dark" shell.

(Fig. 3, upper right corner). Such observations may indicate that
the rays are not composed of straight lines of particles but that
they are tubular. It is currently uncertain whether the short rays
are also tubular. Preliminary histochemical investigations
indicate that only the long rays are tubular.

     Very often the rays of neighbouring concentric bodies are
connected to each other (Figs 3 and 5). The particles of the rays
are rather unstable since after freeze-fracturing they can be
observed scattered around the concentric bodies (Fig. 4b). Such
fragility has already been reported (Peveling, 1969; Boissière,
1982). In particular the long rays tend to break. In some cases,
when the long rays are incomplete, we could observe similar dark
lines in the outer shell as observed earlier (Peveling, 1969;
Griffiths and Greenwood, 1972; Boissière, 1977, 1982).

     The present results indicate that the shells of the concentric
bodies are formed of rays consisting of subunits and not by

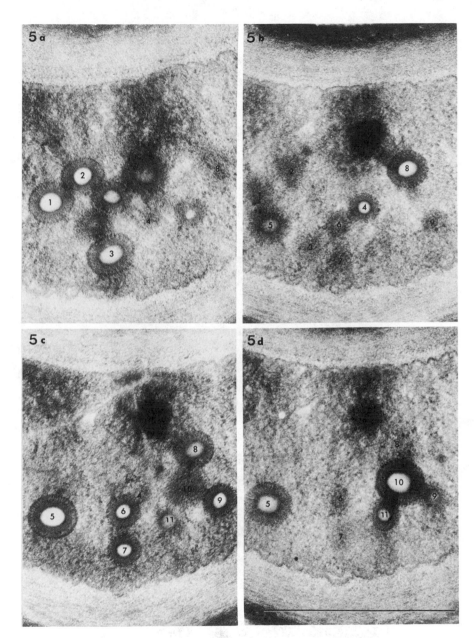

Fig. 5a – d Serial sections through a group of concentric bodies.
For easy identification the single organelles have been
numbered. Between concentric bodies many rays are
connected.

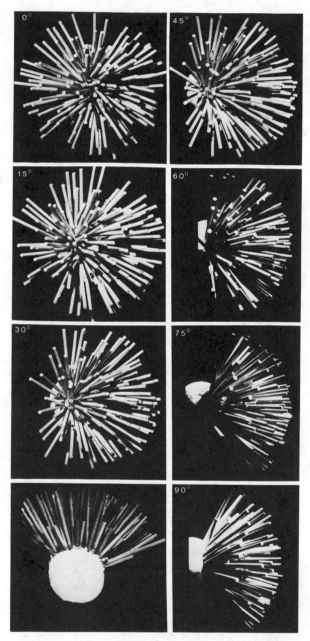

Fig. 6 Photographs of a model of a concentric body taken under the given angles. The photograph in the left corner at the bottom was taken from the halved model.

membranes. This is confirmed by serial sections (Fig. 5). When
densely arranged, the rays may have the appearance of membranes.
This is demonstrated with a scale model produced according to the
size of the components of the concentric bodies and photographed at
different angles (Fig. 6). In this way the diversity of views from
peripheral sections through the concentric bodies becomes obvious
(compare with Fig. 1). Many curved lines can be drawn along the
model sticks. At the same time it becomes clear that the ends of
the rays are only sectioned in some cases; therefore the circles
described are not always visible.

     The tubular system already mentioned, which may appear close
to the concentric bodies (Fig. 2), consists of numerous tubules.
Sometimes the tubules seem to run parallel to each other (Fig. 2a)
but in the following section this orientation may already be
different (Fig. 2b). Therefore a definite tubule length cannot be
determined. The width of a single tubule is about 9 – 10 nm. The
tubules are not continuous structures, but consist of particles
which appear circular in transverse sections (Fig. 2b). The
connecting proteinaceous substance between the tubules which was
described by Boissière (1982) could not be detected. However, a
connection between the tubules and another, undescribed, organelle
is visible. This organelle shows a very uniform grey appearance
(Fig. 2). A surrounding single membrane is not always present
although circles, like in the transverse sections of the tubules,
can be observed inside.

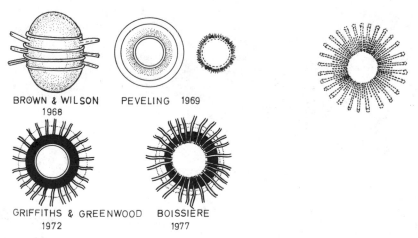

Fig. 7   Different models for the architecture of concentric
           bodies. The model without reference is the new one
           according to the presented observations.

Summarising the results, we propose a model of the architecture of the concenctric bodies which differs from the former ones in the composition of the outer shells (Fig. 7). According to our observations the shells are formed by rays, which are composed of particles. (The distances between the particles is larger in the drawing than in the electron micrographs.) There are indications that these particles, at least in the long rays, are formed as tubules. According to our present knowledge, the concentric dotted lines of earlier models appear only if the constituents of the concentric bodies are incomplete. The sequence of concentric body models demonstrates how much improved preparations and technique lead to a better understanding of structure.

Although we know many details about the architecture of the concentric bodies, we are far from understanding their origin and function. It is known from the observations of Boissière (1979) that they appear mainly in the growing region of the thallus, suggesting that concentric bodies occur in areas of high metabolic activity. The obvious similarity between the rays from the concentric bodies and the tubular system, which occasionally appears close to the concentric bodies, is a starting point for further investigations to find out both the relation between these structures and other organelles and their physiological role.

ACKNOWLEDGEMENT

We are thankful for the financial support by the Deutsche Forschungsgemeinschaft to the first author.

REFERENCES

Boissière, M.C., 1977, Un méchanisme possible d'absorption des glucides d'origine cyanophytique par les hyphes de quelques lichens, Revue Bryologie et Lichénologie, 42: 617-635.
Boissière, M.C., 1979, Cytologie du Peltigera canina (L.) Willd. en microcopie électronique: III.- le mycobionte à l'état végétatif. Revue Mycologie, 43: 1-49.
Boissière, M.C., 1982, Cytochemical ultrastructure of Peltigera canina: Some features related to its symbiosis, The Lichenologist, 14: 1-27.
Brown, R.M., and Wilson, R., 1968, Electron microscopy of the lichen Physcia aipolia (Ehrh.) Nyl., Journal of Phycology, 4: 230-240.
Galun, M., Behr, L., and Ben-Shaul, Y., 1974, Evidence for protein content in concentric bodies of lichenized fungi, Journal de Microscopie, 19: 193-196.
Galun, M., and Bubrick, P., 1984, Physiological interactions between the partners of the lichen symbiosis, Encyclopedia of Plant Physiology, 17: 362-401.

Griffiths, H.B., and Greenwood, A.D., 1972, The concentric bodies
    of lichenized fungi, Archiv für Mikrobiologie, 87: 285-302.
Jacobs, J.B., and Ahmadjian, V., 1971, The ultrastructure of
    lichens. II. Cladonia cristatella. The lichen and its
    isolated symbionts, Journal of Phycology, 7: 71-81.
Luft, J.H., 1961, Improvements in epoxy resin embedding methods,
    Journal of Biophysical and Biochemical Cytology, 9: 409-414.
Moor, H., and Mühlethaler, K., 1963, Fine structure in
    frozen-etched yeast cells, Journal of Cell Biology, 17:
    609-628.
Pease, D.C., 1964, "Histological techniques for electron
    microscopy," Academic Press, London, New York.
Peveling, E., 1969, Elektronenoptische Untersuchungen an Flechten
    III. Cytologische Differenzierungen der Pilzzellen im
    Zusammenhang mit ihrer symbiontischen Lebensweise, Zeitschrift
    für Pflanzenphysiologie, 61: 151-164.
Reynolds, E.S., 1963, The use of lead citrate at high pH as an
    electron opaque stain in electron microscopy, Journal of Cell
    Biology, 17: 208-212.

FINE STRUCTURE OF DIFFERENT TYPES OF SYMBIOTIC RELATIONSHIPS IN
LICHENS

R. Honegger

Cytologie,
Institut für Pflanzenbiologie der Universität Zürich
CH - 8008 Zürich, Switzerland

INTRODUCTION

Many different groups of fungi became independently and at
various times lichenized. As they are symbiotic with a wide range of
distantly related phyco- and cyanobionts with structurally and compositionally distinct cell walls, it is not surprising that many
different types of symbiotic relationships can be found among lichens.
Extensive comparative light microscopic studies of the mycobiont-phyco-
or cyanobiont interface in distantly related lichens were carried out
by Tschermak (1941) and Plessl (1963) who, respectively, first
observed a correlation between the anatomical organisation or evolutionary stage of a lichen and the type of mycobiont-phycobiont
relationship. Their data were largely confirmed by different transmission electron microscopic (TEM) studies (reviewed by Honegger, 1984).

The three commonest and best investigated types of mycobiont-phyco-
or cyanobiont interfaces will be discussed in detail. It must,
however, be pointed out that there are many other types of interactions among lichens which deserve further investigation.

PELTIGERACEAN SPECIES WITH NOSTOC CYANOBIONTS

The fine structure of the mycobiont-cyanobiont interface in
Peltigera species, with either thalline or cephalodial Nostoc
colonies, was studied by various authors (e.g. reviewed by Boissière,
1982; Honegger, 1982a). In both thalline and cephalodial Nostoc,
relatively thin-walled mycobiont hyphae enter the gelatinous sheath of
the colony and are then observed in close proximity to the cyanobacterial cells; neither haustorial structures nor tight cell-to-
cell adhesion have been observed.

The very hydrophilic gelatinous sheath (glycocalyx) of Nostoc
cyanobacteria is mainly built up from fibrous and amorphous glucans
(e.g. Drews and Weckesser, 1982; Honegger, 1982a) and thus represents
a nutrient-rich medium for a growing hyphal tip (Figs 1 & 8).   In
studies on carbohydrate movement between the symbionts of Peltigera
polydactyla, Hill (1972) postulated the breakdown of extracellular
glucans of the cyanobacterium to glucose by mycobiont-derived extracellular glucanase(s); glucose is then used by the mycobiont.   The
cell wall of symbiotic Nostoc is structurally and compositionally
similar to the walls of free-living cyanobacteria.   An outer, often
blebbing lipoprotein membrane with fimbriae covers the murein (peptido-
glucan) layer which, upon isolation and purification, can reveal an
interesting surface pattern (Dick and Stewart, 1980; Honegger, 1982a).

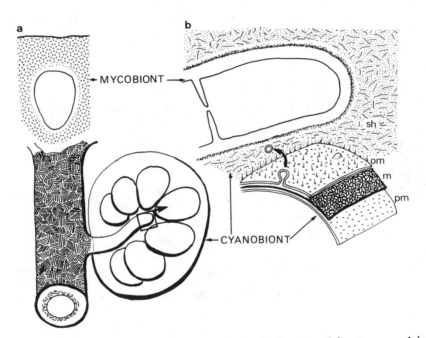

Fig. 1.   Diagrammatic interpretation of the mycobiont-cyanobiont
          (Nostoc) interface in Peltigeraceae.   (a): external
          morphology of the interface.   Thick-walled aerial hypha with
          a thin outermost rodlet layer and a small, relatively
          thin-walled lateral hypha entering the gelatinous sheath of a
          Nostoc colony.   (b): Detail of the contact site.   m, murein
          (peptidoglucan) layer; om, outer membrane with fimbriae and
          blebs which are released into the gelatinous sheath; pm,
          plasma membrane; sh, hydrophilic gelatinous sheath with
          fibrous and amorphous components (mainly glucans).   (Sources:
          Dick and Stewart, 1980; Boissière, 1982; Honegger, 1982a).

The situation at the mycobiont-cyanobiont interface in
Peltigeraceae can, to a certain degree, be compared with lecanoralean
lichens with intraparietal haustoria (see Fig. 3k-m) although the
contact between the symbionts is less complex.

PELTIGERACEAN SPECIES WITH COCCOMYXA PHYCOBIONTS

Neither intracellular, nor intraparietal haustoria but tight wall-
to-wall apposition has so far been detected in Peltigeraceae with
Coccomyxa phycobionts (see Honegger, 1984). This seems to be largely
due to the distinct structure and composition of the phycobiont cell
wall. In all free-living and symbiotic Coccomyxa species so far
investigated, tripartite cell walls were detected with an innermost
amorphous layer of irregular size and shape, an outer, regular layer
with fibrous, probably cellulosic elements, and an outermost trilaminar layer of membrane-like appearance. This trilaminar sheath
contains protein-like particles embedded in an amorphous matrix on its

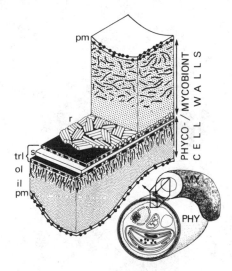

Fig. 2.   Diagrammatic interpretation of the mycobiont-phycobiont
          interface in Peltigeraceae with Coccomyxa phycobionts. il,
          inner, irregular layer of the cell wall (amorphous); ol,
          outer, regular wall layer with fibrous elements (probably
          cellulose-like); pm, plasma membrane; r, outermost wall
          layer of mycobiont with rodlet pattern. This rodlet layer
          tightly adheres to the phycobiont wall surface. s,
          sporopollenin-containing central layer of the trilaminar
          sheath (trl). The electron-dense inner and outer surfaces of
          the trilaminar sheath contain protein-like particles embedded
          in an amorphous matrix. (Sources: Honegger and Brunner,
          1981; Honegger, 1982a, 1984).

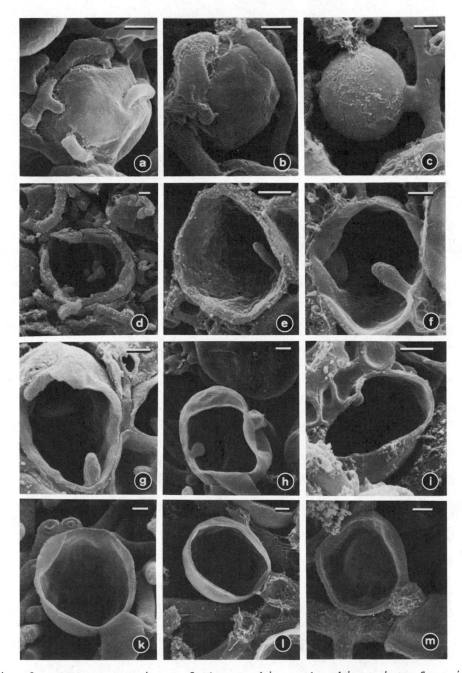

Fig. 3.   SEM preparations of the mycobiont-phycobiont interface in
          different groups of lichens with either <u>Trebouxia</u> s.lat. or
          <u>Trentepohlia</u> phycobionts.
          Bar equals 2 μm.   (Caption continued on opposite page)

electron-dense inner and outer surface and a thin, enzymatically non-
degradable, central layer with the highly resistant biopolymer sporopollenin (Honegger and Brunner, 1981). During autospore formation the
inner two layers of the cell wall are degraded whereas the trilaminar
sheath persists. Several layers of trilaminar, non-degradable mother
cell walls were found in the algal layer of taxonomically distantly
related asco- and basidiolichens (Honegger and Brunner, 1981,
Honegger, 1984).

In all peltigeracean species so far investigated with freeze-
fracture techniques, the mycobiont hyphae of the upper medullary and
algal layer reveal a thin outermost wall layer with a distinct pattern
of semicrystalline rodlets (Figs 1 & 2) which is comparable to the
rodlet layer of aerial hyphae of a great number of non-lichenized
fungi (see Honegger, 1984). In Peltigera and Solorina species this
rodlet layer tightly adheres to the cell wall surface of the Coccomyxa
phycobionts. Due to this peculiar mycobiont-phycobiont relationship
symbiotic Coccomyxa cells can be easily separated from the mycobiont,
whereas no pure phycobiont fractions can be obtained of the
trebouxioid phycobionts of lecanoralean species with either intracellular or intraparietal haustoria (see below).

LECANORALEAN SPECIES WITH TREBOUXIOID PHYCOBIONTS

The external features of the mycobiont-phycobiont contact sites
in crustose, placodioid and squamulose forms of different lecanoralean
families, including thallus squamules and podetia of Cladoniaceae with
trebouxioid phycobionts, are different from those of foliose and
fruticose groups: these external differences correlate with diver-
gences in the internal interface. In the more primitive crustose,
placodioid or squamulose forms, each phycobiont cell is contacted by
various mycobiont hyphae (Fig. 3b), whereas in evolved foliose and
fruticose species usually only one highly specialised hypha, when

---

Fig. 3. Caption continued.
          a-c, external, d-m, internal aspects of the contact site.
          a, d-g, crustose species with 1 or more intracellular
          haustoria.
          b, h-i, placodioid or squamulose species with 1 or more
          intraparietal haustoria.
          c, k-m, foliose and fruticose species with usually 1
          specialised intraparietal haustorium.
          a and e, Caliciales: Chaenotheca chrysocephala. b and i,
          Lecanorales: Cladonia macrophylla. c, Lecanorales: Cetraria
          islandica. d, Opegraphales: Lecanactis abietina. f,
          Lecanorales: Buellia punctata. g, Lecanorales: Lecanora
          subfuscata. h, Lecanorales: Lecanora muralis. k,
          Lecanorales: Parmelia acetabulum. l, Lecanorales: Anaptychia
          ciliaris. m, Teloschistales: Teloschistes flavicans.
          All Trebouxia except d, Trentepohlia.

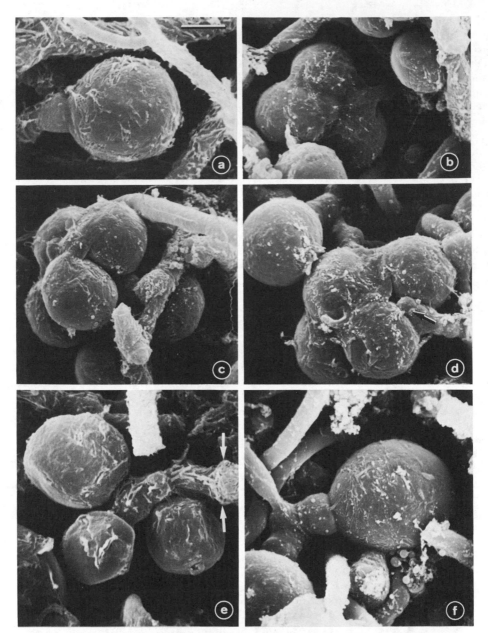

Fig. 4. SEM preparations of successive stages of autospore and
         intraparietal haustorium formation in <u>Cetrelia olivetorum</u>.
         (a): mature phycobiont cell with intraparietal haustorium.
         (b): group of autospores ensheathed by the mother cell wall
         and contacted by a mycobiont hypha. (c-d): group of
         autospores partly ensheathed by the degrading mother cell
                              (Caption continued on opposite page)

fully developed, is found in close contact with the algal cell (Figs 3c & 4a-f).

Correlations between the type of external contact and the haustorial type are obvious. In crustose thalli of lecanoralean and some other lichens with Trebouxia s.lat. or Trentepohlia phycobionts, prominent intracellular fungal haustoria are found in the majority of the algal cells (Fig. 3d-g). In placodioid and squamulose forms (including Cladoniaceae) a special type of intraparietal haustorium can be seen where the phycobiont cell wall is not pierced by the mycobiont but grows around the invading hyphal plug (Fig. 3h-i). Very interesting appressorium-like intraparietal haustoria (Figs 3c, k-m & 4a-f) are characteristic of evolved foliose and fruticose lecanoralean lichen species. All these different haustorial types have been recognized by Tschermak (1941) and Plessl (1963) and are reviewed by Honegger (1984).

Only the most primitive type of interaction with numerous hyphae in contact with single algal cells and prominent intracellular haustoria has so far been successfully resynthesised in pure culture (see Ahmadjian and Jacobs, 1982, 1983). According to these authors intracellular haustoria are characteristic of the axenically resynthesised Cladonia species, whereas intraparietal, but not intracellular, haustoria were detected in TEM and SEM preparations of naturally grown thalli of Cladonia macrophylla and C. caespiticia (Honegger, unpublished). The more complex type of intraparietal haustoria characteristic of evolved foliose and fruticose lichens has so far not been resynthesised (e.g. Bubrick et al., this volume). Its formation in the thallus follows special rules.

In all squamulose, foliose and fruticose lichens so far investigated the growing hyphal tips of the mycobiont tend to contact young autospores at an early stage of their development when they are still enclosed in the degrading mother cell wall. In their studies on Parmelia saxatilis, Greenhalgh and Anglesea (1979) suggested that leakage of nutrients within the autospore packets might stimulate the mycobiont to penetrate the mother cell wall. In Parmeliaceae and probably also in other evolved foliose Lecanorales and Teloschistales, only those mycobiont hyphae which had been contacting the young auto-

---

Fig. 4.   Caption continued.
          wall and contacted by a mycobiont hypha which, with high
          probability, was the haustorial hypha of the mother cell.
          The arrow points to a tightly adhering hypha which will not
          form an intraparietal haustorium. (e): later developmental
          stage after dissolution of the mother cell wall. The arrows
          point to the former intraparietal haustorial structure. (f):
          fully developed intraparietal haustoria.
          Bar equals 2 μm.

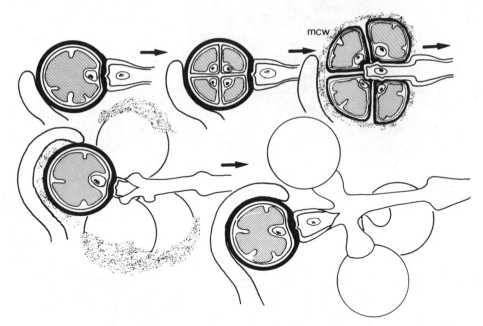

Fig. 5. Diagrammatic interpretation of the development of
         intraparietal haustoria and the shifting of algal cells by
         contacting mycobiont hyphae as observed in SEM preparations
         (Fig. 4a–f). mcw, degrading mother cell wall. (Sources:
         Honegger, 1982b, 1984).

spores within the mother cell wall will form the peculiar intraparietal haustoria (Figs 4 & 5). Hyphae contacting the phycobiont
cells at a later developmental stage will tightly adhere but not form
the typical intraparietal haustoria. In phycobiont strains and
species which usually form large numbers of autospores, several mycobiont hyphae penetrate the degrading mother cell wall, whereas often
only one, namely the hypha which was forming an intraparietal
haustorium in the mother cell, will contact the young autospores in
strains and species with few ($\pm$ 4) autospores (Figs 4 & 5).

Greenhalgh and Anglesea (1979) concluded from their ultrastructural observations that the establishment of a symbiotic relationship between growing mycobiont hyphae and young, not fully developed,
phycobiont cells might be essential for nutrient transfer between the
symbionts. With regard to the peculiar properties of the growing
hyphal tip (Fig. 8) and the situation observed at the interface of
other, equally successful fungus–host relationships such as vesicular-
arbuscular and ecto-mycorrhizae (reviewed by Harley and Smith, 1983;
Figs 9 & 10), the hypothesis of Greenhalgh and Anglesea (1979) merits
further attention. In addition, these authors claimed that the

features of the early mycobiont-phycobiont relationship are also
important for the movement of algal cells in a highly structured heteromerous thallus, a process which undoubtedly requires peculiar and very
precise regulatory mechanisms.

Another interesting aspect of the early stages of the mycobiont-
phycobiont relationship in Lecanorales with trebouxioid phycobionts
was noted in freeze-fracture preparations and ultrathin sections of
species of Cladoniaceae and Parmeliaceae (Fig. 6) in conjunction with
comparative ultrastructural studies of the cell wall surface of cultured and symbiotic myco- and phycobionts (Honegger, 1984). As demonstrated in a series of biochemical studies on the properties of phycobiont cell wall surfaces, such as antigenicity, staining properties or
lectin-binding pattern, cultured phycobiont cells differ from symbiotic ones (Bubrick and Galun, 1980; Bubrick et al., 1981, 1982; Hersoug,

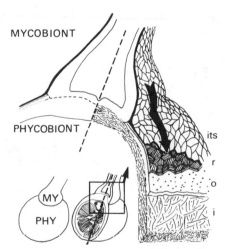

Fig. 6.   Diagrammatic interpretation of the mycobiont-phycobiont
          interface in lecanoralean lichen species with Trebouxia
          s.lat. phycobiont as observed in ultrathin sections and
          freeze-fracture preparations (Honegger, 1984). i, inner cell
          wall layer with cellulose-like fibres embedded in an
          amorphous matrix; o, outer layer with protein-like particles
          embedded in an amorphous matrix; r, rodlet layer, so far
          detected on the wall surface of young autospores of
          Cladoniaceae only; its origin (algal or fungal) remains
          unclear. its, irregularly tessellated surface layer, with
          high probability of fungal origin. This layer is covering
          the surface of myco- and phycobiont cells.
          Inner (i) and outer (o) wall layers are found in the cultured
          state of the phycobiont as well, whereas the irregularly
          tessellated surface layer is a peculiarity of the symbiotic
          state.

1983). Cytological studies indicate that there are also significant
ultrastructural differences between the cell wall surfaces of cultured
and symbiotic myco- and phycobionts (Honegger, 1984). In freeze-fracture preparations of the algal layer of Cladoniaceae and Parmeliaceae
it was noted that both myco- and phycobiont cells are covered by a continuous surface layer which reveals an irregularly tessellated pattern,
whereas no comparable structures were noted in the cultured stage of
either symbiont. In Cladoniaceae it was possible to obtain fractures
of the cell wall surface of young autospores which were still en-
sheathed by the mother cell wall but were already contacted by mycobiont hyphae. These preparations strongly suggest a spreading of the
irregularly tessellated outermost wall layer of the mycobiont over the
wall surface of the young phycobiont cell, which lacks this surface
layer in young developmental stages. It is possible that the
irregularly tessellated surface layer, of unknown chemical composition, is responsible for the water repellent nature of of the upper
medullary and algal layer and thus plays a rôle in the translocation
of water and/or metabolites between the symbionts.

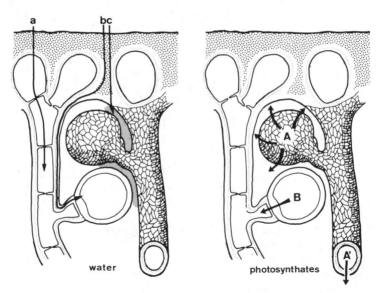

Fig. 7   Hypotheses on translocation of water and photosynthates in
          lichen thalli. Water: a, symplastic (e.g. Sievers, 1908); b,
          apoplastic within the outer part of the cell walls (von
          Goebel, 1926a, b; Stocker, 1956); c, apoplastic, along the
          cell wall surfaces by capillary forces acting between hyphae
          and/or algal cells (e.g. Stocker, 1927). Photosynthates: A,
          leaking out of phycobiont cells; working hypothesis of the
          "inhibition technique" (Richardson, 1973; Hill, 1976). A',
          leaking out of cut ends of mycobiont hyphae is also possible
          and can not be excluded in the "inhibition technique". B,
          haustorial.

TRANSLOCATION OF WATER AND PHOTOSYNTHATES BETWEEN THE SYMBIONTS

As pointed out by Blum (1973), it is not clear how water is translocated within heteromerous lichen thalli. All investigators agree
that the hydrophilic extracellular material of the cortical layer
plays an important rôle in water uptake and storage. From the cortex
water is supposed to move symplastically through the cytoplasm (e.g.
Sievers, 1908), apoplastically along the cell wall surfaces by
capillary forces acting between fungal hyphae and/or algal cells (e.g.
Stocker, 1927) or apoplastically within the cell walls, probably also
by means of capillary forces (von Goebel, 1926a,b; Stocker, 1956)
(Fig. 7). The morphologist von Goebel clearly distinguished between
swelling hyphae ("Schwellhyphen") of the cortical layer(s) and water-
repellent aerial hyphae ("Lufthyphen") of the gas-filled algal and
medullary layers. On the basis of simple experiments, he demonstrated
that apoplastic water translocation along the cell wall surfaces is
doubtful, since the aerial hyphae are almost unwettable unless their
surface properties are changed by treatment with solvents. Von Goebel
supposed the water repellency to be due to lichen substances located
at the cell wall surfaces (see Green et al., Richardson et al., this
volume). Further studies have to show whether the peculiar surface
layers such as the irregularly tessellated layer of Lecanorales or the
rodlet layer of Peltigerales are, together with crystalline extra-
parietal lichen products, responsible for the water repellency of the
algal and medullary layers. Structurally comparable cell wall surface
layers have been detected in aerial hyphae of non-lichenized fungi
such as conidia and conidiophores in Hyphomycetes or the hymenial
surface of basidiocarps (reviewed by Honegger, 1984). It is possible
that all of these particular surface layers have a regulatory function
in water uptake and/or water loss and might be functionally analogous
to the cuticle of higher plants.

Photosynthates are supposed to leak out of the phycobiont cells
and thus to be detectable in the incubation medium on which lichen
discs, whose medullary layer was cut off, have been placed. In this
model, which is the basis of a large and interesting series of experiments on the movement of carbohydrates between the symbionts with the
so-called inhibition technique according to Smith and his co-workers
(see Richardson, 1973; Hill, 1976), one cannot, however, exclude the
possibility that the (labelled) compounds detected in the incubation
medium were leaking out of the cut ends of mycobiont hyphae and thus
do not necessarily represent the type of carbohydrate which was moving
from phycobiont to mycobiont. In Peltigeraceae with <u>Nostoc</u> cyanobionts it is possible that products (glucose) of the glucan sheath,
which had been hydrolysed by mycobiont-derived hydrolases, are
detectable in the incubation medium. The very low and slow rates of
translocation of photosynthates from trebouxioid phycobionts to
lecanoralean mycobionts measured with the inhibition technique
(Richardson, 1973) might refer to different modes of carbohydrate
mobilisation and movement in this system. The cytological data,

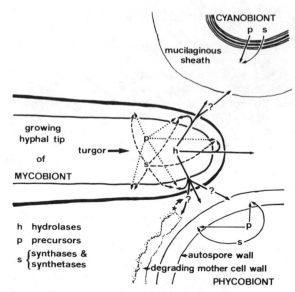

Fig. 8.   Properties of the growing hyphal tip as reviewed by Farkas
          (1979): a delicately balanced equilibrium exists between cell
          wall synthesis and hydrolysis. Hydrolases are also released
          into the substratum (e.g. host cell walls in pathogenic and
          symbiotic systems). The production of hydrolases can be
          stimulated by components of the substratum (* products of
          degrading mother cell walls might have a stimulatory effect
          on mycobiont hydrolases).

mainly the presence of a continuous surface layer on myco- and phycobiont cells, do not support the idea of carbohydrate leakage in the
surrounding area but rather emphasises the significance of haustorial
structures in translocation.

     The rôle of haustorial and appressoria-like structures in lichens
may have been underestimated (e.g. Collins and Farrar, 1978). As the
most highly evolved types of haustoria in lecanoralean lichens are
formed by growing hyphal tips with their peculiar properties (reviewed
by Farkas, 1979) and not fully developed algal cells, the events
taking place within the cell walls at the contact site are particularly interesting (Fig. 8). This situation strongly resembles the interface in ectomycorrhizae (Figs 9 & 10) where growing mycobiont
hyphae contact young, not fully differentiated, root cells (Harley and
Smith, 1983). Passive leakage of carbohydrates may be achieved as a
final step in the fully developed symbiotic relationship in Lecanorales
but at the early stages of the mycobiont interaction with young, not
fully differentiated phycobiont cells the mobilisation of cell wall
components is, with high probability, an equally important event.

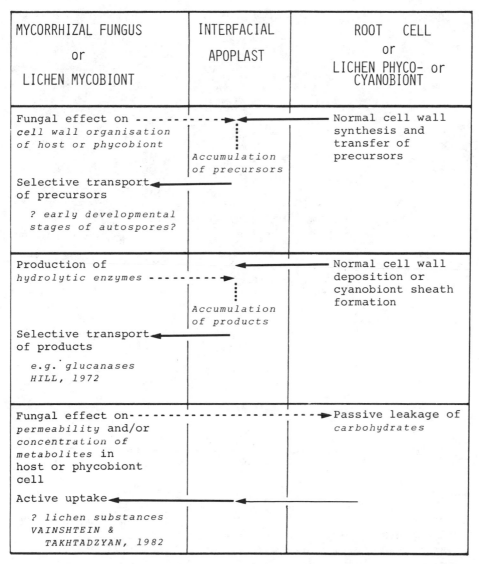

MYCORRHIZAL FUNGUS or LICHEN MYCOBIONT	INTERFACIAL APOPLAST	ROOT CELL or LICHEN PHYCO- or CYANOBIONT
Fungal effect on *cell wall organisation of host or phycobiont* *Accumulation of precursors* Selective transport of precursors *? early developmental stages of autospores?*		Normal cell wall synthesis and transfer of precursors
Production of *hydrolytic enzymes* *Accumulation of products* Selective transport of products *e.g. glucanases HILL, 1972*		Normal cell wall deposition or cyanobiont sheath formation
Fungal effect on *permeability* and/or *concentration of metabolites* in host or phycobiont cell Active uptake *? lichen substances VAINSHTEIN & TAKHTADZYAN, 1982*		Passive leakage of *carbohydrates*

Fig. 9.   Three possible modes of interaction at the fungus-host
          interface in mycorrhizae, as reviewed by Harley and Smith
          (1983), which might equally occur in the lichen symbiosis.

     Further cytological, biochemical and physiological investigations
of the mycobiont- phyco- and cyanobiont interfaces are needed to
improve our understanding of the symbiotic relationship in lichens.

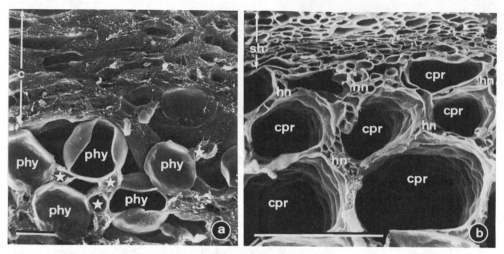

Fig. 10. SEM preparations of parts of cross-sections through (a) the
lichen <u>Anaptychia ciliaris</u> and (b) an undetermined
ectotrophic mycorrhiza of <u>Abies alba</u>. (a): c, cortex; phy,
phycobiont cell; asterisks on mycobiont hyphae forming
intraparietal haustoria in phycobiont cells. (b): hn, Hartig
net (mycobiont) in between the cells of the cortical
parenchyma of the root (cpr); sh, external fungal sheath.
Bar equals 5 μm.

ACKNOWLEDGEMENTS

    My sincere thanks are due to Dr H.R. Hohl for many stimulating
discussions, to Dr D.H. Brown for assistance with the manuscript and
to Mr Ueli Lüthi for excellent technical assistance.

REFERENCES

Ahmadjian, V., and Jacobs, J.B., 1982, Algal/fungal symbioses, <u>in</u>:
    "Progress in Phycological Research," F.E. Round and D.J. Chapman,
    eds, Volume 1, pp. 179-233, Elsevier Biomedical Press, Amsterdam.
Ahmadjian, V., and Jacobs, J.B., 1983, Algal-fungal relationships in
    lichens: recognition, synthesis and development, <u>in</u>: "Algal
    symbiosis," L.J. Goff, ed., pp. 147-172, Cambridge University
    Press, Cambridge.
Blum, O.B., 1973, Water relations, <u>in</u>: "The Lichens," V. Ahmadjian and
    M.E. Hale, eds, pp. 381-400, Academic Press, New York.
Boissière, M.-C., 1982, Cytochemical ultrastructure of <u>Peltigera</u>
    <u>canina</u>: some features related to its symbiosis, <u>The Lichenologist</u>
    14: 1-28.
Bubrick, P., and Galun, M., 1980, Symbiosis in lichens: differences in
    cell wall properties of freshly isolated and cultured phyco-
    bionts, <u>FEMS Microbiology Letters</u>, 7: 311-313.

Bubrick, P., Galun, M., and Frensdorff, A., 1981, Proteins from the
    lichen Xanthoria parietina which bind to phycobiont cell walls.
    Localization in the intact lichen and cultured mycobiont,
    Protoplasma, 105: 207-211.
Bubrick, P., Galun, M., Ben-Yaacoov, M., and Frensdorff, A., 1982,
    Antigenic similarities and differences between symbiotic and
    cultured phycobionts from the lichen Xanthoria parietina, FEMS
    Microbiology Letters, 13: 435-438.
Collins, C.R., and Farrar, J.F., 1978, Structural resistance to mass
    transfer in the lichen Xanthoria parietina, New Phytologist, 81:
    71-83.
Dick, H., and Stewart, W.D.P., 1980, The occurrence of fimbriae on a
    N_2-fixing cyanobacterium which occurs in lichen symbiosis,
    Archives of Microbiology, 124: 107-109.
Drews, G., and Weckesser, J., 1982, Function, structure and
    composition of cell walls and external layers, in: "The Biology
    of Cyanobacteria," N.G. Carr and B.A. Whitton, eds, pp. 333-357,
    Blackwell Scientific Publications, Oxford.
Farkas, V., 1979, Biosynthesis of cell walls in fungi, Microbiological
    Reviews, 43: 117-144.
Goebel, K. von, 1926a, Die Wasseraufnahme der Flechten, Berichte der
    Deutschen Botanischen Gesellschaft, 44: 158-161.
Goebel, K. von, 1926b, Morphologische und biologische Studien. VII.
    Ein Beitrag zur Biologie der Flechten, Annales du Jardin
    botanique de Buitenzorg, 36: 1-83.
Greenhalgh, G.N., and Anglesea, D., 1979, The distribution of algal
    cells in lichen thalli, The Lichenologist, 11: 283-292.
Harley, J.L., and Smith, S.E., 1983, "Mycorrhizal symbiosis," Academic
    Press, London.
Hersoug, L.G., 1983, Lichen protein affinity towards walls of cultured
    and freshly isolated phycobionts and its relationship to cell
    wall cytochemistry, FEMS Microbiology Letters, 20: 417-420.
Hill, D.J., 1972, The movement of carbohydrate from the alga to the
    fungus in the lichen Peltigera polydactyla, New Phytologist, 71:
    31-39.
Hill, D.J., 1976, The physiology of lichen symbiosis, in:
    "Lichenology: Progress and Problems," D.H. Brown, D.L. Hawksworth,
    and R.H. Bailey, eds, pp. 457-496, Academic Press, London.
Honegger, R., 1982a, Cytological aspects of the triple symbiosis in
    Peltigera aphthosa, Journal of the Hattori Botanical Laboratory,
    52: 379-399.
Honegger, R., 1982b, Ascus structure and function, ascospore
    delimitation and phycobiont cell wall types associated with the
    Lecanorales (lichenized ascomycetes), Journal of the Hattori
    Botanical Laboratory, 52: 417-429.
Honegger, R., 1984, Cytological aspects of the mycobiont-phycobiont
    relationship in lichens, The Lichenologist, 16: 111-127.
Honegger, R., and Brunner, U., 1981, Sporopollenin in the cell walls
    of Coccomyxa and Myrmecia phycobionts of various lichens: an
    ultrastructural and chemical investigation, Canadian Journal of

Botany, 59: 2713-2734.

Plessl, A., 1963, Ueber die Beziehungen von Pilz und Alge im
Flechtenthallus, Österreichische Botanische Zeitschrift, 110:
194-269.

Richardson, D.H.S., 1973, Photosynthesis and carbohydrate movement,
in: "The Lichens," V. Ahmadjian, and M.E. Hale, eds, pp. 249-288,
Academic Press, New York.

Sievers, F., 1908, Ueber die Wasserversorgung bei Flechten.
Wissenschaftliche Beilage 38. Jahresbericht der Landwirtschaft-
lichen Schule Marienberg zu Helmstedt.

Stocker, O., 1927, Physiologische und oekologische Untersuchungen an
Laub- und Strauchflechten, Flora, N.F. 21: 334-415.

Stocker, O., 1956, Wasseraufnahme und Wasserspeicherung bei
Thallophyten, in: "Handbuch der Pflanzenphysiologie," W. Ruhland,
ed., Vol. 3, pp. 160-172, Springer, Berlin.

Tschermak, E., 1941, Untersuchungen über die Beziehung von Pilz und
Alge im Flechtenthallus, Österreichische botanische Zeitschrift,
90: 233-307.

Vainshtein, E.A., and Takhtadzhyan, E.A., 1982, Physiological changes
in the lichen alga Trebouxia during cultivation. Translated from
Fiziologiya Rastenii, 28 (5): 1037-1044 (1981) in Soviet Plant
Physiology 28: 763-769.

CHANGES IN PHOTOBIONT DIMENSIONS AND NUMBERS DURING CO-DEVELOPMENT OF

LICHEN SYMBIONTS

D.J. Hill

Department of Botany
The University
Bristol, BS8 1UG, UK

INTRODUCTION

The development of the changing interactions between symbionts
that occurs in a growing lichen, from the lobe tips to the more mature
regions of the thallus, has not been specified precisely nor has it
been investigated in detail. Such work that has been done has been
reviewed by various authors, e.g. Nienburg (1926), Jahns (1973),
Henssen and Jahns (1974), Greenhalgh and Anglesea (1979), Hill (1981).
These developmental processes, which are of key importance to the
lichen symbiosis, have two major features. Firstly, the overall
structure is primarily composed of fungal tissues originating from
hyphae which aggregate in some kind of process of continuous morphogenesis. The vegetative thallus that has evolved, at least in the
more advanced species, does not appear to have any obvious counterpart
in non-lichenized fungi. Secondly, the developmental sequence only
leads to a functionally competent lichen thallus if both the fungus
and the photobiont develop in a co-ordinated tandem, termed here
balanced co-development. The balance of the co-development involves
the co-ordination of the expression of two separate genomes. Anatomi-
cally, it involves growth of the fungus by hyphal aggregation and of
the photobiont, which is usually unicellular or in some cases filamen-
tous, by aplanospore formation or cell fission.

This paper deals with some of the morphological and anatomical
manifestations of these processes with special reference to photobiont
cell size and cell density. The lichen species chosen were those that
contained different photobionts but which had well developed lobes
that might have a developmental sequence along them.

MATERIALS AND METHODS

Lobes that were apparently in an active state of vegetative growth were carefully selected. Hydrated lichen thalli were dissected under a x10 binocular microscope to remove debris and a small portion, approximately 1 mm^2, was cut out of the thallus. This was then dissected using a hypodermic needle and a fine pointed pair of forceps under a x75 binocular microscope, to remove medullary tissue, and then photobiont cells were teased into a drop of water. The dissection was done as thoroughly as possible so that the free cells constituted a representative sample. The drop of water containing the free cells was then transferred, in a pasteur pipette with a finely drawn out point, to a microscope slide and covered with a coverslip. Excess water was squeezed out and the coverslip ringed with sealant (Gurr).

Photobiont cell diameter was measured to the nearest half unit of the eye-piece graticule at x1000 magnification. The histograms showing cell size distribution are set out in the original eye-piece unit size-classes to avoid any artefacts created by conversion to absolute unit size-classes. Photobiont cells were for the most part selected at random on the microscope slide by measuring every cell that coincided with the graticule scale during scans of the microscope slide. Techniques that have been used to estimate photobiont cell numbers by Drew and Smith (1967), Millbank and Kershaw (1969) and Kershaw and Millbank (1970) have been considered unreliable (Millbank 1976). The method adopted here is the Millbank (1972) modification of the Hill and Woolhouse (1966) technique using a chromium trioxide solution for tissue maceration. To obtain pieces of thallus of standard area from Peltigera and Lobaria, 3.5 mm discs were punched out with a cork borer while for Xanthoria and Hypogymnia pieces were cut with a scalpel on graph paper. The concentration of cells in the macerate was measured using a haemocytometer. Care was taken to distinguish between photobiont cells and other spherical cells of similar size e.g. cortical cells, since cell contents were no longer discernible following the chromium trioxide treatment.

Thallus thickness was measured to 0.05 mm with vernier-scale calipers on dry thalli. Thalli were dried overnight at 60°C for dry weight determination.

RESULTS

Photobiont Dimensions

Measurement of photobiont cell diameter indicates that photobiont cell size may increase as the tissues age in the growing lobe. In Hypogymnia physodes (photobiont: Trebouxia) the mean diameter of the photobiont cells was greater 5 mm from the lobe end (9.18 $\pm$ 0.18 μm) than at the growing point (7.45 $\pm$ 0.10 μm) (Fig. 1). The range of cell size was however about the same. In Xanthoria parietina (photo-

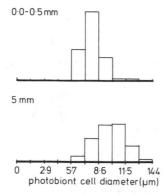

Fig. 1. Photobiont cell diameter distribution in <u>Hypogymnia physodes</u>
(February).
94 cells measured per sample. Note: vertical axis represents
number of cells in each size class; when range of two
distances from the lobe end is given, sample was taken from
between these points; when one is given, sample was taken at
that distance.

biont: <u>Trebouxia</u>) a similar pattern was observed in measurements taken
in February and July (Figs. 2 & 3). Although the mean cell diameter
in the two samples was slightly different, it is not clear whether

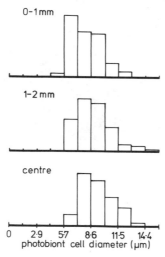

Fig. 2. Photobiont cell diameter distribution in <u>Xanthoria</u>
(February).
100 cells measured per sample. See notes in Fig. 1; centre =
> 5mm from lobe end.

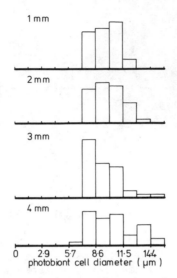

Fig. 3.   Photobiont cell diameter distribution in <u>Xanthoria</u> (July).
100 cells measured per sample; see notes in Fig. 1.

this was due to seasonal variation. In both samples the proportion of
larger cells increased and the proportion of smaller cells decreased
with increasing distance from the growing point. The February data
showed a statistically significant increase in the mean (t-test, P <
0.05). In <u>Peltigera horizontalis</u> (photobiont: <u>Nostoc</u>) the cyanobiont
showed a substantial increase in cell diameter from 6.41 + 0.10 µm at
the growing point to 8.65 ± 0.14 µm about 10 mm from it (Fig. 4).   A
size difference of this magnitude is noticeable by eye without measurement and was also conspicuous in <u>Peltigera praetextata</u> (photobiont:

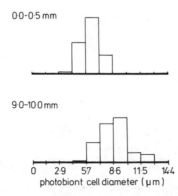

Fig. 4.   Photobiont cell diameter distribution in <u>Peltigera
horizontalis</u> (February).
80 cells measured per sample; see notes in Fig. 1.

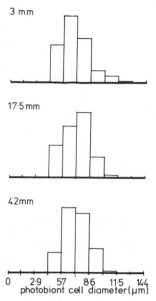

Fig. 5.   Photobiont cell diameter distribution in <u>Lobaria pulmonaria</u>
(February).
(lobe A; see Table 1).   100 cells measured per sample; see
notes in Fig. 1.

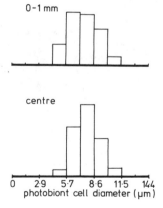

Fig. 6.   Photobiont cell diameter distribution in <u>Lobaria pulmonaria</u>
(February).
(lobe B; see Table 1).   100 cells measured per sample; see
notes in Fig. 1.

Table 1. Ratio of Diameter of Large and Small Photobiont Cells (1)

Distance from lobe end (mm)	Size/Number Score (2)		

Lobaria pulmonaria

	< 7.2 μm	> 7.2 μm	Ratio
LOBE A			
0.0-3.5	84	56	0.67
17.5	90	65	0.72
42	62	66	1.06
LOBE B			
Growing point of lobe	55	87	1.58
Central region (3)	37	95	2.57

Xanthoria parietina

	< 10 μm	> 10 μm	Ratio
FEBRUARY			
0-1	104	50	0.48
1-2	53	83	1.56
centre (4)	45	102	2.27
JULY			
0-1	34	20	0.59
1-2	32	26	0.81
2-3	44	20	0.45
3-4	31	41	1.32

Notes: (1) see Figures 2, 3, 5 & 6
       (2) number of cells in an eye-piece unit size-class multiplied
           by the number of classes that class is away from the mean
       (3) > 4 cm from lobe end
       (4) > 5 mm from lobe end

Nostoc) collected in July. In *Lobaria pulmonaria* (photobiont:
Myrmecia), little change in the size-class distribution of the cells
was immediately obvious (Fig. 5). Scoring the numbers of smaller and
larger cells and comparing the scores for different parts of the lobe
indicated that there was some change in the size distribution of the
cells (Table 1). In another lobe of *L. pulmonaria* measurements of
photobiont cell diameter from the growing point were compared with
those from the centre of the thallus (> 4 cm from lobe end) (Fig. 6
and Table 1) with a similar suggestion of an increase in cell size.

Table 2. Density of Photobiont Cells

Distance from lobe end (mm)	Number of cells (10^6 cm^{-2})	Thickness of photobiont layer (μm)	% cells empty
Xanthoria parietina			
0-1	229 + 34	-	-
1-2	167 + 15	-	-
centre	196 + 14	-	-
Lobaria pulmonaria			
0-3.5	120 + 22	37 + 3	15.6 + 2.2
17.5	202 + 32	59 + 3	16.2 + 4.3
42	128 + 20	56 + 3	28.0 + 3.8

Density of Photobiont Cells in the Thallus

Estimates were made of the number of photobiont cells per unit area of thallus. Cell density appeared to be remarkably constant along the growing lobe indicating that there was no substantial increase in the number of photobiont cells during lobe growth. This was evident in the data obtained for **X. parietina** (Table 2),

Table 3. Photobiont Cell Density in **Peltigera praetextata** in February

Distance from lobe end (mm)	Filament density (10^6 cm^{-2})	Cells per filament (1)	Cell density calculated (10^6 cm^{-2})	Cell density observed (10^6 cm^{-2})
0-1	0.28 + 0.02(2)	36.1 + 4.4	10.1 + 3.7	5.04 + 0.81
2-4	0.61 + 0.11(3)	57.6 + 5.4	35.0 + 19.6	-
4-6	0.31 + 0.02(2)	20.3 + 1.3	6.3 + 1.6	-
6-8	0.29 + 0.02(2)	53.1 + 3.3	15.4 + 3.6	-
8-10	0.47 + 0.12(3)	55.2 + 9.4	26.1 + 2.7	-
10-12	0.27 + 0.06(3)	31.3 + 2.6	8.5 + 5.0	-
12-14	0.29 + 0.02(2)	30.9 + 3.5	9.1 + 3.4	4.70 + 0.65

Notes: (1) cells of 11-30 filaments counted
       (2) > 130 filaments counted
       (3) < 50 filaments counted
       Standard error of means is that between estimates not samples

Table 4.   Photobiont Cell Density in <u>Peltigera praetextata</u> in July

Distance from lobe end (mm)	Filament density $(10^6 \text{ cm}^{-2})(1)$	Cells per filament (2)	Calculated cell density (10^6 cm^{-2})	Heterocyst frequency (%)
LOBE A				
0– 3.5	0.16 + 0.01	69 + 5	11.0	2.1 + 0.4
3.5– 7.0	0.17 + 0.02	–	–	–
7.0–10.5	0.16 + 0.02	–	–	–
10.5–14.0	0.09 + 0.02	–	–	–
14.0–17.5	0.13 + 0.02	65 + 10	6.5	2.9 + 0.3
17.5–21.0	0.09 + 0.01	–	–	–
21.0–24.5	0.19 + 0.02	–	–	–
24.5–28.0	0.18 + 0.01	–	–	–
28.0–31.5	0.12 + 0.02	–	–	–
31.5–35.0	0.17 + 0.02	52 + 7	8.8	6.9 + 0.7
35.0–38.5	0.14 + 0.01	–	–	–
38.5–42.0	0.12 + 0.01	–	–	–
42.0–45.5	0.18 + 0.02	–	–	–
45.5–49.0	0.14 + 0.02	–	–	–
49.0–52.5	0.17 + 0.02	29 + 3	4.9	7.6 + 0.9
LOBE B				
0.0– 3.5	0.10 + 0.01	72 + 12	7.2	4.3 + 0.8
14.0–17.5	–	33 + 3	–	4.5 + 0.6
31.5–35.0	–	37 + 3	–	5.1 + 0.7
49.0–52.5	0.21 + 0.02	36 + 5	7.6	9.0 + 1.0

Notes: (1) 9 estimates
       (2) 7 or > 7 estimates
       Standard error of mean is that between estimates not samples

<u>L. pulmonaria</u> (Table 3) and <u>P. praetextata</u> (Tables 3 and 4). In
addition, other characteristics were measured. The photobiont layer
of <u>L. pulmonaria</u> proved very difficult to measure reliably owing to
the diffuse nature of its boundaries. The number of empty cells
appeared to be somewhat greater at 42 mm from the growing point than
at the growing point (using t-test the difference being significant at
$P < 0.05$).

    In <u>P. praetextata</u> the photobiont cell number was difficult to
estimate accurately in the haemocytometer since the folding of

filaments obscured some of the cells. Therefore the density of the
filaments rather than the cells was estimated. The number of cells,
and heterocysts, per filament was readily counted when the filaments
were mounted on a microscope slide and the coverslip pressed down to
spread the filament. The estimated number of cells per unit area was
very variable since it involved the multiplication of the errors of
the means. To confirm that the estimate of the numbers of cells per
unit area was close to the number actually in the photobiont layer,
attempts were made to count the number of cells per unit area by
direct observation of the photobiont layer after removal of the
medulla. Careful focusing of the microscope through the layers of
cells in the photobiont layer permitted a reasonably thorough count.
The estimates of the photobiont cell density by these two methods were
reasonably close and of the same order of magnitude (Table 3).
Moreover, both methods indicate that cell density is similar whether
near the growing point or in more mature regions of the thallus; there
was also no significant increase in the number of cells per filament.
A similar estimate of the photobiont density made in July (Table 4)
confirmed these findings. However, the density of the filaments was
somewhat lower and the numbers of cells per filament greater but it is
not known whether these differences were due to seasonal variation.
Heterocyst frequency increased significantly from the growing point in
accord with other findings of Hill (unpublished).

Fungal Development

     Although it is not the purpose of this paper to include fungal
differentiation, it is important to relate the development of the
photobiont to that of the fungus. In both P. horizontalis and
L. pulmonaria thallus thickness and thallus dry weight per unit area
increased with distance from the growing point (Figs 7 & 8).

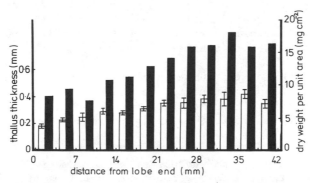

Fig. 7. Dry weight per unit area and thallus thickness along lobe of
         Peltigera horizontalis (February).
         Mean of data from 15 discs. Bars with standard error lines
         are thickness; black bars are dry weight.

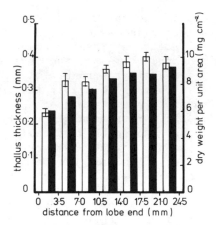

Fig. 8.   Dry weight per unit area and thallus thickness along lobe of
          Lobaria pulmonaria (February).
          Mean of data from five lobes. Bars with standard error lines
          are thickness; black bars are dry weight.

DISCUSSION

     It is well known that the size of algal cells is very variable
and can be affected by many environmental factors. In the lichen
thallus changes in size distribution were studied along the length of
single lobes and were presumed to indicate developmental patterns on
the assumption that distance from the lobe tip is directly related to
age. This assumption is also made for cellular development and
differentiation in the roots of higher plants. Since the thalli were
collected from the field and not grown under constant and controlled
conditions it may be argued that the differences in cell size may be
the result of environmental fluctuations. However, this seems
unlikely since similar patterns were found in different species and at
different times of year. Anglesea et al. (1983) also found that photobiont cells in the thallus tip of Usnea subfloridana were smaller
(7.85 $\pm$ 0.02 μm) than in the rest of the thallus (9.80 $\pm$ 0.02 μm).

     Clarification of the question of how control of the morphological
development of the photobiont is exercised, presumably by the fungus,
would be assisted by information on quantitative characteristics of
the population of photobiont cells in the growing thallus. Evidence
to date suggests that:-

     a) The density of the photobiont cells number per unit surface
area of thallus remains fairly constant from 1 mm of the lobe tip to
senescent parts of the thallus. This has been observed in Xanthoria,
Lobaria and Peltigera (Tables 2, 3 & 4) but it is not known if it is
also a feature of crustose species. In Xanthoria a slightly higher

density in the thallus tip and a higher density reported in podetia tips in <u>Cladonia</u> (Plummer and Gray, 1972) suggests that density distribution requires further investigation.

b) Cell division occurs at a greater frequency at the thallus tip than in older regions (Greenhalgh and Anglesea, 1979; Boissière, 1982; Anglesea et al., 1983).

c) Some cell division was observed in all parts of thallus, suggesting a turnover of cells to maintain constant cell number (Galun et al., 1970; Greenhalgh and Anglesea, 1979; Anglesea et al., 1983).

d) Senescent or dead cells are present in numbers that increase with distance from the thallus tip (Galun et al., 1970; Greenhalgh and Anglesea, 1979; Anglesea et al., 1983; Table 2).

e) Photobiont cell density varies with habitat (Hill and Woolhouse, 1966; Harris, 1971) and apparently with season (Harris, 1969; Plummer and Gray, 1972) although this requires further investigation.

f) The proportion of larger photobiont cells increases and the proportion of smaller cells decreases, resulting in an increase in their mean diameter, with distance from the thallus tip (this paper; Anglesea et al., 1983).

g) The space for the photobiont is created by fungal growth mainly at the thallus tip (Hale, 1970; Boissière, 1972; Fisher and Proctor, 1978) but also to a lesser and decreasing extent within the first few millimetres from the tip.

h) In cyanobionts, heterocyst frequency increases with distance from the growing point of the lobe (Englund, 1977; Table 4).

These features suggest that the photobiont population increases rapidly at and near the growing point but, with a certain amount of cell turnover, remains more or less constant in the rest of the thallus. It appears, therefore, that the photobiont cell population must be under some kind of limiting control involving changed rates of progress through the cycle of cell growth and division.

The cell cycle in micro-algae, as in other micro-organisms, is associated with cell diameter changes such that G1 cells are small and cell enlargement takes place in the S phase leading to G2 in which the largest cells occur. Progress through the cell cycle can be observed by monitoring cell diameter. In a non-synchronous culture, mean cell diameter remains constant but if the cell cycle is slowed or arrested the point at which it is impeded or halted will be indicated by accumulation of cells at a particular size.

The histograms of photobiont cell diameters at the growing point (Figs 1-6), show size distributions that include G1, S and G2 phase cells. In H. physodes (Fig. 1) and P. horizontalis (Fig. 4) the size distribution in the more mature regions of the thallus indicate that some cells have exceeded the diameter one would expect for the largest cells near the growing point and hence the diameter of cells normally in G2 in an increasing population. Cell enlargement to this extent has also been observed in the cyanobiont of Azolla in which the vegetative cells of Anabaena in the cavities of the mature leaves reach up to 7x the volume of the dividing cells at the growing point of the stem (Hill, 1977). Cell enlargement in cyanobiont vegetative cells in other plant symbioses has also been reported (Stewart et al., 1980) and is well known in bacteroid development in legume root nodules. These cell enlargements suggest that control of the cell cycle may occur by interruption at the S phase such that the cells no longer exit from S and continue growing beyond the size required to enter G2 and cell division.

The size distribution in the mature regions of Xanthoria and Lobaria did not show a marked increase beyond the maximum diameter of the cells at the growing point. The increase in the proportion of larger cells and decrease in the proportion of smaller cells seems to suggest that the cell cycle may be controlled between wall synthesis and cell division. It is possible also that the rate of cell turnover in these species was greater and both Greenhalgh and Anglesea (1979) and Anglesea et al. (1983) have observed substantial proportions of cells in a state of division and the associated empty cell walls of dead cells. Empty cell walls have been observed in lichens by numerous authors (e.g. Galun et al., 1970; Hill, 1976; Greenhalgh and Anglesea, 1979; Anglesea et al., 1983). There is a question as to whether these cell walls in lichens containing Chlorococcalean photobionts such as Trebouxia, result from autospore formation, therefore representing the discarded mother cell wall, or whether they result from cell death. About as many authors have described senescent photobiont cells in lichen thalli so it would be expected that at least some of the empty cell walls are indeed indicative of cell turnover.

There are features of the Peltigera photobiont that are relevant to cyanobiont development. It is not clear whether filaments observed after maceration in Cr_2O_3 have been extracted from the thallus intact. Their constant length in macerates might be due to the method of tissue homogenization although greater homogenization with the pasteur pipette failed to reduce their length further. With special attention to maintaining a standard technqiue (Table 5) a reduction in filament length was observed with increasing distance from the lobe tip but further investigation is required to find out if this is correlated with ageing in the thallus. In Azolla, filament fragmentation of the cyanobiont appears to be correlated with ageing and leaf senescence (Hill, 1977).

Although the dimensions of the photobiont layer remain remarkably constant throughout lobe development, fungal growth continues in mature and seemingly differentiated parts of the thallus. In X. parietina the thallus becomes more corrugated and ascocarps develop, while in Peltigera and Lobaria the fungus continues to grow with an accumulation of dry matter and a greater medulla thickness. Such fungal growth is presumably a sink for the carbohydrate produced by the photobiont in the mature thallus, since it appears, from studying growth patterns, that carbohydrate available in these central parts of the thallus cannot be translocated to the growing point in any significant amounts (Hill, 1981).

Further work must be done to find out if the patterns of photobiont co-development described here are generally observed in growing lichen thalli and whether the conclusions that they suggest can be substantiated by other experimental evidence, e.g. enzyme activities (Lallemant and Savoye, this volume) and possibly lectin interaction (Bubrick et al., this volume). Anglesea et al. (1983) made an excellent study of the structural relationship between the hyphae of the mycobiont and the Trebouxia cells in the growing point of U. subfloridana. In Azolla, cytological and biochemical investigation of the sequence of the co-development in both symbionts (Peters and Calvert, 1983) has provided considerable insight into the symbiosis. As suitable material for research both lichens and Azolla have the advantage that algal symbiont co-development is displaced spatially along the thallus or axis as a result of the continuous morphogenesis of these organisms. In other algal symbioses, e.g. Hydra, symbiont regulation may also be mediated by control of the cell cycle (Douglas and Smith, 1984) without there being any spatial separation of the different stages of co-development. Results that can be obtained from further research using lichens could therefore have a wider application in symbiotic studies in general.

ACKNOWLEDGEMENTS

I should like to thank Professor A.E. Walsby for allowing me to use the facilities of the Botany Department and Dr D.H. Brown for helpful discussion and criticism of the manuscript.

REFERENCES

Anglesea, D., Greenhalgh, G.N., and Veltkamp, C.J., 1983, The structure of the thallus tip in Usnea sublforidana, The Lichenologist, 15: 73-80.
Boissière, M.-C., 1972, Cytologie du Peltigera canina (L.) Willd. en microscopie électronique: 1. Premières observations, Revue génerale de Botanique, 79: 167-185.
Boissière, M.-C., 1982. Cytochemical ultrastructure of Peltigera canina: some features related to its symbiosis, The Lichenologist, 14: 1-27.

Douglas, A., and Smith, D.C., 1984, The green hydra symbiosis VIII.
    Mechanisms in symbiont regulation, Proceedings of the Royal
    Society, London B, 221: 291-319.
Drew, E.A., and Smith, D.C., 1967, Studies in the physiology of
    lichens VII. The physiology of the Nostoc symbiont of Peltigera
    polydactyla compared with cultured and free living forms, New
    Phytologist, 66: 379-388.
Englund, B., 1977, The physiology of the lichen Peltigera aphthosa,
    with special reference to the blue-green phycobiont (Nostoc sp.),
    Physiologia Plantarum, 41: 298-304.
Fisher, P.J., and Proctor, M.C.F., 1978, Observations on a season's
    growth in Parmelia caperata and P. sulcata in South Devon, The
    Lichenologist, 10: 81-89.
Galun, M., Paran, N., and Ben-Shaul, Y., 1970, Structural
    modifications of the phycobiont in the lichen thallus,
    Protoplasma, 69: 85-96.
Greenhalgh, G.N., and Anglesea, D., 1979, The distribution of algal
    cells in lichen thalli, The Lichenologist, 11: 283-292.
Harris, G.P., 1969 "A study of the ecology of corticolous lichens,"
    Ph.D. Thesis, University of London, England.
Harris, G.P., 1971, The ecology of corticolous lichens II. The
    relationship between physiology and the environment, Journal of
    Ecology, 59: 441-452.
Hale, M.E., 1970, Single-lobe growth-rate patterns in the lichen
    Parmelia caperata, The Bryologist, 73: 72-81.
Henssen, A., and Jahns, H.M., 1974, "Lichenes. Eine Einführung
    in die Flectenkunde," G. Thieme, Stuttgart.
Hill, D.J., 1976, The physiology of the lichen symbiosis, in
    "Lichenology: Progress and Problems," D.H. Brown,
    D.L. Hawksworth, and R.H. Bailey, eds, pp. 457-496, Academic
    Press, London and New York.
Hill, D.J., 1977, The role of Anabaena in the Azolla-Anabaena
    symbiosis, New Phytologist, 78: 611-616.
Hill, D.J., 1981, The growth of lichens with special reference to the
    modelling of circular thalli, The Lichenologist, 13: 265-287.
Hill, D.J., and Woolhouse, H.W., 1966, Aspects of the autecology of
    Xanthoria parietina (agg.), The Lichenologist, 3: 207-214.
Jahns, H.M., 1973, Anatomy, morphology and development, in:
    "The Lichens," V. Ahmadjian, and M.E. Hale, eds, pp. 3-58,
    Academic Press, New York.
Kershaw, K.A., and Millbank, J.W., 1970, Nitrogen metabolism of
    lichens II. The partition of cephalodial-fixed nitrogen between
    the mycobiont and phycobionts of Peltigera aphthosa, New
    Phytologist, 69: 75-79.
Millbank, J.W., 1972, Nitrogen metabolism in lichens IV. The
    nitrogenase activity of the Nostoc phycobiont in Peltigera
    canina, New Phytologist, 71: 1-10.
Millbank, J.W., 1976, Aspects of nitrogen metabolism in lichens, in:
    "Lichenology: Progress and Problems," D.H. Brown,
    D.L. Hawksworth, and R.H. Bailey, eds, pp. 441-445, Academic

Press, London and New York.

Millbank, J.W., and Kershaw, K.A., 1969, Nitrogen metabolism of
    lichens I. Nitrogen fixation in the cephalodia of Peltigera
    aphthosa, New Phytologist, 68: 721-729.

Nienburg, W., 1926, "Anatomie der Flechten", Handbuch der
    Pflanzenanatomie II Abteilung I Tiel Thallophyten, Band VI,
    Borntraeger, Berlin.

Peters, G.A., and Calvert, H.E., 1983, The Azolla-Anabaena azollae
    symbiosis, in: "Algal Symbiosis," L.J. Goff, ed. pp. 109-145,
    Cambridge University Press, Cambridge.

Plummer, G.L., and Gray, B.D., 1972, Numerical densities of algal
    cells and growth in the lichen genus Cladonia, American Midland
    Naturalist, 87: 355-365.

Stewart, W.D.P., Rowell, P., and Rai, A.N., 1980, Symbiotic
    nitrogen-fixing cyanobacteria, in: "Nitrogen Fixation," W.D.P.
    Stewart, and J.R. Gallon, eds, pp. 239-277, Academic Press,
    London and New York.

SELECTIVITY IN THE LICHEN SYMBIOSIS

P. Bubrick[a], A. Frensdorff[b] and M. Galun[a]

[a] Department of Botany, [b] Department of Microbiology
Tel Aviv University 69978
Tel Aviv
Israel

INTRODUCTION

General

The terms specificity and selectivity, which have been used to describe distinct types of cellular behaviour during cell-cell adhesion between animal cells (Garrod and Nicol, 1981), will be used here to describe cellular interactions between symbionts. In our view, specificity describes an interaction in which absolute exclusiveness is expressed; two types of bionts associate only with one another and no other potential combinations between them are observed. In contrast, selectivity describes a situation where bionts interact preferentially with one another. If a host (or vice versa) is presented with a choice of bionts, it will preferentially associate with one over the others.

We wish to stress the distinction between these two terms for several reasons. There has been a tendency to use these terms interchangeably which can lead to misunderstanding (Garrod and Nicol, 1981). The misunderstanding may be further compounded when, for instance, both terms are used as criteria in the definition of recognition (e.g. Heslop-Harrison, 1978a; Bauer, 1981; Smith, 1981a).

We also wish to distinguish between cellular behaviour during an interaction and the underlying molecular events which ultimately govern that behaviour. In several examples of the Rhizobium-legume symbiosis, lectin binding to the bacterium is postulated to be an important molecular event during the early stages of the interaction (Dazzo and Truchet, 1983). While lectin binding is

probably a specific interaction at the molecular level, lectin
binding per se does not guarantee that subsequent cellular events
will be biologically meaningful. Lectin binding in conjunction
with other molecular events is needed before biologically
meaningful responses are observed.

At the cellular level, these terms distinguish between
cellular interactions where exclusiveness may be a biological
requirement (e.g. fertilization) and interactions where it may not.
In many symbioses there is often a range of _potential_ biont
combinations. However, only one or a few of all potential
combinations are _preferred_, as expressed in the overall biological
fitness of the symbiosis (e.g. Trench, 1979).

Lichens

In keeping with these distinctions, lichens are considered, on
the whole, to be non-specific. In general, however, lichens are
selective with respect to their potential biont combinations. A
more detailed discussion of the selective nature of lichens has
been presented by Galun and Bubrick (1984).

When evaluating selectivity in lichens, it must be recalled
that lichens are a heterogeneous assemblage of organisms derived
polyphyletically from several orders of ascomycetes and
basidiomycetes (Hawksworth, 1973), as well as phycobionts from both
the prokaryotes and eukaryotes. It is also probable that lichen
associations arose at different times during the evolution of
different groups of fungi (Smith et al., 1969; James and Henssen,
1976). Therefore, we should expect that the degree of selectivity,
as well as the mechanisms by which it is achieved, may differ
between phylogenetically or evolutionally distinct groups of
lichens (also see Honegger, 1983).

The process of lichen resynthesis may be divided into a number
of phases, including pre-contact, contact, envelopment of algal
cells, incorporation of both symbionts into a common matrix,
morphological, physiological and biochemical modifications of both
symbionts from the free-living to the lichenized state, and
integration into regulatory mechanisms. While still tentative,
this division attempts to take into account the observed
developmental stages of lichen resynthesis. Undoubtedly, as
resynthesis is further studied, additional phases will be
recognised. It is possible that some degree of selectivity is
expressed during each of these phases of resynthesis. One goal of
investigators in this area will be to assess the contribution of
each phase to the overall selective interaction between lichen
symbionts.

CONTACT PHASE STUDIES

Our main interest has been in the contact phase. This phase
is potentially an important one, since contact between symbionts is
a prerequisite for continued development. In several symbioses
where contact has been studied in detail, it has been suggested
that this phase plays an important rôle in determining overall
selectivity between bionts (Pool and Muscatine, 1980; Pool, 1981:
Reisser, 1981; Dazzo and Truchet, 1983; Reisser and Wiessner,
1984). However, it has also been recognised that overall
selectivity is not solely determined by events during contact
(Jolley and Smith, 1980; Bauer, 1981; Fitt and Trench, 1981). This
has also been emphasised by investigators of the lichen symbiosis
(Smith, 1981b; Honegger, 1982; Galun and Bubrick, 1984).

It is frequently assumed that the expression of selectivity at
the contact phase is the result of straightforward molecular
recognition events at the cell surface. However, often such a view
cannot adequately explain selective contact behaviour between
animal cells (Jones, 1980; Thomas et al., 1982) or between the
bionts of symbiotic associations (Bauer, 1981; Dazzo, 1981).
Instead, contact itself should be viewed as a multi-step process
which requires co-ordinated efforts between one or both symbionts.

We have been studying the contact phase in the lichen
Xanthoria parietina. This lichen was chosen for study because it
is not known to produce any types of asexual reproductive
structures which contain both symbionts. The only observed
reproductive strategy is the production of ascospores by the
fungus. Thus, the colonisation of a substrate by this lichen is
dependent upon the ejected ascospore of the mycobiont coming in
contact with suitable phycobionts. The extent to which resynthesis
of X. parietina takes places in nature is not known, although it is
probably more frequent than has been previously suspected (Bubrick
et al., 1984).

One method for studying the contact phase in lichens is by in
vitro resynthesis experiments (Ahmadjian and Jacobs, 1983).
Unfortunately, the conditions necessary for the in vitro
resynthesis of X. parietina are not known. Therefore, we have
adopted a somewhat different approach. It is not well appreciated
that cell walls play more than just a structural role. Cell walls
are viewed as structures capable of complex communication, not only
with their own plasma membranes, but also with other walled or
wall-less organisms in the environment (Heslop-Harrison, 1978b;
Calleja et al., 1981; Reisert, 1981; Albersheim et al., 1983;
Yeoman, 1984). Thus, our initial studies focused on the partial
characterisation of the cell walls of the phycobiont and mycobiont.
It was assumed that if there was sufficient molecular diversity in
their walls, then this may form the basis for a biological

mechanism(s) by which symbionts could distinguish between one another. The results from a variety of approaches and from several laboratories have indeed shown that the cell walls of both phycobionts and mycobionts are not chemically uniform; they exhibit a reasonable degree of molecular diversity (Galun et al., 1976; Takahashi et al., 1979; Bubrick and Galun, 1980a; König and Peveling, 1980; Bubrick et al., 1982, 1984; Honegger, 1982; Honegger and Brunner, 1981; Robenek et al., 1982; Marx and Peveling, 1983).

In addition, we have isolated and purified to apparent homogeneity a glycoprotein from X. parietina which binds to cells of the normal cultured phycobiont of this lichen (partially described in Galun et al., 1984). This algal-binding protein (ABP) is also able to distinguish between a variety of potential phycobionts. We have suggested that ABP may play a rôle during the contact phase between resynthesising X. parietina bionts (Galun et al., 1984). A comparison is presented here between ABP and other cell-adhesion molecules (CAMs) which have been described in the literature. The criteria of CAMs used for comparison have been taken from Barondes (1980), rephrased into lichenological terminology and are discussed below.

Production of ABP by Fungal Symbiont. Indirect evidence suggests that ABP is produced by the fungal component of the intact lichen (Bubrick and Galun, 1980a; Bubrick et al., 1981a). There is also evidence that ABP is produced by the mycobiont grown in culture. Crude protein extracts have been isolated from the in vitro grown mycobiont which, when fluorescently labelled, were observed to bind to the cell wall of the appropriate cultured phycobiont (Bubrick et al., 1981b).

Crude cultured-mycobiont extracts, labelled with radioactive iodine (^{125}I), have been analysed by sodium dodecyl sulphate-polyacrylamide gel electrophoresis (SDS-PAGE) followed by gel autoradiography. It was found that a protein from the crude extract had the same electrophoretic mobility as purified ABP.

Crude mycobiont extracts were also partially characterized with antibody directed against purified ABP (α ABP). In these experiments, serial dilutions of αABP were titrated with ABP and crude mycobiont extracts (Fig. 1). The results demonstrate that a component of this extract is highly cross-reactive with αABP. It still remains to be shown that this cross-reactive molecule and the protein detected by SDS-PAGE are the same.

These data suggest that ABP is produced by the mycobiont grown in culture. Thus, it would appear that ABP is neither dependent upon symbiosis, nor dependent upon the presence of potential phycobionts, for its synthesis and/or expression.

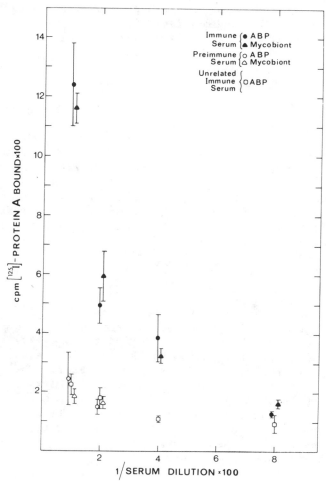

Fig. 1. Titration of ABP and crude protein extracts from the in
vitro cultured Xanthoria parietina mycobiont.
Immunoglobulin binding detected with ^{125}I-Protein A (input
was 100,000 cpm/incubation vessel). Unrelated immune
serum was directed against human serum albumin.

Cellular Location of ABP. To date, there is no direct
evidence that ABP is present at the cell surface of the mycobiont
in symbiosis or in culture. With the isolation and purification of
ABP, localisation studies have now become possible.

Occurrence of ABP when Adhesiveness Develops. If adhesiveness
is viewed in a broad sense as the ability of bionts to bind
together, then mycobiont hyphae are clearly adhesive. However,
little attention has been paid to the types of adhesive behaviour
expressed in lichens. Observations on the in situ acquisition of

secondary phycobionts by intact lichens show that fungal hyphae can adhere to, and envelop, appropriate phycobionts (e.g. James and Henssen, 1976; Brodo and Richardson, 1978; Renner and Galloway, 1982). It is not known whether adhesion was due to an inherent property of the fungal hyphae, or developed as a result of the interaction with an appropriate phycobiont.

An interesting form of fungal adhesion has been observed in lichens with so-called hymenial phycobionts (summarised in Pyatt, 1973). In these cases, phycobionts are present in the fungal reproductive structures and adhere to ascospores. The ejected ascospore thus carries its normal phycobiont into the environment.

The study of resynthesis in vitro clearly demonstrates that fungal hyphae adhere to phycobionts (Ahmadjian and Jacobs, 1983). These authors have also observed that the Cladonia cristatella mycobionts produced two types of hyphae in culture; one type grew within or on the substrate, while the second exhibited aerial growth. Phycobiont cells preferentially adhered to the aerial hyphae.

During attempts to resynthesise X. parietina in vitro, we noticed that the normal phycobiont adhered to fungal hyphae prior to envelopment. Phycobiont cells adhered to young (less than 3 months) and old (more than 6 months) hyphae; this adherence was strong enough to resist a gentle stream of water. When a Coccomyxa phycobiont was introduced into the culture, it aligned itself along the fungal hyphae, but was easily dislodged by a gentle water stream. This observation leads us to the tentative conclusion that living X. parietina mycobiont hyphae were inherently adhesive under the conditions used in that study. It remains to be demonstrated that the adhesion was due to ABP.

Appropriate ABP Binding Molecules on the Phycobiont Surface. We now have considerable evidence that ABP binds to some molecular structures on the phycobiont cell wall.

Radioiodinated ABP (^{125}I-ABP) has been used as a probe to study binding to phycobiont cells in both non-competitive and competitive binding assays. For non-competitive assays, cells were incubated with various concentrations of ^{125}I-ABP, excess unbound ligand removed and bound ^{125}I-ABP detected by gamma counting. Data from one such assay, using the culture phycobionts from X. parietina and Ramalina pollinaria, are shown in Table 1. There were both quantitative and qualitative differences in ^{125}I-ABP binding to the two phycobionts. The X. parietina phycobiont had a much greater binding capacity for ^{125}I-ABP than did the R. pollinaria phycobiont. The qualitative aspects of the observed binding are not yet fully understood.

Table 1. Binding of ^{125}I-ABP to Phycobionts from <u>Xanthoria</u>
          <u>parietina</u> and <u>Ramalina pollinaria</u>[a,b]

^{125}I-ABP added (ng)[c]	^{125}I-ABP bound (ng)	
	X. parietina	R. pollinaria
1611 ± 56	21.3 ± 1.7	3.9 ± 0.0
857 ± 23	14.0 ± 0.3	2.9 ± 0.5
454 ± 7	6.2 ± 0.3	0.8 ± 0.3
226 ± 8	3.0 ± 0.4	0.3 ± 0.1
115 ± 3	1.9 ± 0.4	0.3 ± 0.1
62 ± 1	0.8 ± 0.1	0.2 ± 0.1
33 ± 2	0.6 ± 0.1	0.1 ± 0.1

[a]Nanograms calculated from specific activity of ^{125}I-ABP. Data
rounded to the nearest whole number.
[b]Data corrected for non-specific binding of ^{125}I-ABP to incubation
vessel only.
[c]All tests in triplicate

     To confirm that the radioactivity detected was due to
^{125}I-ABP, bound ligand was eluted in SDS-PAGE sample buffer and
subjected to electrophoresis. Autoradiographs of the gels
demonstrated that cell-bound radioactivity was predominantly due to
^{125}I-ABP.

     Competitive binding assays were also used to characterize
^{125}I-ABP binding. Various concentrations of unlabelled ABP were
first incubated with cells, unbound ligand removed and cells then
incubated with a constant concentration of ^{125}I-ABP. The results
(Table 2) demonstrate that unlabelled ABP competes with ^{125}I-ABP
for available binding sites on the phycobiont. However, as in the
non-competitive studies, precise quantitative aspects of
competition are not easily interpreted.

     The nature of the binding forces have been examined by
attempting to elute bound ^{125}I-ABP from phycobiont cells (Table 3).
Potentially non-destructive elution has been accomplished with urea
or guanidine HCl, suggesting that hydrogen bonds are involved in
binding. Preliminary attempts to elute ^{125}I-ABP from cells with a
variety of sugars were unsuccessful.

     Attempts have been made to partially characterise the
molecules of the phycobiont wall which bind ^{125}I-ABP. The
<u>X. parietina</u> phycobiont was exposed to a number of chemical and

Table 2. Competitve Inhibition of ^{125}I-ABP Binding to Phycobiont
Cells by Unlabelled ABP[a,b]

Unlabelled ABP added (ng)	^{125}I-ABP bound (%)[c]
0	100 ± 9
100	90 ± 10
200	83 ± 3
500	78 ± 10
1,000	76 ± 6
2,000	66 ± 8
5,000	61 ± 16
10,000	59 ± 2

[a]Input of ^{125}I-ABP was 250 ng
[b]Data corrected for non-specific binding of ^{125}I-ABP to incubation
 vessel only
[c]Calculated as ng ^{125}I-ABP bound in the presence of unlabelled
 ABP/ng bound in the absence of unlabelled ABP x 100. Rounded to
 nearest percentage. All tests in triplicate

Table 3. Elution of Bound ^{125}I-ABP from Phycobiont Cells[a]

Eluent[b]	^{125}I-ABP in eluate (%)[c]
NaCl (1.5 M)	2 ± 9
EDTA (0.1 M, pH 2)	2 ± 8
Formic acid (1.0 M, pH2)	8 ± 4
Boric acid (0.1 M, pH 10)	17 ± 2
NaOH (4.0 N, pH 12)	72 ± 10
Triton X-100 (1% v/v)	24 ± 7
Dioxane (10% v/v)	7 ± 5
Ethylene glycol (5% v/v)	6 ± 4
KI (2.0 M)	21 ± 4
Urea (8.0 M)	46 ± 3
Guanidine HCl (6.0 M)	42 ± 7

[a]Cells were incubated with ^{125}I-ABP for 1 hr at room temperature
 (RT), unbound ligand removed and cells incubated with eluent for 1
 hr at RT. Eluent was collected and counted in a gamma counter.
[b]Eluent at pH 7.4 unless otherwise stated
[c]Relative to % in eluate after phosphate buffered saline (0.14 M
 NaCl) elution. All tests done in triplicate; data pooled from 2
 experiments and standard deviation is between experiments.

physical treatments and then tested for its ability to bind ^{125}I-ABP (Table 4). Pretreatment of phycobiont with EDTA, urea or heat had no effect on ^{125}I-ABP binding. Interestingly, both SDS and non-specific protease pretreatment greatly increased binding to cells, suggesting that these reagents expose masked binding sites. ^{125}I-ABP binding was reduced after pretreatment with either trypsin or periodate. Although it is tempting to speculate about the nature of the binding structures based on these findings, further work is needed before meaningful conclusions can be drawn.

Collectively, these data indicate that there are molecular structures on the phycobiont cells to which ABP binds. As of yet, there is no direct evidence for the existence of a discrete cellular "receptor" for ABP. This lack of evidence for a "receptor" is based on the quantitative and qualitative nature of the observed ^{125}I-ABP binding to cells. It is not completely clear whether the presented data accurately reflects the binding of ABP to phycobiont cells, or is in part due to the known effects of radioiodination on the biological activities of many types of molecules (Bolton, 1977). Further studies are needed to clarify this problem.

Table 4. Various Pretreatments on Phycobiont Cells: Their Effect on ^{125}I-ABP Binding[a]

Pretreatment	^{125}I-ABP bound to cells after treatment (%)[b]
EDTA (0.1 M, 1 hr)[c]	100 ± 8
Urea (8.0 M, 1 hr)[c]	100 ± 11
Heat (100°, 10 min)[c]	100 ± 4
SDS (1% w/v, 1 hr)[c]	976 ± 16
Protease (0.1% w/v, 1 hr)[d]	345 ± 20
Trypsin (0.2% w/v, 2 hr)[e]	74 ± 7
Periodate (0.03 M, 4 hr)	80 ± 4

[a]Cells were pretreated, washed and incubated with ^{125}I-ABP for 1 hr at RT. Following incubation, unbound ligand was removed and cells counted in a gamma counter

[b]Relative to ^{125}I-ABP binding after pretreatment in PBS or appropriate inhibited reaction mixture. All tests done in triplicate; data pooled from 2 experiments and standard deviation is between experiments

[c-f]Reactions inhibited or stopped by: [c]repeated centrifugation, [d]repeated centrifugation in the presence of bovine serum albumin, [e]phenylmethylsulphonyl fluoride, [f]ethylene glycol

Table 5. Binding of ^{125}I-ABP to Heterologous Phycobionts[a,b]

Cultured phycobiont from	ng bound[c] per g added	% bound[d]
Xanthoria parietina	21.00 + 0.31	100
X. steineri	2.16 + 0.02	10
Physconia pulvenulenta	6.24 + 0.06	30
Ramalina pollinaria	1.00 + 0.11	5
Peltigera aphthosa (Coccomyxa)	0.90 + 0.11	4

[a]Assay done as described for non-competitive binding assays
[b]Nanograms calculated from specific activity of ^{125}I-ABP. Data
rounded to nearest hundredth.
[c]Input of ^{125}I-ABP was 250 ng. Data expressed as ng bound/g added.
All tests done in triplicate.
[d]% bound calculated relative to binding to the X. parietina
phycobiont. Rounded to the nearest percent.

Selectivity of ABP. The selective binding of ABP to potential
phycobionts is an important consideration if ABP is to play a
significant role during the contact phase of resynthesising lichen
bionts. As summarised in Table 5, ^{125}I-ABP exhibits quantitatively
selective binding to the culture X. parietina phycobiont.

Existence of Appropriate ABP-Binding. The question of whether
normally symbiotic phycobionts can be found free-living in nature
is a controversial one, not only in lichens, but also in other
algal symbioses. (Free-living is used here to indicate that the
normal phycobiont is found outside symbiosis. The extent to which
they can actively multiply is not fully known.) One approach to
this question is from the point of view of resynthesis under
natural conditions. It is well known that in certain symbioses,
algal bionts are transmitted from generation to generation during
sexual reproduction of the host. However, there are also numerous
examples in which offspring are completely devoid of algal bionts
and must acquire them from the environment; in most of these cases
reinfection is essential for the completion of the life history of
the host (e.g. Trench, 1979). Thus, at some point in time, or
during some stage in their life cycle, algal bionts must be present
free-living in nature. Their apparent rarity in situ has been
interpreted to mean that they have a far greater fitness in
symbiosis than when free-living (Law and Lewis, 1983). In some of
the cases where algal bionts have not been reported free-living,
the problem may be in detection, rather than presence or absence.

The same argument can be applied to the lichen symbiosis. Many lichens produce asexual reproductive structures (diaspores) which disperse units containing both mycobiont and phycobiont; in these cases, resynthesis is a _fait accompli_. However, many lichens do not produce diaspores; instead, the mycobiont ejects phycobiont-free ascospores into the environment. In order to complete their life cycles, hyphae from germinated ascospores must make contact with and resynthesize suitable phycobionts. Thus, in those lichens which rely solely on sexual reproduction as a strategy for dispersal, appropriate phycobionts must be, to some extent, available in nature.

We have provided evidence that both the mycobiont and normal phycobiont of X. parietina can be found free-living in nature (Bubrick et al., 1984). Free-living Trebouxia and/or Pseudotrebouxia were isolated, purified and grown in culture. Free-living isolates were compared with the cultured X. parietina phycobiont using both morphological and immunological criteria. At least one of these free-living isolates was found to be indistin-guishable from the cultured X. parietina phycobiont. The immuno-logical approach employed in this study was particularly conclusive since the antiserum used was known to be highly sensitive for the X. parietina phycobiont (unpublished observations). Thus, it appears that the normal X. parietina phycobiont can be found free-living in nature. It is probable that other Trebouxia and Pseudotrebouxia phycobionts can also be found growing outside the lichen symbiosis (Tschermak-Woess, 1978).

Is ABP a CAM? Does it contribute in some way to selectivity during the contact phase in X. parietina? The answer to both of these questions is - maybe. Certainly ABP exhibits a number of the theoretical characteristics required of a CAM. However, it remains to be shown that the binding of ABP to the normal phycobiont plays a biologically meaningful rôle during the contact phase in X. parietina.

COMPARISON WITH OTHER STUDIES

The possible participation of lectins during the contact phase in the lichen genus Peltigera has been reviewed by Galun and Bubrick (1984) (see also Petit et al., 1983; Lallemant and Savoye, this volume). In general, lectins from Peltigera are produced by the fungal symbiont and bind selectively to potential Nostoc phycobionts. The possible participation of lectins during the initial interactions of the Peltigera bionts has been suggested by Petit et al., (1983). The presence of a lectin(s) in X. parietina has been reported by Ingram (1982), although no attempt has been made to correlate the lectin(s) with the biology of this lichen.

To our knowledge, there has been only one other report on the

presence of ABP-like molecules in lichens with eukaryotic
phycobionts (Hersoug, 1983). Hersoug confirmed that a component(s)
of crude X. parietina protein extracts binds to the cells of its
normal culture phycobiont. This report also demonstrated that
ABP-like molecules may not be present in other genera of lichens.
Of five lichens tested, only crude extracts from X. parietina
possessed a binding component(s) for its normal phycobiont. Thus,
ABP-like molecules may be restricted to certain groups of lichen.

The contact phase has also been studied during the in vitro
resynthesis of lichens. In resynthesis experiments between the
Cladonia cristatella mycobiont and a variety of phycobionts and
non-symbiotic algae, a fairly high degree of selectivity was
observed in the overall pattern of resynthesis (Ahmadjian and
Jacobs, 1981). In the authors' opinion, this selectivity was not
expressed during the contact phase between symbionts; all tested
algae were enveloped by fungal hyphae. However, there appeared to
be more than one type of fungal behaviour during the contact phase.
Following initial contacts, but prior to envelopment, mycobiont
hyphae and phycobiont cells were said to be firmly bound together,
apparently by a phycobiont-produced extracellular matrix. Such
behaviour has been observed in at least two cases (Ahmadjian et
al., 1978; Ahmadjian and Jacobs, 1981). It is not clear whether
this matrix bound all tested mycobiont-phycobiont combinations, or
only those in which resynthesis proceeded beyond the stage of
soredia-development to squamule formation. It was also observed
that when glass beads were substituted for phycobiont cells, hyphae
grew along the contour of the beads but did not envelop them.
According to Ahmadjian and Jacobs (1981), the hyphae could be
"easily dislodged" from the beads, in contrast to phycobiont cells.
These different types of contact behaviour emphasize the need for
caution when attempting to generalize about selectivity during the
contact phase between lichen bionts.

In our studies the resynthesis of the X. parietina bionts has
not proceeded beyond the phase of envelopment of algal cells.
Under conditions where the mycobiont was enveloping the normal
phycobiont, it did not interact with glass beads, or with
cationized or hydrophobic derivatives of the beads. It may be that
the X. parietina mycobiont displays a higher degree of selectivity
during the contact phase than does the C. cristatella mycobiont.

REFERENCES

Ahmadjian, V., and Jacobs, J.B., 1981, Relationship between fungus
    and alga in the lichen Cladonia cristatella Tuck., Nature,
    289: 169-172.
Ahmadjian, V., and Jacobs, J.B., 1983, Algal-fungal relationships
    in lichens : recognition, synthesis, and development, in:
    "Algal Symbiosis," L.J. Goff, ed., pp. 147-172, Cambridge

University Press, Cambridge.

Ahmadjian, V., Jacobs, J.B., and Russell, L.A., 1978, Scanning
electron microscope study of early lichen synthesis, Sciences,
200: 1062-1064.

Albersheim, P., Darvill, A.G., McNeil, M., Valent, B.S.,
Sharp, J.K., Nothnagel, E.A., Davis, K.R., Yamazaki, N.,
Gollin, D.J., York, W.S., Dudman, W.F., Darvill, J.E., and
Dell, A., 1983, Oligosaccharins : naturally occurring
carbohydrates with biological regulatory functions, in:
"Structure and Function of Plant Genomes," O. Ciferri, and
L. Dure III, eds, pp. 293-312, Plenum Press, New York.

Barondes, S.H., 1980, Endogenous cell-surface lectins: Evidence
that they are cell adhesion molecules, in: "Cell Surface.
Mediator of Developmental Processes," S. Subtelney, and
A.K. Wessels, eds, pp. 349-363, Academic Press, New York.

Bauer, W.D., 1981, Infection of legumes by rhizobia, Annual Review
of Plant Physiology, 32: 407-449.

Bolton, A.E., 1977, "Radioiodination Techniques", Radiochemical
Centre Review 18, Amersham International Limited, Printarium
Limited, Wembley.

Brodo, I.M., and Richardson, D.H.S., 1978, Chimeroid associations
in the genus Peltigera, The Lichenologist, 10: 157-170.

Bubrick, P., and Galun, M., 1980a, Proteins from the lichen
Xanthoria parietina which bind to phycobiont cell walls.
Correlation between binding patterns and cell wall
cytochemistry, Protoplasma, 104: 167-173.

Bubrick, P., and Galun, M., 1980b, Symbiosis in lichens:
differences in cell wall properties of freshly isolated and
cultured phycobionts, FEMS Microbiology Letters, 7: 311-313.

Bubrick, P., Galun, M., and Frensdorff, A., 1981a, Proteins from
the lichen Xanthoria parietina which bind to phycobiont cell
walls. Localization in the intact lichen and cultured
mycobiont, Protoplasma, 105: 207-211.

Bubrick, P., Frensdorff, A., and Galun, M., 1981b, Differences
in the cell wall properties between freshly isolated and
cultured phycobionts from the lichen, Xanthoria parietina,
XIII International Botanical Congress, Sydney, Australia.

Bubrick, P., Galun, M., Ben-Yaacov, M., and Frensdorff, A., 1982,
Antigenic similarities and differences between symbiotic and
cultured phycobionts from the lichen Xanthoria parietina,
FEMS Microbiology Letters, 13: 435-438.

Bubrick, P., Galun, M., and Frensdorff, A., 1984, Observations on
free-living Trebouxia de Puymaly and Pseudotrebouxia
Archibald, and evidence that both symbionts from Xanthoria
parietina (L.) Th. Fr. can be found free-living in nature, New
Phytologist, 97: 455-462.

Calleja, G.B., Johnson, B.F., and Yoo, B.Y., 1981, The cell wall as
a sex organelle in fission yeast, in: "Sexual interactions in
Eukaryotic Microbes," D.H. O'Day, and P.A. Horgen, eds,
pp. 225-259, Academic Press, New York.

Dazzo, F.B., 1981, Bacterial attachment as related to cellular recognition in the Rhizobium-legume symbiosis, Journal of Supramolecular Structure and Cell Biochemistry, 16: 29-41.

Dazzo, F.B., and Truchet, G.L., 1983, Interactions of lectins and their saccharide receptors in the Rhizobium-legume symbiosis, Journal of Membrane Biology, 73: 1-16.

Fitt, W.K., and Trench, R.K., 1981, Spawning, development, and acquisition of zooxanthellae by Tridacna squamosa (Mollusca, Bivalvia), Biological Bulletin, 161: 213-235.

Galun, M., and Bubrick, P., 1984, Physiological interactions between the partners of the lichen symbiosis, in: "Encyclopedia of Plant Physiology. New series Vol. 17, Cellular Interactions," H.F. Linskens, and J. Heslop-Harrison, eds, pp. 362-401, Springer-Verlag, Berlin, Heidelberg, New York and Tokyo.

Galun, M., Braun, A., Frensdorff, A., and Galun, E., 1976, Hyphal walls of isolated lichen fungi. Autoradiographic localization of precursor incorporation and binding of fluorescein-conjugated lectins, Archives of Microbiology, 108: 9-16.

Galun, M., Bubrick, P., and Frensdorff, A., 1984, Initial stages in fungus-alga interaction, The Lichenologist, 16: 103-110.

Garrod, D.R., and Nicol, A., 1981, Cell behaviour and molecular mechanisms of cell-cell adhesion, Biological Reviews, 56: 199-242.

Hawksworth, D.L., 1973, Some advances in the study of lichens since the time of E.M. Holmes, Botanical Journal of the Linnean Society, 67: 3-31.

Hersoug, L.G., 1983, Lichen protein affinity toward walls of cultured and freshly isolated phycobionts and its relationship to cell wall cytochemistry, FEMS Microbiology Letters, 20: 417-420.

Heslop-Harrison, J., 1978a, "Cellular Recognition Systems in Plants", Edward Arnold (Publishers) Limited, London.

Heslop-Harrison, J., 1978b, Genetics and Physiology of angiosperm incompatibility systems, Proceedings of the Royal Society of London series B, 202: 73-92.

Honegger, R., 1982, Ascus structure and function, ascospore delimitation and phycobiont cell wall types associated with the Lecanorales (lichenized Ascomycetes), Journal of the Hattori Botanical Laboratory, 52: 417-429.

Honegger, R., 1983, The ascus apex in lichenized fungi. IV. Baeomyces and Icmadophila in comparison with Cladonia (Lecanorales) and the non-lichenized Leotia (Helotiales), The Lichenologist, 15: 57-71.

Honegger, R., and Brunner, U., 1981, Sporopollenin in the cell walls of Coccomyxa and Myrmecia phycobionts of various lichens: an ultrastructural and chemical investigation, Canadian Journal of Botany, 59: 2713-2734.

Ingram, G.A., 1982, Haemagglutinins and haemolysins in selected lichen species, The Bryologist, 85: 389-393.

James, P.W., and Henssen, A., 1976, The morphological and taxonomic significance of cephalodia, in: "Lichenology: Progress and Problems," D.H. Brown, D.L. Hawksworth, and R.H. Bailey, eds, pp. 27-77, Academic Press, London and New York.

Jolley, E., and Smith, D.C., 1980, The green Hydra symbiosis. II. The biology of the establishment of the association, Proceedings of The Royal Society of London, series B, 207: 311-333.

Jones, B.M., 1980, Regulation of the contact behaviour of cells, Biological Reviews, 55, 207-235.

König, J., und Peveling, E., 1980, Vorkommen von Sporopollenin in der Zellwand des Phycobionten Trebouxia, Zeitschrift fur Pflanzenphysiologie, 98: 459-464.

Law, R., and Lewis, D.H., 1983, Biotic environments and the maintenance of sex - some evidence from mutualistic symbioses, Biological Journal of the Linnean Society, 29, 249-276.

Marx, M., and Peveling, E., 1983, Surface receptors in lichen symbionts visualized by fluorescence microscopy after use of lectins, Protoplasma, 114: 52-61.

Petit, P., Lallemant, R., and Savoye, D., 1983, Purified phytolectin from the lichen Peltigera canina var. canina which binds to the phycobiont cell walls and its use as a cytochemical marker in situ, New Phytologist, 94: 103-110.

Pool, R.R. Jr., 1981, The establishment of the Hydra - Chlorella symbiosis : recognition of potential algal symbionts. Berichte der Deutschen Botanischen Gesellschaft, 94: 565-569.

Pool, R.R. Jr., and Muscatine, L., 1980, Phagocytic recognition and the establishment of the Hydra viridis - Chlorella symbiosis in: "Endocytobiology. Endosymbiosis and Cell Biology, Volume l," W. Schwemmler, and H.E.A. Schenk, eds, pp. 223-238, Walter de Gruyter and Company, Berlin and New York.

Pyatt, F.B., 1973, Lichen propagules, in: "The Lichens," V. Ahmadjian and M.E. Hale, eds, pp. 117-145, Academic Press, New York and London.

Reisert, P.S., 1981, Plant cell surface structure and recognition phenomena with reference to symbioses, International Review of Cytology, Supplement 12: 71-111.

Reisser, W., 1981, Host-symbiont interaction in Paramecium bursaria: physiological and morphological features and their evolutionary significance, Berichte der Deutschen Botanischen Gesellschaft, 94: 557-563.

Reisser, W., and Wiessner, 1984, Autotrophic eukaryotic freshwater symbionts, in: "Encyclopedia of Plant Physiology, New Series, Volume 17, Cellular interactions," H.F. Linskens, and J. Heslop-Harrison, eds, pp. 59-74, Springer-Verlag, Berlin, Heidelberg, New York and Tokyo.

Renner, B., and Galloway, D.J., 1982, Phycosymbiodemes in Pseudocyphellaria in New Zealand, Mycotaxon, 16: 197-231.

Robenek, H., Marx, M., and Peveling, E., 1982, Gold-labelled concanavalin A binding sites at the cell surface of two

phycobionts visualized by deep-etching, Zeitschrift fur
     Pflanzenphysiologie, 106: 63-68.
Smith, D.C., 1981a, The role of nutrient exchange in recognition
     between symbionts, Berichte der Deutschen Botanischen
     Gesellschaft, 94: 517-528.
Smith, D.C., 1981b, The symbiotic way of life, Transactions of the
     British Mycological Society, 77: 1-8.
Smith, D., Muscatine, L., and Lewis, D., 1969, Carbohydrate
     movement from autotrophs to heterotrophs in parasitic and
     mutalistic symbiosis, Biological Reviews, 44: 17-90.
Takahashi, K., Takeda, T., and Shibata, S., 1979, Polysaccharides
     of lichen symbionts, Chemical and Pharmaceutical Bulletin, 27:
     238-241.
Thomas, W.A., Edelman, B.A., Lobel, S.M., Breitbart, A.S., and
     Steinberg, M.S., 1982, Two chick embryonic adhesion systems:
     molecular vs. tissue specificity, in: "Cellular Recognition,"
     W.A. Frazier, L. Glaser, and D.I. Gottllieb, eds, pp. 77-89.
     Alan R. Liss, Inc., New York.
Trench, R.K., 1979, The cell biology of plant-animal symbiosis,
     Annual Review of Plant Physiology, 30: 485-531.
Tschermak-Woess, E., 1978, Myrmecia reticulata as a phycobiont and
     free-living - free-living Trebouxia - the problem of Stenocybe
     septata, The Lichenologist, 10: 69-79.
Yeoman, M.M., 1984, Cellular recognition systems in grafting, in:
     "Encyclopedia of Plant Physiology, New Series, Volume 17,
     Cellular Interactions," H.F. Linskens, and J. Heslop-Harrison,
     eds, pp. 453-472, Springer-Verlag, Berlin, Heidelberg, New
     York and Tokyo.

LECTINS AND MORPHOGENESIS: FACTS AND OUTLOOKS

R. Lallemant and D. Savoye

Laboratoire de Biologie et Cytophysiologie Végétales
U.E.R. des Sciences de la Nature, Université de Nantes
2, Chemin de la Houssinière, F-44072 Nantes Cedex, France

INTRODUCTION

Lectins are often defined as proteins, or glycoproteins, with
special binding sites allowing them to "recognize", i.e. to form a
complex with, the carbohydrates present at the surface of red blood
cells. For that reason, they are also known as hemagglutinins. More
recently, the term "lectin" has been extended to molecules other than
proteins, which have the same binding properties, and also to any
sugar-specific molecule. We will confine this review to those lectins
corresponding to the narrowest, older definition.

Lectins are known to be produced by a very great variety of
organisms ranging from microorganisms to plant tissues. They seem to
be present, or at least formed more frequently, in plants and over
1000 phytolectins have now been detected.

THE ROLE OF LECTINS

The biological rôle of lectins is far from being thoroughly
understood, and has more often been deduced from biochemical properties rather than actually determined. While more specialised books
can be consulted for a detailed review, certain points should be
emphasised here:

1. From their affinity for carbohydrates, it has been suggested that
lectins might play a role in the transport and storage of saccharides
(Ensgraber, 1958);

2. Lectins have often been found associated with cell walls,
particularly in the seeds and roots of Leguminosae, where they have

335

been postulated to be involved in protection against fungal infections by inhibiting fungal saccharases (Lotan et al., 1975a,b).

3.   Lectins have been found to have mitogenic properties (for example in Toyoshima et al., 1971; Tokuyama, 1973; Kawaguchi et al., 1974; Waxdal, 1974; Toyoshima and Osawa, 1975), but it is not known whether this is of biological significance;

4.   Malignant cells are preferentially agglutinated (Aub et al., 1963; Cohen, 1968; Burger, 1969; Bezkorovainy et al., 1971; Sharon and Lis, 1972), but again the significance of this property has not been elucidated.

    In addition to these possible rôles, cell-to-cell binding has been demonstrated in many instances and, whatever the exact meaning might be, cell recognition is the most precisely established function of lectins. Cell recognition has been postulated, though sometimes with insufficient experimental evidence in:

    a.   Rhizobium-legume symbioses (see Graham, 1981)

    b.   aggregation of cells of slime moulds (Rosen et al., 1973, 1974, 1976, 1977; Chang et al., 1975, 1977; Reitherman et al., 1975; Barondes and Rosen, 1976; Siu et al., 1976)

    c.   invasion by or protection against phytopathogens (Kojima and Uritani, 1974, Albersheim and Anderson, 1975; Doke et al., 1975; Mirelman et al., 1975; Goodman et al., 1976; Graham and Sequeira, 1977; Graham et al., 1977; Sequeira and Graham, 1977; Sequeira et al., 1977; Sing and Schroth, 1977; Anderson and Jasalavich, 1979)

    d.   compatible reproductive and vegetative structures (Crandall and Brock, 1968; Li and Hubbel, 1968; Bowles and Kauss, 1975)

    e.   recognition and adhesion of Escherichia coli on human mucosal cells (Gibbons and Van Houte, 1975; Ofek et al., 1977; Ofek and Sharon, 1983).

In vertebrate tissues a similar rôle, though not clearly established, is suspected in a number of cases (brain : Barondes et al., 1974; Simpson et al., 1978; Barondes, 1980; muscle : Nowak et al, 1976; Den and Malinzak, 1976), so that it has been speculated that lectins might be involved in a "general mechanism, which regulates the recognition and subsequent adhesion of cells in development and ageing processes" (Simpson et al., 1978).

HOW COULD LECTINS REGULATE DEVELOPMENT PROCESSES?

If lectins regulate developmental processes, some mechanism to regulate their production or activity would be required. Their presence is so ubiquitous that it is difficult, though not impossible, to imagine one simple system. The available data are still scarce but a number of results, depending on the system studied, have already been obtained. Recognition phenomena appear to be very complex and are but very poorly understood. What was considered as a "fairly simple lock-and-key phenomenon is obviously a complex array of dynamic and active developmental conditions, influences, and responses involving the interacting partners" (Graham, 1981) and the various responses obtained in the field of lectin activity might be only aspects of a more complicated general system.

In a number of cases, the production of lectins has been shown to be developmentally regulated. This is the case in slime moulds (Rosen, 1972; Rosen et al., 1973, 1975; Barondes and Rosen, 1976), in chick thigh muscle (Den and Malinzak, 1976) and in rat brain (Simpson et al., 1978). A complicated and still unique system of regulation was found with the hepatic-binding glycoprotein, a lectin involved in the recognition and clearance of glycoproteins in the blood stream (Stockert et al., 1974; Ashwell and Morell, 1977).

Of particular interest are the data obtained on the regulation of lectin-mediated cell recognition in the Rhizobium-legume systems. Bhuvaneswari and Bauer (1978), using cultures under various conditions, found that the environment provided by plant roots is an important factor in the development of receptors specific for the lectin of Glycine max on the cell surface of Rhizobium japonicum but they did not attempt to identify the environmental factor involved. At the same time, Dazzo and Brill (1978) showed that nitrate and ammonium substantially reduced the concentration of lectins involved in the binding of R. trifolii to clover root hairs. Furthermore, Dazzo et al., (1979) found that the occurrence of lectin receptors on R. trifolii changed with culture time on synthetic media. In those studies, nitrogen compounds seem to be one of the elements in the regulation of lectin-mediated cell-recognition.

LECTINS IN LICHENS

Lectins have been reported and extracted from a number of lichen species (Estola and Vartia, 1955; Howe and Barrett, 1967, 1970; Barrett and Howe, 1968; Xavier Filho et al., 1971, 1980; Lockhart et al., 1978; Rogers et al., 1980; Petit, 1982) but very little is known of their physiological rôle. The extracts were shown in three instances (Peltigera canina : Lockhart et al., 1978; Petit, 1982; P. polydactyla : Lockhart et al., 1978; P. horizontalis : Petit et al., 1983) to be of fungal origin and to bind to the appropriate cyanobiont. In P. praetextata, the binding of lectins have been localised within the lichen thallus (Lallemant, 1983; Petit et al., 1983).

The wide variety of species in which lectins have been found
(including basidiolichens, Xavier Filho et al., 1980) suggests that
their presence may be universal. A number of studies, still incomplete, tend to show that cell surfaces are important factors in the
mycobiont-photobiont recognition processes (Ahmadjian et al., 1978;
Lockhart et al., 1978; Bubrick and Galun, 1980a; Bubrick et al., 1981,
1983; Marx and Peveling, 1983), and it could reasonably be thought
that lectins are involved. However, in one case a non-lectin molecule,
at least in the narrow sense, has been shown to have similar recognition properties (Bubrick and Galun, 1980a; Bubrick et al., 1981, 1983,
this volume).

Another area of research has been the exploration of the ability
of lichens or lichen symbionts to bind commercially available, non-
lichen-extracted lectins. In most cases, the ability of the photobiont to bind lectins depended on whether it was freshly isolated or
not (Bubrick and Galun, 1980b; Marx and Peveling, 1983). From the
binding patterns of a number of mycobionts, photobionts and free-
living algae, Peveling and Marx proposed that the ability to bind
various lectins could be related to the possibility of having various
symbiont partners in the formation of lichens.

A STUDY OF LECTINS IN THE COURSE OF THE DEVELOPMENT OF A LICHEN ORGAN:
THE FORMATION OF REGENERATION LOBULES OF PELTIGERA PRAETEXTATA

We have studied the development of the isidia-like structures of
P. praetextata. We refer to them as regeneration lobules, as their
formation is induced by wounding, and they appear to be the result of
a "repair" or "regeneration" process. This choice was determined by
the fact that within a few weeks, which is very quick in comparison
with the usual slow growth of lichens, lobules acquire the typical
structure of an adult lichen thallus lobe. Furthermore we have grown
in vitro cultures of the lichen thallus and, following Scott (1956),
we have developed an experimental technique whereby the formation of
lobules can be induced by wounds such as scarifications or by cutting
small discs out of the thallus.

We simultaneously studied, at the cellular level, the fixation of
exogenous lectins and the presence of key enzymes of ammonia assimilation. The lectins were extracted and purified by Dr. P. Petit (Paris)
from P. canina. The enzymes studied were glutamate dehydrogenase
(GDH) and glutamine synthetase (GS), two enzymes characteristic of the
two main pathways of nitrogen assimilation. We chose them for two
reasons. 1. Morphogenesis implies cell divisions and cell movements,
i.e. not only the necessity for coordination and recognition between
cells but also changes in the symbiotic relationships between
partners. It is known that the nitrogen assimilating enzyme activity
of Nostoc differs between the free and symbiotic state (Stewart and

Rowell, 1977; Rai et al., 1981; Rowell et al., this volume). 2. As
reviewed above, the concentrations of nitrogen derivatives might, in
some cases, interfere with the activity of lectins.

Material and Methods

     Thalli of P. canina and P. praetextata were collected from the
Fontainebleau forest (Seine-et-Marne, France). Thallus sections were
made from fresh material with a cryotome, as described in Petit et
al., (1983).

     The lectin was extracted and purified from P. canina (Petit et
al., 1983). We have shown that it binds well to the Nostoc from P.
praetextata, though slightly less efficiently than to those of P.
canina. The lectin was labelled with fluorescein isothiocyanate
(FITC) using the technique described by Lockhart et al., (1978). The
in situ localisation of the FITC-labelled lectin fixation sites was a
modification (Lallemant, 1983) of the method of Tkacz, et al., (1981).

     Localisation of glutamate dehydrogenase activity, involving the
precipitation of formazan blue, was by the method of Hitzeman (1963)
and Lillie (1965) modified by Lallemant (1983).

Fig. 1. In situ localisation of free sites available for the
         fixation of lectins (adult thallus). Arrows indicate
         trichomes labelled with FITC-lectin.

Glutamine synthetase activity was localised by the method of
Lallemant and Savoye (Lallemant, 1983) as a brown-red precipitate of
γ - glutamyl-hydroxamate.

The Adult Thallus

The structure of the adult thallus is that of a pseudo-cladome,
e.g. multiaxial cladome (Chadefaud, 1975), with a thick algal layer
showing many spirally wound trichomes and hyphae coming from the
medulla, passing through it and being incorporated into the upper
cortex. The medulla is composed of the central pseudo-cladomian axis,
the pseudo-pleuridia producing the upper and lower cortex.

About 20 per cent of the trichomes were labelled by the FITC-
lectin, indicating they had free sites, while the other 80 per cent
had their sites occupied by endogenous lectins (Fig. 1). The labelled
trichomes were basal in the algal layer, spirally wound or beginning
to unfold and in the course of or at the end of cell division
(Fig. 2). GDH activity was also found at the level of the lowest
Nostoc of the algal layer (Fig. 3), particularly at the level of
hyphal ramifications involved in the intercalary growth of the cortex
(Fig. 4) and at the level of hyphae surrounding the unwinding tri-

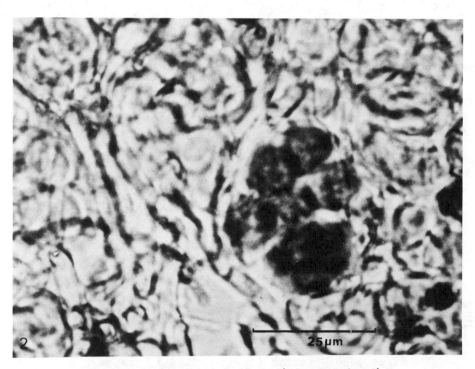

Fig. 2.   Trichome labelled with FITC-lectin.

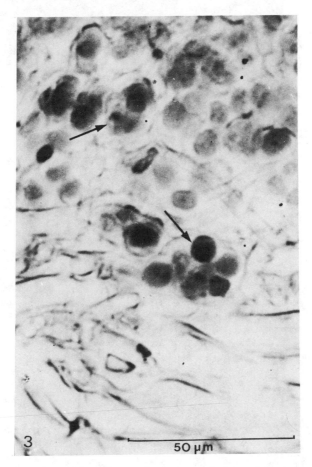

Fig. 3.   In situ localisation of GDH activity (adult thallus): basal
          part of the algal layer.

chomes. This GDH activity is most probably fungal but it might be
mistaken for Nostoc activity, as it appears mainly at the fungus-
Nostoc interface. GS activity was not located in any particular part
of the algal layer; it was found in uniformly distributed trichomes
which are tightly contained in a thick, hard, mucilaginous shell
(Fig. 5). It is obvious that the GS activity and the free sites for
the fixation of lectins are different; but the trichomes with many
free sites and those in contact with GDH-active hyphae might be the
same ones. There is no definite information on this point as it would
require performing both reactions on the same sections, which does not
seem technically possible. However, from the detailed observation of
both types of slides we think that there is good evidence to support
this suggestion. Thus it seems that there is an intense fungal GDH
activity, mainly at the junction of the medulla/ algal layer and more

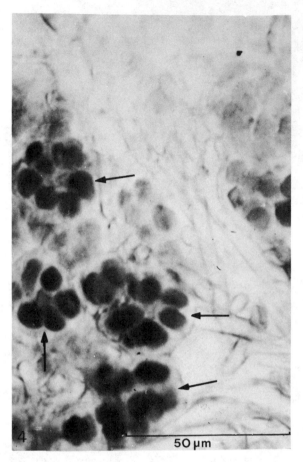

Fig. 4. In situ localisation of GDH activity (adult thallus):
trichomes adjacent to hyphae from the medulla passing through
the algal layer to the upper cortex.

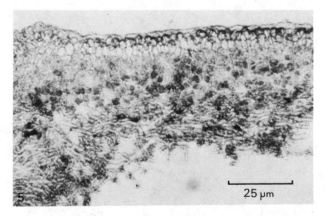

Fig. 5. In situ localisation of GS activity (adult thallus).

precisely in the zones of fungal growth, which, combined with the
inhibition of Nostoc GS activity in those zones, tends to suggest that
atmospheric nitrogen fixed by the Nostoc cells, is used mainly by the
fungus.

In the adult thallus we can conclude that each bundle of hyphae,
either pseudo-cladomian axis or pseudo-pleuridia, is fed by adjacent
trichomes with high fungal GDH activity and low Nostoc GS activity.
It follows that nutrient flows are organized in small production units
and the concentration of ammonia must be more important in these areas
of the thallus. The free sites available for the fixation of lectins
are numerous but we do not know whether they are due to the cyanobacteria being young (trichomes beginning to elongate) and the slowly
released lectins not being fixed yet, or whether the presence of free
sites is linked to the high concentrations of ammonia, as in other
symbioses. This might mean that lectins, which permit the recognition
of symbiotic partners, have a role in the control of nitrogen flow
between symbionts.

The formation of young lobules

Stage 1. The lobules are initiated by wounding, which changes
the quality and the intensity of the light reaching the cyanobiont,
cuts hyphae and results in a modification in the circulation of

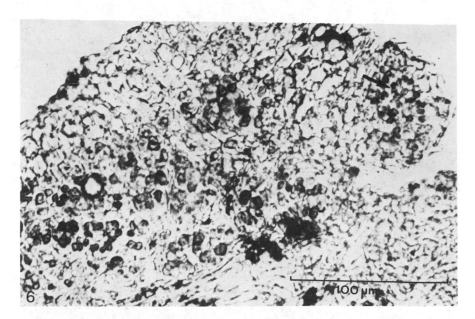

Fig. 6.   Localisation of GS activity (formation of lobule, stage 2).
          Arrow indicates numerous sub-terminal Nostoc cells showing
          GS activity.

nutrients and thus in the metabolism of the partners. The first stage
is characterised by medullary hyphae thickening in the subalgal zone
and lengthening through the algal layer to form a subcortical swelling
at the border of the wound.

Stage 2. Trichomes unfold, lengthen and then migrate towards the
subcortical swelling. At this time, Nostoc GS activity is very high
(Fig. 6) and GDH activity is very low. This means that the Nostoc
cells do not provide nitrogen to the hyphae at this stage of development; the hyphae are fed by the parental thallus from which they
originate.

Stage 3. Hyphae lengthen and become organised into a pseudo-
cladomian structure. Some trichomes lengthen in parallel to the
hyphae of the central axis of this new pseudo-cladome; the affinity of
these for the exogenous FITC-labelled lectin is more significant than
at any other stage. There is no contact yet between hyphae and
Nostoc. There appears to be GDH activity in the Nostoc, though it is

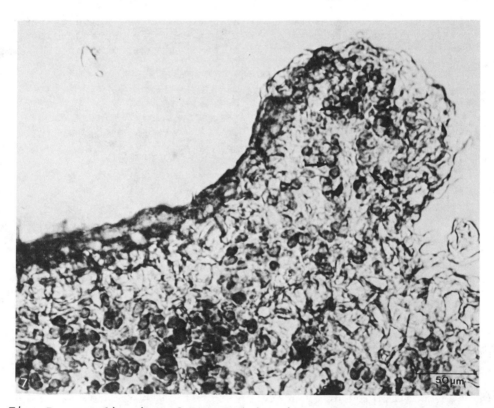

Fig. 7.   Localisation of GDH activity (formation of lobule, stage 3).
          Arrows indicate unwound trichomes showing (?) GDH activity.

usually considered that there is none in the cyanobacteria (Stewart, 1974; Rowell et al, this volume); further work is needed to confirm this.

The lack of contacts between the symbionts results in the absence of fixation of endogenous lectins (the free sites are numerous, thereby allowing fixation of the exogenously-labelled lectin). Ammonia might be excreted but not assimilated by the hyphae, which show no GDH activity. This would confirm the hypothesis that lectins might control the nitrogen flow between partners; with a high concentration of ammonia being generated, the Nostoc trichomes would induce their own GDH.

Stage 4. The structure of the adult thallus is rapidly reached; GS activity, GDH activity and sites available for the fixation of exogenous lectins are distributed as in the adult thallus.

CONCLUSIONS

From earlier work (Petit et al., 1983) we had already proposed that lectins play a role in the recognition process between symbiotic partners. From our study of the development of the lobules of P. praetextata, there is also good evidence that they have a direct role in the changes of the types of contact between mycobiont and photobiont. It also seems that the number of free sites which can bind lectins is related to the concentration of ammonia. However, further investigations are required to establish the precise relationship between those two factors and to determine which one influences the other. We also suspect, on the basis of preliminary data, that, in response to such simple factors as the concentration of ammonia, lectins modify the behaviour of groups of cells. Lectins might be key molecules not only for morphogenesis but also for the circulation of nutrients. Consequently it is possible that lectins play a major role in the conservation of the fragile symbiotic equilibrium.

Another, more general, fact to be emphasised, is that during morphogenesis the lichen is organised in small units related to nutrient flows. Thus, the concentration of ammonia may vary locally. What is true of ammonia is probably also true of many other metabolites. Lichen physiologists should remember that the lichen thallus is not an homogenous entity and, because a lichen is a composite object, it is difficult to talk of the physiology of a lichen as a whole regardless of anatomical and histological structures.

REFERENCES

Ahmadjian, V., Jacobs, J.B., and Russell, L.A., 1978, Scanning
    electron microscopic study of early lichen synthesis, Science,
    200: 1062-1064.
Albersheim, P., and Anderson, A.J., 1975, Carbohydrates, proteins and
    the biochemistry of pathogenesis, Annual Review of Plant
    Physiology, 26: 31-52.
Anderson, A.J., and Jasalavich, C., 1979, Agglutination of Pseudomonas
    cells by plant products, Physiological Plant Pathology, 15:
    149-159.
Ashwell, G., and Morrell, A.G., 1977, Membrane glycoproteins and
    recognition phenomena, Trends in Biochemical Science, 2: 76-78.
Aub, J.C., Tieslau, C., and Lankester, A., 1963, Reaction of normal
    and tumor cell surfaces to enzymes. I. Wheat-germ lipase and
    associated mucupolysaccharides, Proceedings of the National
    Academy of Sciences of the U.S.A., 50: 613-626.
Barondes, S.H., 1980, Developmentally regulated lectins in slime
    moulds and chick tissue - are they cell adhesion molecules? in:
    "Cell adhesion and motility," A.S.G. Curtis and J. Pitts, eds,
    pp. 309-328. Cambridge University Press, Cambridge.
Barondes, S.H., and Rosen, S.D., 1976, Cell surface carbohydrate-
    binding proteins: role in cell recognition, in: "Neuronal
    recognition," S.H. Barondes, ed, pp.331-356, Chapman and Hall,
    London.
Barondes, S.H., Rosen, S.D., Simpson, D.L., and Kafka, J.A., 1974,
    Agglutinins of formalinized erythrocytes: changes of activity
    with development of Dictyostelium discoideum and embryonic chick
    brain, in: "Dynamics of degeneration and growth in neurons." K.
    Fuxe, L. Olson and Y. Zotterman, eds, pp. 449-454, Pergamon
    Press, Oxford.
Barrett, J.T., and Howe, M.L., 1968, Haemagglutination and haemolysis
    by lichen extracts, Applied Microbiology, 16: 1137-1139.
Bezkorovainy, A., Springer, G.F., and Desai, P.R., 1971, Physical
    properties of the Eel anti-human blood-group H(O) antibody,
    Biochemistry, 10: 3761-
Bhuvaneswari, T.V., and Bauer, W.D., 1978, Role of lectins in plant-
    microorganism interactions. III. Influence of rhizosphere-
    rhizoplane culture conditions on the soybean lectin-binding
    properties of Rhizobia, Plant Physiology, 62: 71-74.
Bowles, D.J., and Kauss, H., 1975, Carbohydrate-binding proteins from
    cellular membranes of plant tissues, Plant Science Letters, 4:
    411-418.
Bubrick, P., Ben-Yaakov, M., Frensdorff, A., and Galun, M., 1983,
    Lichen symbiosis: does the surface of the phycobiont's cell wall
    play a role in the discrimination between compatible and
    incompatible symbionts? Israel Journal of Botany, 32: 47-48.
Bubrick, P., and Galun, M., 1980a, Proteins from the lichen Xanthoria
    parietina which bind to phycobiont cell walls. Correlation
    between binding patterns and cell wall cytochemistry,

Protoplasma, 104: 167-173.

Bubrick, P., and Galun, M., 1980b, Symbiosis in lichens: differences
    in cell wall properties of freshly isolated and cultured
    phycobionts, FEMS Microbiology Letters, 7: 311-313.

Bubrick, P., Galun, M., and Frensdorff, A., 1981, Proteins from the
    lichen Xanthoria parietina which bind to phycobiont cell walls.
    Localization in the intact lichen and cultured mycobiont,
    Protoplasma, 105: 207-211.

Burger, M.M., 1969, A difference in the architecture of the surface
    membrane of normal and virally transformed cells, Proceedings of
    the National Academy of Sciences of the U.S.A., 62: 994-1001.

Chadefaud, M., 1975, L'origine "para-floridéenne" des eumycètes et
    l'archétype ancestral de ces champignons, Annales des Sciences
    naturelles, 12ème série, Botanique et Biologie Vegetale, 16:
    217-247.

Chang, C.M., Reitherman, R.W., Rosen, S.D., and Barondes, S.H., 1975,
    Cell-surface location of discoidin, a developmentally regulated
    carbohydrate-binding protein from Dictyostelium discoideum,
    Experimental Cell Research, 95: 136-159.

Chang, C.M., Rosen, S.D., and Barondes, S.H., 1977, Cell surface
    location of an endogenous lectin and its receptor in Polysphondy-
    lium pallidum, Experimental Cell Research, 104: 101-109.

Cohen, E., 1968, Immunologic observations of the agglutinins of the
    hemolymph of Limulus polyphemus and Birgus latro, Transactions of
    the New York Academy of Sciences, 30: 427.

Crandall, M.A., and Brock, T.D., 1968, Molecular basis of mating in
    the yeast Hansenula wingei, Bacteriological Reviews, 32: 139-163.

Dazzo, F.B., and Brill, W.J., 1978, Regulation by fixed nitrogen of
    host-symbiont recognition in the Rhizobium-clover symbiosis,
    Plant Physiology, 62: 18-21.

Dazzo, F.B., Urbano, M.R., and Brill, W.J., 1979, Transient
    appearance of lectin receptors on Rhizobium trifolii, Current
    Microbiology, 2: 15-20.

Den, H., and Malinzak, D.A., 1976, β-galactoside-specific lectin and
    fusion of chick myoblasts, Federation Proceedings, 35: 1409.

Doke, N., Tomiyama, K., Nishimura, N., and Lee, H.S., 1975, In vitro
    interactions between components of Phytophora infestans zoospores
    and components of potato tissue, Annals of the Phytopathology
    Society of Japan, 41: 425-433.

Ensgraber, A., 1958, Die Phytohämagglutine und ihre Funktion in der
    Pflanze als Kohlenhydrat-Transportsubstanzen, Berichte der
    Deutschen Botanischen Gesselschaft, 71: 349-361.

Estola, E., and Vartia, K.O., 1955, Phytoagglutinins in lichens,
    Annales Medicinae Experimentalis et Biologie Fenniae, 33:
    392-395.

Gibbons, R.J., and Van Houte, J., 1975, Bacterial adherence in oral
    microbial ecology, Annual Review of Microbiology, 29: 19-44.

Goodman, R.N., Huang, P.Y., and White, J.A., 1976, Ultrastructural
    evidence for immobilization of an incompatible bacterium, Pseudo-
    monas pisi, in tobacco leaf tissue, Phytopathology, 66: 754-764.

Graham, T.L., 1981, Recognition in Rhizobium-Legume symbioses. International Review of Cytology, Supplement 13: 127-148.

Graham, T.L., and Sequeira, L., 1977, in: "Cell wall biochemistry related to specificity in host-plant pathogen interactions," B. Solheim and J. Raa, eds, pp. 417-422, Columbia University Press, New York.

Graham, T.L., Sequeira, L., and Huang, P;., 1977, Bacterial liposaccharides as inducers of disease resistance in tobacco, Applied and Environmental Microbiology, 34: 424-432.

Hitzeman, J.W., 1963, Observations on the subcellular localization of oxidative enzymes with nitro blue tetrazolium, Journal of Histochemistry, 11: 62-70.

Howe, M.L., and Barrett, J.T., 1967, Phytohaemagglutinins from lichens, Federation Proceedings, 26: 756.

Howe, M.C., and Barrett, J.T., 1970, Studies on haemagglutinin from the lichen Parmelia michauxiana, Biochimica et Biophysica Acta, 215: 97-104.

Kawaguchi, T., Matsumoto, I., and Osawa, T., 1974, Studies on haemagglutins from Maackia amurensis seeds, The Journal of Biological Chemistry, 249: 2786-2792.

Kojima, N., and Uritani, I., 1974, The possible involvement of a spore agglutinating factor(s) in various plants in establishing host specificity by various strains of black-rot fungus Ceratocystis fimbriata, Plant Cell Physiology, 15: 733-737.

Lallemant, R., 1983, Quelques problèmes de morphogénèse dans la symbiose lichénique: étude descriptive et expérimentale, Thèse de Doctorat ès Sciences Naturelles, Université Pierre et Marie Curie, Paris.

Li, D., and Hubbel, D.H., 1969, Infection thread formation as a basis of nodulation specificity in Rhizobium-strawberry clover associations, Canadian Journal of Microbiology, 15: 1133-1136.

Lillie, R.D., 1965, "Histopathological technic and practical histochemistry," 3rd Edition. McGraw Hill, New York.

Lockhart, C.M., Rowell, P., and Stewart, W.D.P., 1978, Phytohaemagglutinins from the nitrogen-fixing lichens Peltigera canina and Peltigera polydactyla, FEMS Microbiology Letters, 3: 127-130.

Lotan, R., Galun, E., Sharon, N., and Mirelman, D., 1975a, Abstracts of the 10th meeting of the Federation of European biochemical Societies, Paris, Abstract 1001.

Lotan, R., Sharon, N., and Mirelman, D., 1975b, Interaction of wheat-germ agglutinin with bacterial cells and cell-wall polymers, European Journal of Biochemistry, 55: 257-262.

Marx, M., and Peveling, E., 1983, Surface receptors in lichen symbionts visualized by fluorescence microscopy after use of lectins, Protoplasma, 114: 52-61.

Mirelman, D., Galun, E., Sharon, N., and Lotan, R., 1975, Inhibition of fungal growth by wheat germ agglutinin, Nature, 256: 414-416.

Nowak, T.P., Haywood, P.L., and Barondes, S.H., 1976, Developmentally regulated lectin in embryonic chick muscle and a myogenic cell line, Biochemical Biophysical Research Communications, 68: 650-657.

Ofek, I., Mirelman, D., and Sharon, N., 1977, Adherence of Escherichia coli to human mucosal cells mediated by mannose receptors, Nature, 265: 623-625.

Ofek, I., and Sharon, N., 1983, Comment les bactéries adhèrent aux cellules, La Recherche, 14: 376-378.

Petit, P., 1982, Phytolectins from the nitrogen-fixing lichen Peltigera horizontalis: the binding pattern of primary protein extract, New Phytologist, 91: 705-710.

Petit, P., Lallemant, R., and Savoye, D., 1983, Purified phytolectin from the lichen Peltigera canina var. canina which binds to the phycobiont cell walls and its use as cytochemical marker in situ, New Phytologist, 94: 103-110.

Rai, A.N., Rowell, P., and Stewart, W.D.P., 1981, Glutamate synthase activity in symbiotic cyanobacteria, Journal of General Microbiology, 126: 515-518.

Reitherman, R.W., Rosen, S.D., Frazier, W.A., and Barondes, S.H., 1975, Cell surface species-specific high affinity receptors for discoidin: developmental regulation in Dictyostelium discoideum, Proceedings of the National Academy of Sciences of the U.S.A., 72: 3541-3546.

Rogers, D.J., Blunden, G., Topliss, J.A., and Guiry, M.D., 1980, A survey of some marine organisms for haemagglutinins, Botanica Marina, 23: 569-577.

Rosen, S.D., 1972, A possible assay for intercellular adhesion molecules. Ph.D. thesis, Cornell University.

Rosen, S.D. Chang, C.M., and Barondes, S.H., 1977, Intercellular adhesion in the cellular slime mold Polysphondylium pallidum inhibited by interaction of asiolofetuin or specific univalent antibody with endogenous cell surface lectin, Developmental Biology, 61: 202-213

Rosen, S.D., Haywood, P., and Barondes, S.H., 1976, Inhibition of intercellular adhesion in a cellular slime mold by univalent antibody against a cell surface lectin, Nature, 263: 425-427.

Rosen, S.D., Kafka, J., Simpson, D.L., and Barondes, S.H., 1973, Developmentally-regulated, carbohydrate binding protein in Dictyostelium discoideum, Proceedings of the National Academy of Sciences of the U.S.A., 70: 2554-2557.

Rosen, S.D., Simpson, D.L., Rose, J.E., and Barondes, S.H., 1974, Carbohydrate-binding protein from Polysphondylium pallidum implicated in intercellular adhesion, Nature, 252: 149-151.

Scott, G.D., 1956, Further investigation of some lichens for fixation of nitrogen, New Phytologist, 55: 111-117.

Sequeira, L., Gaard, G., and De Zoeten, G.A., 1977, Attachment of bacteria to host cell walls: its relation to mechanism of induced resistance, Physiological Plant Pathology, 10: 43-50.

Sequeira, L., and Graham, T.L., 1977, Agglutination of avirulent strains of Pseudomonas solanacearum by potato lectin, Plant Pathology, 11: 43-54.

Sharon, N., and Lis, H., 1972, Lectins: cell-agglutinating and sugar-specific proteins, Science, 177: 949-959.

Simpson, D.L., Thorne, D.R., and Loh, H.H., 1978, Lectins: endogenous carbohydrate-binding proteins from vertebrate tissues: functional role in recognition processes? Life Sciences, 22: 727-748.

Sing, V.O., and Schroth, M.N., 1977, Bacteria-plant cell surface interactions: active immobilization of saprophytic bacteria in plant leaves, Science, 197: 759-761.

Siu, C.H., Lerner, R.A., Ma, G., Firtel, R.A., and Loomis, W.F., 1975, Developmentally regulated proteins of the plasma membrane of Dictyostelium discoideum, Journal of Molecular Biology, 100:157-178.

Stewart, W.D.P., 1974, Prerequisites for biological nitrogen fixation in blue-green algae, in: "The biology of nitrogen fixation," A. Quispel, ed., pp. 696-718, North Holland, Amsterdam.

Stewart, W.D.P., and Rowell, P., 1977, Modifications of nitrogen-fixing algae in symbiosis, Nature, 265: 371-372.

Stockert, R.J., Morell, A.G., and Scheinberg, I.H., 1974, Mammalian hepatic lectin, Science, 186: 365-366.

Tkacz, J.S., Cybulska, E.B., and Lampen, J.O., 1971, Specific staining of wall mannan with fluorescein-conjugated concanavalin A, Journal of Bacteriology, 105: 1-5.

Tokuyama, H., 1973, Isolation and characterization of pokewood mitogen-like phytomitogens from shoriku, Phytolacca esculenta, Biochimica et Biophysica Acta, 317: 338-350.

Toyoshima, S., Akiyama, Y., Nakano, K., Tonomura, A., and Osawa, T., 1971, Phytomitogen from Wistaria floribunda seeds and its interaction with human peripheral lymphocytes, Biochemistry, 10: 4457-4463.

Toyoshima, S., and Osawa, T., 1975, Lectins from Wistaria floribunda seeds and their effect on membrane fluidity of human peripheral lymphocytes, The Journal of Biological Chemistry, 250: 1655-1660.

Waxdal, M., 1974, Isolation, characterization and biological activities of five mitogens from pokewood, Biochemistry, 13: 3671-3676.

Xavier Filho, L., Mendes, C.G., and Vasconcelo, C.A.F., 1971, Fitohemaglutinina em alguns criptogamos, Estudoe Pesquisas, serie B, N_2: 1-8.

Xavier Filho, L., Mendes, C.G., Vasconcelo, C.A.F., and Costa, A.C., 1980, Fitohemaglutinina (lectinas) en basidioliquens, Boletin da Sociedade Broteriana, 4: 41-46.